Techniques
of
Optimization

Techniques of Optimization

edited by

A. V. Balakrishnan

University of California
Los Angeles, California

4th IFIP Colloquium on Optimization Techniques
Held at Los Angeles, California
on October 19-22, 1971
Sponsored by the Technical Committee
on Optimization (TC 7)

Academic Press   New York and London   1972

ACADEMIC PRESS, INC.
111 Fifth Avenue, New York, New York 10003

United Kingdom Edition published by
ACADEMIC PRESS, INC. (LONDON) LTD.
24/28 Oval Road, London NW1 7DD

LIBRARY OF CONGRESS CATALOG CARD NUMBER: 72-77726

PRINTED IN THE UNITED STATES OF AMERICA

CONTENTS

CONTENTS

STOCHASTIC CONTROL

ECONOMICS

LINEAR AND NONLINEAR PROGRAMMING

vii

PREFACE

This volume is based on papers presented at the 4th IFIP Colloquium on Optimization Techniques held in Los Angeles, October 19-22, 1971. The Colloquium was sponsored by the Technical Committee on Optimization [TC 7], and was devoted to computational aspects of optimization problems arising in a variety of disciplines from the more traditional areas such as Aerospace to the more recent areas such as Public Systems. Although interdisciplinary in nature, a high mathematical level was emphasized. A noteworthy feature of the Colloquium was the mixing of mathematical economists and specialists in optimal control-calculus of variations.

For convenience of the reader, papers have been grouped according to subject matter, although of course, the classification is not precise.

I am indebted to Professors J. Marschak, E. Blum, and L. W. Neustadt for their continual help and advice in all phases of the Colloquium activity. Thanks are also due to the IFIP Colloquium Committee: R. Conti, J. L. Lions, G. Marchuk, L. S. Pontrjagin, F. de Veubeke, and A. A. Dorodnicyn. Financial help for the Colloquium was provided by a grant from the National Science Foundation, Applied Mathematics Division.

PATTERN CLASSIFICATION–IDENTIFICATION

APPLICATION OF PATTERN RECOGNITION TO SOME PROBLEMS
IN ECONOMICS

Jean-Marie Blin, Northwestern University

King-Sun Fu and Andrew B. Whinston, Purdue University

1. Introduction

Economic theory, as it stands today, displays a rather
appalling contrast between the level of sophistication
attained by micro economic theory on the one hand and
"public economics" on the other hand. As a matter of fact
the very term "public economics" is vague enough to warrant
some further specification. Micro economic theory has
given us a very general mathematical model of the production and consumption of private goods. These goods are
such that their consumption (or the consumption of their
services) by one individual excludes the consumption of the
same goods by another individual. Generally speaking they
are divisible (or, at least, their services are) and they
are privately appropriated -- at least in most western
countries.

A moment's reflection will remind us of the existence
of a very different kind of commodity, viz, "public goods."
These are usually indivisible commodities, jointly consumed
by a large number of individuals who may or may not pay an
equal price for their services, depending upon the nature
of the good and its particular financing. The provision of
public goods is not an individual decision but rather a
<u>collective</u> one. The elaborate models of production and con-
sumption of private goods are no longer applicable or, to
put it another way, the decision rules of utility maximization for the individual consumer and profit maximization
for the individual producer are of no help since what we
need is a <u>collective</u> decision rule. The study of such
collective decision rules is actually two-fold: on the one
hand, one may ask how such a rule happens to be - or ought

3

to be - chosen. In other words what rule can we use to decide upon a rule? This problem, which has the flavor of an infinite regress, is known in the literature as the "constitutional choice problem." On the other hand, although the concepts to be outlined below have been quite useful in its solution (see [1]), we shall concentrate here on a logically posterior problem, viz. the choice of a collective decision rule by some outside "advisor" on the basis of various rationality requirements. This is the classical problem of "individual preference aggregation" in Economics. The basic aim of this study is to show how one can use a single underlying structure, viz. that of a preference pattern, and its various mathematical representations to tackle the problem of social choice. The use of classical methods of pattern recognition leads to (1) a completely general concept of collective decision rule and (2) actual processes for collective decision-making. Our study will proceed as follows. First of all, in order to motivate our discussion, we shall briefly review the aggregation problem for consumer preferences and then translate its formulation in terms of pattern recognition which will give us a very clear understanding of the so-called "voting paradox." Then we will relate some classical methods of pattern recognition with the notion of a collective decision rule. And finally, we will present two possible aggregation algorithms for binary preference patterns.

2. The aggregation problem in Economics and the notion of consumer preference patterns

2.1. The "voting paradox": For at least two hundred years it has been known that such a widely used decision rule as majority voting could lead to some peculiar results. More specifically let us consider a group of 3 individuals who are to decide between 3 alternatives (a, b, c), which one they collectively prefer. When we talk about collective preference we must, of course, have decided upon a rule for combining individual preferences. In this case let us suppose majority voting has been chosen. Let us now consider that each individual is able to rank order the alternatives (a, b, c) and that their respective preference orderings are

$$
\left\{
\begin{array}{rcl}
1 & : & b \underset{1}{>} c \underset{1}{>} a \\[2ex]
2 & : & a \underset{2}{>} b \underset{2}{>} c \\[2ex]
3 & : & c \underset{3}{>} a \underset{3}{>} b
\end{array}
\right.
\tag{1}
$$

(Conventionally we shall denote the preference relation for any individual h by $\underset{h}{>}$ which is read"... is preferred by (h) to..."). These individual preference relations are strict complete orderings; and as such they satisfy the three basic properties: irreflexivity, asymmetry and **transitivity**. In other words if A denotes the set of alternatives,

$$
A = \{a,\ b,\ c,\ \ldots,\ m\}
\tag{2}
$$

and S the set of individual consumers who are to make a collective decision

$$
S = \{h \,|\, h = 1,2,\ldots,\ell\}
\tag{3}
$$

the individual preference orderings are all:

(i)   Irreflexive $<\Rightarrow \forall\, i \in A,\ \forall\, h \in S:\ \sim (i \underset{h}{>} i)$

(where $\sim$ is read "it is not the case that"...).

(ii)   Asymmetric $<\Rightarrow \forall\, (i,j) \in \{A \times A\},\ \forall\, h \in S:$
$$
(i \underset{h}{>} j) \Rightarrow \sim (j \underset{h}{>} i)
$$

(iii)   Transitive $<\Rightarrow \forall\, i,\ j,\ k \in A,\ \forall\, h \in S:$
$$
[\,(i \underset{h}{>} j)(j \underset{h}{>} k)\,] => (i \underset{h}{>} k)
$$

Let us now return to the preference system shown in equation (1) above. The problem is: can we obtain a collective preference ordering in such a society using majority rule. There are $c_3^2 = 3$ paired comparisons to consider: (a,b); (b,c); and (a,c).

We have $a > b$ by a 2/3 majority

$$b > c \text{ by a } 2/3 \text{ majority}$$

$$c > a \text{ by a } 2/3 \text{ majority}$$

(where $>$ without a subscript denotes collective preference).
Now if this collective preference is to be an ordering we
must have $a > c$ for transitivity to hold. But here we
obtain

$$a > b > c > a$$

which violates transitivity.

It can be verified quite easily that for 3 alternatives
another preference system would also result in an intransitive collective preference, namely (b a c), (c b a) and
(a c b). This is known as the voting paradox and forms
the initial basis of the aggregation problem for consumer
preference. Its implication is so obvious that it needs
little elaboration: in effect, if a group of individuals,
a society, makes its social decisions by majority rule it
will often be the case that the final winning alternative
will depend upon the order of voting - if each alternative
defeated is then excluded from further consideration, as
it is the case in most legislative bodies! Had we decided
to adopt a random choice mechanism for a social decision,
we would not have been any worse off than under majority
voting. But the above presentation of such a paradox fails
to illustrate the exact nature of the problem and this is
where pattern recognition will first be helpful.

2.2. <u>The concept of consumer preference pattern.</u>

In order to discuss the logic of aggregation of individual preferences, we must first characterize each consumer in terms of its <u>own</u> most desired social state which
<u>he</u> would choose, if he happened to be the sole decision-
maker. Put another way, each consumer can be uniquely
characterized by an "<u>opinion pattern</u>," in some appropriate
feature space. All such opinion pattern can then be compared with each other to detect similarities that will
allow us to partition the feature space accordingly into
preference classes.

The feature space here will be the space of social

states, an N-dimensional space where each dimension represents some "issue" in a general sense. For instance it may be the various levels of production of a given public good; as a first approximation each dimension may take on a continuum of values and we may identify our feature space F_N with R^N. On the other hand it may also take on a finite number of values and in the limit it may even be a two element set $\{0,1\}$ for instance. The next step consists in recognizing explicitly the similarity, or dissimilarity that exists between any two preference patterns. To formalize this concept of <u>similarity</u>, we should require that patterns that are "similar" should lie close to each other in some appropriate coordinate system. This clustering property requires the adoption of some appropriate <u>metric</u> on the feature space F_N.

We can now summarize the pattern recognition approach to social choice as follows: a set S of ℓ consumers is mapped into specified regions of the feature space F_N - where the features are the public issues at stake. These regions can be viewed as equivalence classes partitioning the F_N space - when the equivalence relation is defined as a similarity index between any two members of a given class. As an illustration for a two-dimensional feature space F_2 let $\{X_1\}$ and $\{X_2\}$ be two such clusters of patterns. This means that all the patterns in $\{X_i\}$ are clustering around the cluster center X_i^* which may be considered as a representative pattern of the class ω_i. The cluster center X_i^*, for example, can be interpreted as an "average" of the patterns in the class ω_i. The separability of these two clusters is guaranteed by the fact that they are non-overlapping. (See Figure 1.)

On the other hand we may find that some preference patterns are as likely to belong to class ω_i as they are to belong to class ω_j. What is needed then is some sort of discriminating procedure that will classify any given pattern into one of the classes in F_N. Thus we will use a set of discriminant functions $g_i(X)$ which will map an arbitrary preference pattern X into the class ω_i if and only if

$$g_i(X) > g_j(X) \; \forall \; j \neq i \; ; \; i,j=1,2,\ldots$$

The decision boundary between class ω_i and class ω_j will then be defined by the equation

$$g_i(X) - g_j(X) = 0.$$

We are then led to search for a class of functions $g_i(X)$ which will reach a maximum if and only if pattern X belongs to the class ω_i. Let us now see how these familiar concepts of pattern recognition relate to the notion of a collective decision rule.

3. Discriminant functions as social decision rules

As we stated at the beginning of this study one of our ultimate goals is to show how the aggregation problem in Economics is best understood and solved through the use of a pattern recognition approach. The most startling result, in this respect, consists in showing that the simple majority voting rule can be interpreted as a simple threshold logic unit (TLU).

3.1. Majority voting as a threshold logic unit

Let us consider a vote on a single clear-cut issue first. In such a case the feature space F_N is one-dimensional (N=1) and this unique feature can take on only two values $\{Yes, No\}$, $\{-1; +1\}$ say.

For each preference pattern the discriminant function is linear:

$$g(X) = (W_h \cdot X_h) \text{ where } h=1,\ldots,\ell.$$
$$(\ell = \text{number of consumers})$$
$$X_h = \begin{cases} -1 \\ +1 \end{cases}$$

and
$$W_h = +1$$

The social decision rule is then:

$$G(X) = \text{sign} \sum_{h=1}^{\ell} W_h X_h$$

where the sign function can take on two values only +1 and -1, and results in an indeterminacy if ℓ is an even number and the two pattern classes are of equal size.

The majority voting technique can thus be viewed as a threshold logic unit (TLU) with two possible responses corresponding to the two courses of collective action opened to society in this case.

Needless to say this one-dimensional equal weighting system is but the most primitive and elementary form of decision one could think of. Its limitations are many. Generally speaking the one-dimensional restriction will more often than not lead to a "misrepresentation of consumer preferences" for it is seldom the case that the issues are independent and can be decided upon sequentially one dimension at a time. The most common case is that a given issue is actually viewed by each consumer as part of a complex of issues where he takes a global stand by allowing some trade-off between various features of his preference. In cutting down the dimensionality of the feature space to one dimension, we lose some crucial information for pattern classification. We are thus forced to conclude that one of the most fundamental and commonly accepted method of democratic choice, viz. that of proceeding by binary comparisons in order to attain a global multi-valued social decision suffers from a basic drawback: often it will lead to some pathological collective preference pattern - pathological in the sense that such a social pattern will fail to behave in the same way as the individual patterns. The possible intransitivity of a social pattern obtained by majority voting constitutes such a pathological case.

A new presentation of the paradox of voting can now be achieved to cast some light on this problem. As we know the paradox arises from the sequential application of the majority voting rule to binary issues. In terms of our feature space each feature is a paired comparison of the form
(a vs. b) and it has only two values $\{$Yes or No$\}$, $\{1;0\}$ say. Thus each consumer preference pattern is binary and we are supposed to combine these boolean vectors into some social pattern - which is itself binary.

As an example let there be 3 alternatives, $A=\{a,b,c\}$ and 3 consumers $S=\{1,2,3\}$. Hence we need $C_3^2=3$ dimensions to compare three alternatives pairwise. If the individual patterns are restricted to complete strict orderings there are only $3!=6$ such patterns available, namely

$$
\begin{array}{ll}
\text{(a b c)} & \text{(c b a)} \\
\text{(b a c)} & \text{(c a b)} \\
\text{(b c a)} & \text{(a c b)}
\end{array}
$$

In terms of boolean vectors, if the paired comparisons are made in the following order

a vs. b   c vs. a   b vs. c

we can also write these patterns as

(a b c) <=> (1 0 1)   (c b a) <=> (0 1 0)

(b a c) <=> (0 0 1)   (c a b) <=> (1 1 0)

(b c a) <=> (0 1 1)   (a c b) <=> (1 0 0)

These are the only admissible preference patterns, under the transitivity requirement. But there are exactly 2^N i.e. $2^3=8$ possible binary patterns that can arise, the eight vertices of the unit cube in 3 dimensions.

The two excluded patterns are

$$(0\ 0\ 0) <\Rightarrow (a < b < c < a)$$

and $$(1\ 1\ 1) <\Rightarrow (a > b > c > a) \ .$$

They are the origin vertex of the unit cube, and its symmetric along the main diagonal of this cube.

As we know the voting paradox will arise in only two cases:

(i) If the preference patterns are all equally distributed in three classes

$$X_1 = (100) <=> \quad a \underset{h}{>} c \underset{h}{>} b$$

$$X_2 = (010) <=> \quad c \underset{h}{>} b \underset{h}{>} a$$

$$X_3 = (001) <=> \quad b \underset{h}{>} a \underset{h}{>} c \qquad .$$

Under majority voting the social preference pattern obtained is

$$X* = (000) <=> \quad (a < b < c < a).$$

(ii) If the preference patterns are equally distribued in the three classes

$$X_1' = (011) <=> \quad b \underset{h}{>} c \underset{h}{>} a$$

$$X'_2 = (101) <=> \quad \underset{h}{a >} \quad \underset{h}{b >} \quad c$$

$$X'_3 = (110) <=> \quad \underset{h}{c >} \quad \underset{h}{a >} \quad b$$

Under majority voting the social preference pattern is

$$X^{**} = (111) = a > b > c > a \ .$$

The interpretation of this paradox in terms of linear discriminant functions is clear. Let us consider Figure 2: each issue, say a vs. b is decided upon by drawing a vertical hyperplane H_1 parallel to the plane (b vs. c;c vs. a). The decision rule tells us to choose whichever side of the half space determined by this hyperplane which "contains" most patterns. Similarly we draw hyperplanes H_2 parallel to the (b vs. c;a vs. b) plane to decide between c and a; and H_3 parallel to the horizontal plane to decide between b and c. The intersection of these 3 half spaces is a convex region viz. the lower cell on this figure and the resulting pattern which is the solution to this sequential threshold logic procedure is the point (0 0 0), the origin which is ruled out by the transitivity axiom.

Similarly if society had been of the form (ii), majority voting would have yielded an intransitive pattern, (1 1 1).

Thus we see that six out of eight patterns (vertices of this cube) are acceptable as social preference patterns and majority voting viewed as a sequential threshold logic unit (hyperplane decision boundaries) may yield any one of the eight patterns.

3.2. <u>The general case</u>. The preference patterns displayed by the individuals need not be binary. It is, of course, true that such a presentation more closely conforms with the actual voting techniques encountered in most societies today. But we can, for the sake of generality, assume that each feature displays a continuum of values; and, for instance, we can take as our feature space the N-dimensional Euclidean space.

3.2.1. Let us consider a 2-dimensional feature space F_2 with two pattern classes ω_1 and ω_2 . Let us agree that the sample means $\bar{X}_1$ and $\bar{X}_2$ constitute a representative pattern for each of the two classes. One possible classifier would be a hyperplane normal to the line joining $\overline{X}_1$ and $\overline{X}_2$ and intersecting at mid-distance from $\overline{X}_1$ and $\overline{X}_2$. In Euclidean space this hyperplane would be represented by the equation

$$g(X) = (\overline{X}_1 - \overline{X}_2) \cdot X + \frac{1}{2} |\overline{X}_2|^2 - \frac{1}{2} |\overline{X}_1|^2 = 0$$

which is of the general form

$$g(X) = W \cdot X - K,$$

where K is a constant.

Here
$$W = (\overline{X}_1 - \overline{X}_2)$$
$$K = -\frac{1}{2} |\overline{X}_2|^2 + \frac{1}{2} |\overline{X}_1|^2 \quad .$$

W the <u>weight vector</u> lies in the weight space, the dual of the feature space where X, the pattern vector lies.

Intuitively we can give a general policy recommendation for the determination of a best social pattern to represent the two classes of opinion encountered in such a society. If, for instance, the two classes are of the same size, a best pattern would be P*, the point of intersection of the $g(X)$ hyperplane with its weight vector $W = (\overline{X}_1 - \overline{X}_2)$ since of all the points on $g(x)$, only P* minimizes the sum total (Euclidean) distance from $g(X)$ to each element in pattern class ω_1 (and ω_2).

3.2.2. A special case of a linear classifier is afforded by the <u>minimum distance</u> classifier whereby we use m reference patterns $P_1, P_2, \ldots, P_m$ representing the m classes to classify each consumer in one (and only one) of the m pattern classes. For each class ω_i the Euclidean distance between an arbitrary pattern X and P_i is of the form

$$d(X) = |X - P_i|^2 = X \cdot X - 2XP_i + P_i \cdot P_i \quad .$$

To minimize the distance $d_i(X)$ is equivalent to maximizing the new discriminant function

$$d_i^*(X) = XP_i - \frac{1}{2} P_i \cdot P_i \quad .$$

The decision rule then says

$$X \in \text{Class } \omega_i <=> d_i^*(X) > d_j^*(X)$$

$$\forall \ i, j = 1, 2, \ldots, m; \ i \neq j$$

or equivalently: Maximize $\quad d_i^*(X)$

$\quad i \in \{1,2,\ldots,m\}$.

There are exactly $C_m^2 = \dfrac{m(m-1)}{2}$ decision surfaces (hyperplanes here) needed to partition our feature space into m classes of opinions. Each decision region being formed by the intersection of a finite number of half spaces is itself <u>convex</u> and open - since any pattern on the decision boundary between two regions is classifiable in either one of the two classes.

For two public issues (features) and three classes of opinion ω_1, ω_2 and ω_3 an illustration of such a linear classification procedure is given in Figure 3.

The economic interpretation of such classification procedures as the hyperplane technique is clear: it can be viewed as a vote on the public issues at stake - where the vote would actually be a rating and not only a Yes-or-No type of answer.

It is clear that many other types of decision rules can be used for classifying preference patterns. For instance we may wish to consider more than a single representative pattern to describe each class: if we use a point set of representative patterns for each class - exactly analogous to a "council" in voting theory - the minimum distance procedure would yield a <u>piecewise linear decision</u> rule. On the other hand polynomial decision rules, quadric or otherwise, could also be used. Whatever type of classifier we care to choose the complete analogy between the notion of a collective decision rule and discriminant functions as used in pattern recognition, still remains.

3.2.3. <u>A characterization of Pareto optimal patterns space</u>. An interesting result follows from this presentation of the social choice problem. For many years welfare economists have used the notion of Pareto optimality as one of their least value-restricted tool to rate various "economic states."

Among the set E of all possible economic states a subset θ is said to form the Pareto optimal region if and only if for any state not in θ, there exists some element in θ which is unanimously preferred to the initial state, i.e.

$$\forall \chi \in (E - \theta) , \exists X' \in \theta : \chi' \underset{u}{\gtrsim} \chi$$

(where $\gtrsim_u$ is read "... unanimously preferred to..."

Thus if we start out from a status quo not in θ, there is always a move that will be unanimously approved. Conditions for achieving Pareto-optimality in a private goods economy have long been known. On the other hand Samuelson [5] has derived conditions for Pareto optimality in the case of public goods. Within our model if we agree to represent each pattern class by its centroid, the following characterization theorem for Pareto optimal patterns can be proved.

<u>Theorem</u>: In a public goods economy represented by the above model, the set of Pareto-optimal patterns is the convex closure of the set of centroids representing each pattern class.

Since there is a finite number of such centroids in general, this convex closure is a polyhedron in N-space whose vertices are the centroids $P_1, P_2, \ldots$. An illustration for five pattern classes in F_2 is afforded by Figure 4.

If <u>the status</u> quo point I is outside of the convex closure $\overline{P_1 P_2 P_3 P_4 P_5}$ then a move to I_0, say, will be unanimously favored as required for Pareto-optimality.* Although such a characterization theorem can be extended and serve as a basis for various taxation schemes, this brief discussion was merely meant to point out how useful the general case of preference patterns in an Euclidean feature space could be. However, it appears that the case of binary preference patterns if it may lack some generality, lends itself more directly to some potential applications.

4. <u>Binary preference pattern aggregation</u>

4.1. <u>Binary patterns as tournament matrices</u>. We have previously encountered binary preference patterns in our discussion of the voting paradox (see 3.1 above). At the time we considered them as boolean vectors but we could also have looked at them as a special class of matrices known as "tournament matrices." If m opponents (alternatives) play each other in a round robin tournament the outcomes can be recorded in matrix form by entering a 1 as the i-jth entry if i defeats j (and a 0 in the j-ith entry). Thus a tournament matrix $T_{m \times m}$ is a boolean matrix which verifies:

---

*For proof and complete discussion of this theorem see [1].

$$T + T^{tr} = E - I \quad \text{where E is an (mxm) matrix of 1's}$$
$$\text{and I is the (mxm) unit matrix.}$$

For m alternatives the set $\mathcal{T}$ of all (mxm) tournament matrices will have cardinality $\exp_2 \binom{m}{2}$.

Also, a <u>generalized tournament matrix P</u> is a non-negative (mxm) matrix such that

$$\forall \ i, j \quad 0 \leq P_{ij} \leq 1 \quad \text{and} \quad P + P^{tr} = E - 1 .$$

Such a matrix can arise, for instance, from the assignment of a probability measure Γ on the set $\mathcal{T}$.

And last we can also note that the upper triangular matrix T obtained from any tournament matrix T is nothing but a binary preference pattern as we defined it earlier since it has $\binom{m}{2}$ boolean entries written in triangular form. For instance if we have the preference pattern (a b c) and we can write it in binary form as (1 1 1) - when the paired comparisons are in the order a vs. b; a vs. c, b vs. c. Its tournament matrix representation T is

$$T = \begin{array}{c} \\ a \\ b \\ c \end{array} \begin{array}{c} a \ b \ c \\ \begin{bmatrix} 0 & 1 & 1 \\ 0 & 0 & 1 \\ 0 & 0 & 0 \end{bmatrix} \end{array} \quad \text{and} \quad T' = \begin{bmatrix} 1 & 1 \\ & 1 \end{bmatrix} .$$

A natural metric on the space of binary patterns is afforded by Hamming distance δ. For instance if two preference patterns T'_1 and T'_2 are such that

$$\begin{array}{c} T'_1 \\ (a{>}b{>}c) \end{array} = \begin{bmatrix} 1 & 1 \\ & 1 \end{bmatrix} \quad \text{and} \quad \begin{array}{c} T'_2 \\ (a{>}c{>}b) \end{array} = \begin{bmatrix} 1 & 1 \\ & 0 \end{bmatrix} \quad \delta (T'_1 \ T'_2) = 1 .$$

In terms of permutation group, it is easy to show (see [1]) that the Hamming distance between any two patterns is equal to the minimal number of transpositions necessary to transform one pattern (ordering) into the other.

Various solutions of the aggregation problem will now be briefly discussed.

4.2. <u>Minimizing the probability of misrepresentation of individual preference patterns</u>. Let us consider ℓ individual preference patterns $\{T_1,\ldots,T_\ell\}$ - a set of ℓ tournament matrices, i.e. points in $\mathbb{R}^{m^2}$. Suppose we define the <u>linear aggregate pattern</u> P_Σ as the generalized tournament matrix whose entries p_{ij}^Σ are the relative frequencies of $i > j$ in this society (other means could be used if we wished, e.g. harmonic, geometric or quadratic means). The fact that this linear aggregate matrix is, in fact a generalized tournament matrix can be shown quite readily (see [1]).

Now it can also be shown that the set θ of all $m \times m$ generalized tournament matrices, a subset of $\mathbb{R}^{m^2}$, forms a convex polyhedron whose vertices are the elementary tournament matrices $T \in \mathcal{T}$ (see [1] and [4]). Define the index set $I(T)$ for any tournament matrix T as:

$$I(T) = \{i,j) \mid t_{ij} = 1\} \ .$$

The probability $\Pi(T)$ that the outcome of a round-robin tournament may be represented by the matrix T_Σ when the a priori probabilities are given by the matrix P_Σ is then written

$$\Pi(T) = \prod_{(i,j) \in I(T)} P_{ij}^\Sigma \ .$$

The error probability (e) is then: $e = 1 - \Pi(T)$.

To minimize e we must maximize $\Pi(T)$. It can be shown (see [1]) that the optimal solution leads to the following decision rule:

$$\text{Choose} \quad \begin{cases} P_{ij} & \text{if and only if } P_{ij} > P_{ji} \\ P_{ji} & \text{if and only if } P_{ji} > P_{ij} \end{cases}$$

(In the case where $P_{ij} = P_{ji}$ we have an indeterminacy.) This result is the maximum likelihood decision rule which is thus seen to be identical to the majority decision rule in voting theory. In this way we can give a decision-theoretic rationale for the majority voting rule.*

However, even though we started, by assumption, from individual transitive patterns, we end up with a social pattern that is optimal in the sense of minimizing the misrepresentation probability but is not necessarily transitive.

*In [1] it is also shown that another algorithm, the "minimal distance algorithm," can be devised, using the geometric properties of the set of binary patterns viewed as tournament matrices.

4.3. <u>Generating an aggregate transitive preference pattern</u>. We can now use the Hamming metric defined on the binary patterns $(P_1, \ldots, P_\ell)$. The aggregation problem can be stated: Find $P*$, a transitive pattern which

$$\text{Min} \quad \sum_{h=1}^{\ell} \quad d(P_h - P*) .$$

This procedure can be justified by noting that if we assume (1) that individual disutility is an increasing function of the distance between the social pattern and the preferred pattern of each individual and (2) that disutilities are interpersonally comparable, then minimizing the total distance index is equivalent to minimizing social loss.

A solution algorithm can be outlined as follows:

1. Find all pairs (i,j) such that $i \underset{h}{\succeq} j \ \forall \ h$ and take the transitive closure of this partial order.

2. If $P*$ is a total order go to 9.

3. List all pairs (i,j) of quasi-agreement <u>in order of decreasing agreement</u>.

4. Include pair #X from this list.

5. Check for transitivity in the resulting partial order $P*$.

6. If $P*$ is intransitive, exclude pair #X.

7. If $P*$ is a total order go to 9.

8. Go to 4.

9. Stop.

<u>Example</u>.

Let $A = \{a, b, c, d\}$ and $\ell = 5$

$P_1 = (b, d, c, a)$

$P_2 = (a, d, b, c)$

$P_3 = (a, c, b, d)$

$P_4 = (c, d, b, a)$

$P_5 = (b, a, d, c)$

The computation can be summarized in the table below.

	a vs. b	a vs. c	a vs. d	b vs. c	b vs. d	c vs. d	$d(P_h\ P^*)$
b d c a	0	0	0	1	1	0	4
a d b c	1	1	1	1	0	0	2
a c b d	1	1	1	1	1	0	1
c d b a	0	0	0	0	0	1	5
b a d c	0	1	1	1	1	0	2
number of patterns in agreement	2	3	3	4	3	1	$\Sigma = 16$
P*	1	1	1	1	1	1	

4.4. Conclusion

This brief discussion of two possible aggregation processes was meant to suggest how useful the concept of binary preference pattern could be. Many other algorithms can and have in fact been devised, starting from other optimality criteria and other algebraic structures (see [1]). Moreover, the search for a class of computationally "economical" algorithms has been quite successful so far - as will be shown elsewhere - and it suffices to note, here, that the results thus obtained cast even more light on the relation between pattern recognition and the aggregation problem.

Bibliography

[1] Blin,J.M., Social decision processes, Ph.D. dissertatation (unpublished), Purdue University, 1971

[2] Condorcet, Marquis de, Essai sur l'application de l'analyse a la probabilité des décisions rendues à la pluralité des voix, Paris, 1785.

[3] Fu,K.S., Sequential methods in Pattern recognition and machine learning, Acad.Press, New York,London, 1968.

[4] Moon,J.W. and N.J.Pullmann, "On generalized tournament matrices," SIAM Review, Vol.12,No.3,July 1970,384-399.

[5] Samuelson,P.A.,"The theory of public expenditures," Review of Economics and Statistics, Vol. XXXVI, 1954 pp. 387-389.

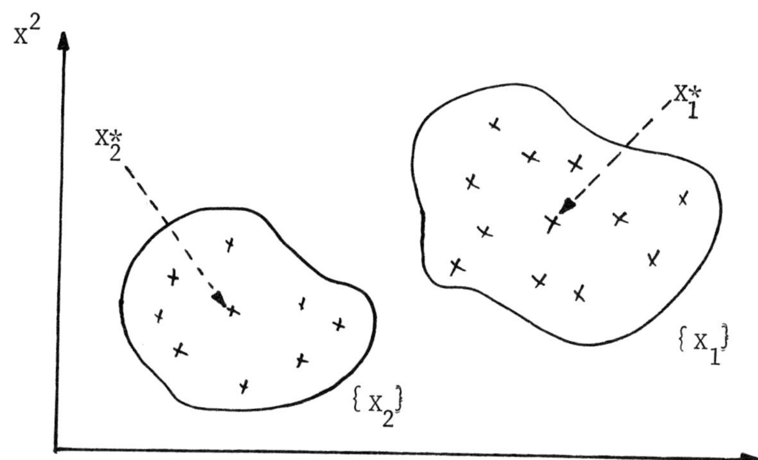

Figure 1. <u>Pattern classes in a two-dimensional feature space.</u>

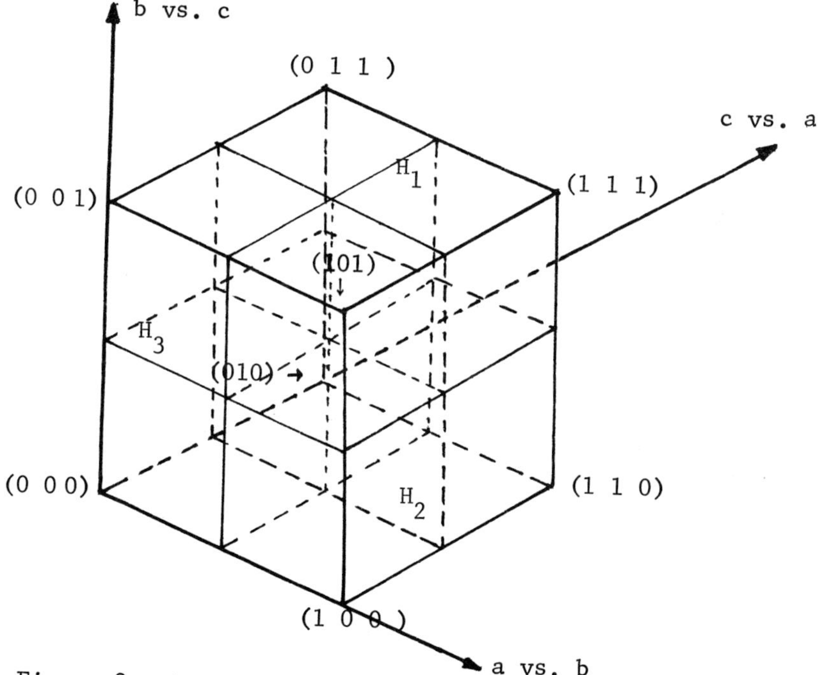

Figure 2. <u>A geometric illustration of the majority voting rule for a set of three alternatives.</u>

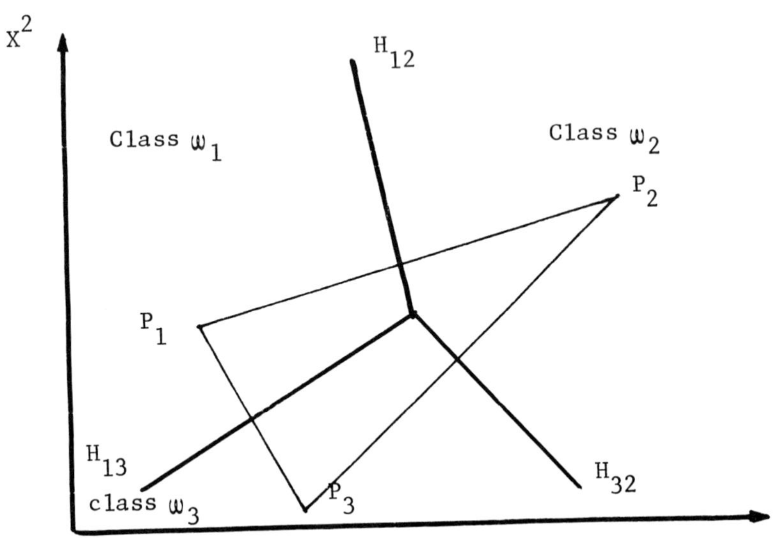

Figure 3. Minimum distance classification for 3 classes.

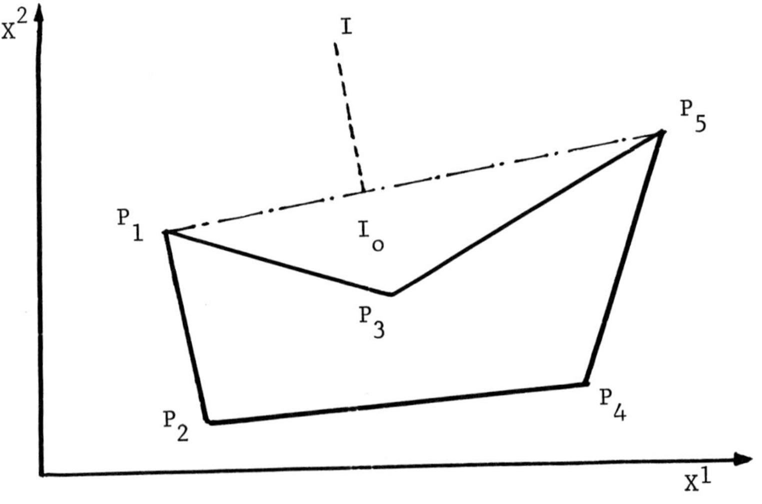

Figure 4. Set of Pareto optimal patterns for a five-class society in F_2.

A DISTRIBUTION-FREE PATTERN CLASSIFICATION PROCEDURE
WITH PERFORMANCE MONITORING CAPABILITY[*]

Harvey L. Kasdan

Recognition Systems, Inc.
Van Nuys, California U.S.A.

Introduction

In pattern recognition problems the underlying distributions are usually unknown. Statistical techniques that are applied to pattern recognition problems must take this into account.

Recently, nonparametric statistical discrimination procedures have been introduced. Cover and Hart (1) showed that the nearest neighbor rule introduced by Fix and Hodges (2) asymptotically has a risk which is less than or equal to twice the Bayes risk independent of the underlying distributions. Fu (3) has recently developed a number of nonparametric sequential methods.

In the past year or so, the need for discrimination procedures with reject or abstention actions has been recognized. Anderson and Benning (4) developed a specific algorithm for a discrimination procedure based on nonparametric tolerance regions suggested by Quesenberry and Gessaman (5). Hellman (6) analyzed a modified nearest neighbor rule with an abstention action.

In this paper, we define a threshold decision rule that has two abstention actions. The statistics or features used to mechanize the rule can be chosen in a number of

____

[*]This paper is based on a dissertation submitted by the author in partial satisfaction of the requirements for the degree of Doctor of Philosophy in Engineering at the University of California, Los Angeles.

ways. When the features used in the rule are $f_i(X) = p(h_i|X)$, the posterior probability of the pattern class given the observable, the threshold rule is an ε-Bayes rule with ε bounded above by a constant times the probability of ambiguous, the second abstention action. Furthermore, an empirical procedure based on nonparametric tolerance regions is developed which asymptotically converges to this ε-Bayes rule.

Another contribution of this work is the idea of using the observed distribution of abstention decisions as an indicator of changes in the underlying distributions. This notion is extremely important in screening systems which make thousands of decisions per hour. Introduction of patterns which are not represented in the learning set or malfunction of a portion of the equipment can cause many costly misclassifications if these changes are not detected.

Finally, the results of a computational study to verify the performance of the empirical rule on three different data sets is presented. The classification accuracy of the rule is studied as well as the extent to which changes in the underlying distributions increase the abstention probabilities. The correlation between an empirical mutual information measure and the performance of the empirical rule is also studied numerically.

Threshold Decision Rule, TDR

To define the Threshold Decision Rule, TDR, in the context of a pattern recognition problem we assume a pattern vector, X, to be a sample from one of M unknown multivariate distributions. A single feature or test statistic, $f_i(X)$, and a corresponding threshold T_i is associated with each pattern class. The threshold rule is then defined as follows.

<u>Definition 1.</u> Let

$$A_i \triangleq \{X: \text{pattern class } i \text{ is rejected}\}, \quad i = 1,\ldots,M \qquad (1)$$

$$\triangleq \{X: f_i(X) \le T_i\}.$$

The decision regions $\{B_i\}_{i=1}^{M}$, B_U and B_A are defined in terms of A_i by the following.

$$B'_i \triangleq \{X: \text{decide in favor of pattern class } i\}, \ i = 1,\ldots,M$$

$$\triangleq \overline{A}_i \bigcap_{\substack{j=1 \\ j\neq i}}^{M} A_j \tag{2a}$$

$$B_U \triangleq \{X: \text{declare an unclassifiable}\} \triangleq \bigcap_{j=1}^{M} A_j \tag{2b}$$

$$B_A \triangleq \{X: \text{declare an ambiguous}\} \triangleq \bigcap_{\substack{2 \text{ or more} \\ j\text{'s}}} \overline{A}_j \tag{2c}$$

From definition 1 we have the following two inequalities using μ_i to denote the measure when pattern class i is the true pattern class.

$$\mu_i(A_i) \geq \mu_i\left(\bigcup_{j\neq i}(\overline{A}_j \bigcap_{k\neq j} A_k)\right) \tag{3a}$$

or

$$P(\text{reject } i|i) \geq P(\text{misclassification}|i) \tag{3b}$$

and

$$\mu_i(A_i) \geq \mu_i\left(\bigcap_{j=1}^{M} A_j\right) \tag{4a}$$

or

$$P(\text{reject } i|i) \geq P(\text{unclassifiable}|i). \tag{4b}$$

Since $P(\text{reject } i|i)$ is an upper bound on both $P(\text{misclassification}|i)$ and $P(\text{unclassifiable}|i)$, the possibility for performance monitoring exists because if

$$\text{Prob}\left[P(\text{unclassifiable}|i) > \alpha_i\right] = \gamma_i \tag{5a}$$

then

$$\text{Prob}\left[P(\text{reject } i|i) > \alpha_i\right] \geq \gamma_i. \tag{5b}$$

Thus if the initial choice of $\{f_i(X)\}$ and $\{T_i\}$ is such that $P(\text{reject } i|i) = \alpha_i$, $i=1,\ldots,M$, one has $P(\text{misclassification}|i) \leq \alpha_i$. If the observed number of unclassifiables implies a change in the unclassifiable probability to

$$P'(\text{unclassifiable}|i) > \alpha_i \tag{6}$$

then by Eqs. (5a,b) the bound on $P(\text{misclassification}|i)$ is greater than α_i.

Threshold Decision Rule with $f_i(X) = p(h_i|X)$

The form of the features $f_i(X)$ is not specified by the definition of the TDR. A particularly "good" choice is $f_i(X) = p(h_i|X)$, the posterior probability density of the pattern class given the observable. To motivate this choice, consider the following hypothesis testing problem. There are M hypothesis and M+1 actions. The ith hypothesis is denoted by h_i and the jth action by a_j. The M+1 st action, a_{M+1}, is to <u>abstain</u> from declaring in favor of any of the hypotheses. Action a_i, i=1,...,M, is to declare in favor of hypothesis i, h_i.

For the purposes of this analysis assume the following cost structure. Let c_{ij}, the cost of action a_j, when h_i is true, be given by

$$c_{ij} = \begin{cases} c_j & \text{if } i \neq j, \quad i,j = 1,\ldots,M \\ 0 & \text{if } i=j \\ c_A & \text{if } j=M+1 \end{cases} \tag{7}$$

We will also use the following notation:

π_i: prior probability of h_i
$\underline{\pi} = (\pi_1,\pi_2,\ldots,\pi_M)^T$
$p_i(X)$: probability of X, the observable, under h_i
$p(h_i|X)$: posterior probability of h_i given the observable,
$B_j \triangleq \{X: \text{the action is } a_j \text{ using the Bayes rule}\}$

The Bayes decision regions for the cost structure defined in Eq. (7) are

$$B_j = \{X: p(h_j|X) > (c_j-c_k)/c_j + (c_k/c_j)p(h_k|X) \text{ for } k \neq j, \text{ and}$$
$$p(h_j|X) > 1 - c_A/c_j\} \qquad j = 1,\ldots,M \tag{8a}$$

$$B_{M+1} = \{X: p(h_j|X) \leq 1 - c_A/c_j \text{ for all } j, j=1,\ldots,M\}. \tag{8b}$$

We can now define a modified rule with M+2 actions in terms of regions $B_j' \triangleq \{X: \text{the action is } a_j \text{ using the modified rule}\}$ by the following.

$$B_j' = \{X: p(h_j|X) > 1 - c_A/c_j \text{ and } p(h_k|X) \leq 1 - c_A/c_j \text{ for } k \neq j\} \quad j = 1,\ldots,M \tag{9a}$$

$$B_{M+1}' = \{X: p(h_j|X) \leq 1 - c_A/c_j \text{ for all } j, j=1,\ldots,M\} \tag{9b}$$

$$B'_{M+2} = \{X: p(h_j|X) > 1 - c_A/c_j \text{ for two or more } j\text{'s},$$
$$j=1,\ldots,M\} \tag{9c}$$

To analyze the risk using this rule, we will assume in addition to the cost structure of Eq. (7) that $c_{iM+2} = c_A$. Although action a_{M+2} is also an abstention action to which we have assigned a cost c_A, we choose to consider it to be distinct from a_{M+1} in order to compare the modified rule to the Bayes rule and to develop properties of the modified rule in the sequel.

We note that this modified rule is a threshold decision rule, TDR, with

$$f_i(X) = p(h_i|X) \quad i = 1,\ldots,M \ , \tag{10a}$$

$$T_i = 1 - c_A/c_i \quad i = 1,\ldots,M \ , \tag{10b}$$

$B'_{M+1} = B_U$ defined in Eq. (2b) and $B'_{M+2} = B_A$ defined in Eq. (2c). Comparing the decision regions $\{B_j\}_{j=1}^{M+1}$ for the Bayes rule to the regions $\{B'_j\}_{j=1}^{M+2}$ for the TDR we note the following relationships:

$$B_{M+1} = B'_{M+1} = B_U \tag{11}$$

$$B_j \supset B'_j \quad j = 1,\ldots,M \tag{12}$$

$$\bigcup_{j=1}^{M} B_j \supset B'_{M+2} = B_A \tag{13}$$

The relationship in Eq. (12) implies that whenever the TDR decision is in favor of h_i so is the Bayes decision. Together Eqs. (12) and (13) imply that the difference between the TDR and the Bayes rule is that in some cases where the Bayes decision is in favor of h_i, the TDR abstains (action a_{M+2}). If an error is defined as a decision in favor of h_j when h_i, $i \neq j$, is true, we also have that the TDR error rate is less than or equal to the Bayes error rate.

Of course, the risk using the TDR, $r(\underline{\pi},d_{TDR})$, is greater than the risk associated with the Bayes rule, $r(\underline{\pi},d_{Bayes})$. The relationship between the two risks is given by

$$r(\underline{\pi},d_{TDR}) = r(\underline{\pi},d_{Bayes}) + \varepsilon \tag{14}$$

where,

$$\varepsilon = \sum_{j=1}^{M} \int_{B_j \cap B'_{M+2}} \left[c_A p(X) - c_j \sum_{\substack{i=1 \\ i \neq j}}^{M} \pi_i p_i(X) \right] d|X|$$

$$\leq \int_{B'_{M+2}} c_A p(X) \, d|X| = c_A P(B'_{M+2}) \qquad (15)$$

and

$p(X) = \sum_{i=1}^{M} p_i(X)$, the mixture distribution. We summarize the discussion above with the following theorem.

Theorem 1. For the cost structure defined in Eq. (7), and assuming $c_{iM+2} = c_A$, the threshold decision rule, TDR, defined in Eq. (9) is an ε-Bayes rule (7) with ε given by Eq. (15). Furthermore,

$$\left(r(\underline{\pi}, d_{TDR}) \right) / \left(r(\underline{\pi}, d_{Bayes}) \right) \leq 1 + \left(c_A / r(\underline{\pi}, d_{Bayes}) \right) P(B'_{M+2}). \quad (16)$$

The significance of this theorem is that since $P(B'_{M+2})$, the unconditional probability of ambiguous, is readily estimated from the observed number of samples falling into B'_{M+2}, one has an empirical estimate of ε.

From Theorem 1 and the discussion preceeding the theorem we see that the smaller $P(B'_{M+2})$, the closer the modified rule given by Eq. (9) is to the Bayes rule, Eq. (8). A result similar to the Neyman-Pearson Lemma is the following.

Theorem 2. For all threshold decision rules (definition 1) with the same P(misclassification|i) and P(unclassifiable|i) $i=1,\ldots,M$, the threshold decision rule with $f_i(X) = p(h_i|X)$, $i=1,\ldots,M$ has the minimum P(ambiguous) and hence the maximum P(correct classification).

The proof of this theorem is similar to the proof of the Neyman-Pearson Lemma and is given in detail in (8).

A Mutual Information Interpretation of the Threshold
Decision Rule with $f_i(X) = p(h_i|X)$

If we let θ be the generic representation for the hypothesis or pattern class then the mutual information between the observable, X, and hypothesis, θ, is given by

$$I(\theta, X) = H(\theta) - H(\theta|X)$$

$$= \sum_{i=1}^{M} \int - \pi_i \log\left(p(X) / p_i(X)\right) \, d\mu_i(X) \qquad (17)$$

where $H(\cdot)$ is the information theoretic entropy. When $f_i(X) = p(h_i|X)$, $i=1,\ldots,M$,

$$A_i = \{X: p(h_i|X) \leq 1 - c_A/c_i\}$$
$$= \{X: p_i(X) / p(X) \leq (1/\pi_i)(1 - c_A/c_i)\} . \qquad (18)$$

From Eq. (18) we may interpret A_i as the critical region for the test of the null hypothesis $X \sim p_i(X)$ versus the alternate hypothesis $X \sim p(X)$. Hence by the Neyman-Pearson Lemma A_i is the R for which

$$\max_{R} \int_{R} \left(p(X) / p_i(X)\right) \, d\mu_i(X) \text{ subj to } \int_{R} d\mu_i(X) = \alpha_i \qquad (19)$$

occurs. Comparing Eq. (19) to Eq. (17) A_i, the region where h_i is rejected, is the fixed size (wrt μ_i) region which makes the least contribution to the mutual information, $I(\theta,X)$.

Empirical Distribution-Free Procedures

Suppose N_i samples from class i are used to form N_i+1 statistically equivalent blocks, SEB's. If each of the regions, A_i, in definition 1 are chosen to be nonparametric tolerance regions formed by a union of a_i of these SEB's, then for each i, $i=1,\ldots,M$, $\mu_i(A_i)$ is a "random measure" distributed according to the Beta distribution independent of the underlying distribution (5),(8). In particular, we have

$$\mathscr{P}\left(\mu_i(A_i)\right) = Be(a_i, N_i-a_i+1). \qquad (20)$$

From Eqs. (3) and (4) since

$$\mu_i(A_i) \geq P(\text{misclassification}|i) \qquad (21a)$$

and

$$\mu_i(A_i) \geq P(\text{unclassifiable}|i), \qquad (21b)$$

when A_i is a nonparametric tolerance region, the upper bound on both $P(\text{misclassification}|i)$ and $P(\text{unclassifiable}|i)$ is a Beta random variable independent of the underlying distribution. Thus if we choose

$$a_i = [\alpha_i(N_i+1)] \quad , \quad [\,] \text{ denotes greatest integer,} \qquad (22)$$

$$\mu_i(A_i) \overset{p}{\to} \alpha_i \text{ as } N_i \to \infty. \qquad (23)$$

For finite N_i we may say $\mu_i(A_i)$ is approximately α_i. Thus to have a bound of approximately α_i on P(misclassification $|i$) and P(unclassifiable$|i$) choose a_i as in Eq. (22).

To develop an empirical rule which converges to the TDR with $f_i(X) = p(h_i|X)$ based on known π_i and $p_i(X)$, $i=1,\ldots,M$ it is sufficient to find the A_i, $i=1,\ldots,M$, described by either Eq. (18) or Eq. (19). If one uses Eq. (18), an empirical estimate of $p_i(X)/p(X)$ is required. This approach would be used if the costs c_A and c_j, $j=1,\ldots,M$ were known. Quite often one may not know the cost structure but is willing to place an upper bound on the misclassification probability. In this case one would use Eq. (19) with α_i equal to the desired upper bound on the misclassification probability.

The following basic algorithm may be used to determine empirical A_i described by either Eq. (18) or Eq. (19). The $\{A_i\}$, $i=1,\ldots,M$ determined using this algorithm will be nonparametric tolerance regions. As the number of samples from each class, N_i, approaches infinity the probability that the action using the empirical rule is the same as the action using the TDR based on known π_i and $p_i(X)$, $i=1,\ldots,M$ approaches 1.

<u>Algorithm 1</u>. 1) Use the coordinates of the pattern vector as ordering functions to form a set of N_i+1 SEB's based on N_i samples from class i. Denote the jth block by $B_j^{(i)}$.
2) Determine the number of samples from all classes, i.e. the mixture distribution, falling in each $B_j^{(i)}$. Denote the number in the jth block by $N_j^{(i)}$.
3) To choose an A_i that approximates the region given by Eq. (18) let

$$A_i = \bigcup_{j \in J} B_j^{(i)}$$

where

$$J = \{j: 1/N_j^{(i)} \leq 1 - c_A/c_i\}$$

As $N_i \to \infty$, $i=1,\ldots,M$, $1/N_j^{(i)} \overset{p}{\to} p(h_i|X)$ in $B_j^{(i)}$. Thus, the empirical regions determined using Algorithm 1 will converge to regions described by Eq. (18).

To use this basic algorithm to determine A_i which approximate the A_i given by Eq. (19) replace step 3) with step 3'):

3') To choose an A_i with approximate size α_i let

$$A_i = \bigcup_{k=1}^{K_i} B_{j_k}^{(i)} \qquad \text{where,} \qquad K_i = [\alpha_i(N_i+1)] \qquad \text{and}$$

$$N_{j_1}^{(i)} \geq N_{j_2}^{(i)} \geq \cdots \quad N_{j_{K_i}}^{(i)} \geq \cdots \geq N_{j_{N_i+1}}^{(i)}$$

Loosely speaking we may think of the convergence of this algorithm in the following terms. As $N_i \to \infty$, $i=1,\ldots,M$ $\mu_i(A_i) \xrightarrow{P} \alpha_i$ as discussed above. Furthermore, as $N_i \to \infty$, $i=1,\ldots,M$

$$\sum_{k=1}^{K_i} N_{j_k}^{(i)}/N_i \xrightarrow{P} \max_R \int_R \left(p(X) / p_i(X) \right) d\mu_i(X) \qquad i=1,\ldots,M.$$

Thus the empirical regions, A_i, determined from Algorithm 1 approach the regions determined using Eq. (19).

To summarize, the empirical rule based on empirical A_i and definition 1 converges to the rule which uses the A_i given by Eq. (18) or Eq. (19) in definition 1. A more detailed discussion of the convergence is presented in (8).

Finally, we make the observation that the $N_j^{(i)}$ determined as a result of applying Algorithm 1 provide the following estimate of the mutual information $I(\theta,X)$ given by Eq. (17).

$$\hat{I}_N(\theta,X) \triangleq \sum_{i=1}^{M} \left(\frac{N_i}{N}\right) \sum_{j=1}^{N_i+1} - \frac{1}{(N_i+1)} \log \left[\frac{N_j^{(i)}(N_i+1)}{N} \right] . \tag{24}$$

In Eq. (24) $N = \sum_{i=1}^{M} N_i$, the total number of learning samples.

Numerical Results

Algorithm 1 was programmed using a binary ranking algorithm suggested by Anderson (9) to determine the statistically equivalent blocks, $B_j^{(i)}$. A number of computational experiments were carried out in order to: 1) evaluate the small sample performance of the empirical rule, 2) study the performance monitoring capability of the empirical rule, 3) study the relationship between the empirical estimate of the mutual information measure (Eq. (24)), the probability of correct classification and the increase in $P(unclass|i)$ when a mixing change occurs.

The three different types of data used in this study were:

1) Iris data used by Kendall (10) and Anderson and Benning (4), 2) Gaussian variates generated by a pseudorandom number generator, and 3) diffraction pattern data from aerial photographs of cloud covered and not cloud covered terrain used by Mead (11).

All of the experiments conducted are described in detail in (8). Three of these will be discussed here to show the general nature of the results.

Iris Data Experiment

Anderson and Benning (4) used the Iris data first used by Fisher (12) in his paper on discriminant functions to evaluate a distribution free method which they developed. Since the Anderson and Benning algorithm is based on non-parametric tolerance regions we repeated their experiments using our empirical rule in order to compare our results with theirs.

The Iris data consists of four measurements on each sample of three different varieties, setosa, virginica and versicolor. Fifty samples from each variety were measured to determine, 1) SL, sepal length, 2) SW, sepal width, 3) PL, petal length, and 4) PW, petal width.

The experiment considered by Anderson and Benning (4) was to discriminate between versicolor, class 1, and virginica, class 2, based on PL and PW. The first 31 samples were used as learning samples for our algorithm, and the remaining 19 samples were used only for recognition. This was done because the binary ranking algorithm requires the number of learning samples to be 2^q-1, where q is an integer. In contrast, Anderson and Benning used all the samples for learning and recognition. Thus our procedure is initially at a disadvantage, but the results indicate no significant degradation in performance.

To compare our results to those of Anderson and Benning (4) consider the results in Table 1. The decision regions for this case were determined by A_i, i=1,2, formed from 2 blocks using PW as the initial ranking component. This is

30

close to their case in which two blocks were chosen based on 50 samples. Using their algorithm for selecting blocks, two cases of versicolor and six cases of virginica were left undecided, while one case of versicolor is misclassified. Using our algorithm with PW as the initial ranking component we have from Table 1, 4 versicolor and 5 virginica undecided but only 1 virginica misclassified. Thus in this case we have but 1 additional misclassification and abstention compared with Anderson and Benning.

Since this is a small sample case and fewer learning samples were used in our algorithm, we conclude that in terms of two class performances our algorithm is probably no worse than that of Anderson and Benning. On the other hand, our algorithm does not require a metric to be defined on the observation space, nor is a separate clustering algorithm and component scaling required. In fact, the results of our algorithm will be invariant under any monotonic transformation of individual coordinates since these transformations do not change the rank ordering.

Experiments Using Gaussian Random Vectors

The case of Gaussian random vectors provides an opportunity to study several properties of the empirical rule in a quantitative way. As noted in (8), in the case of different mean vectors and equal covariance matrices, the ambiguous region is the null set, and the TDR with $f_i(X) = p(h_i|X)$ is in fact the Bayes Rule with an abstention action. Thus for this case, given a covariance matrix and specified mean vectors it is possible to compute exactly the probability of a sample falling in the various regions under either pattern class. We can therefore compare the performance of the empirical rule directly to computed values.

In particular, we study the convergence of the empirical rule by comparing the observed probabilities of correct, ambiguous, and unclassifiable as a function of the number of blocks in A_i to the computed values for the Bayes Rule for a particular case. We then study the decision regions formed by the empirical rule in greater detail by simulating a change in the underlying distribution.

For this experiment, the mean vector for class 1, $\mu_1 = (10,10)$ while the mean vector for class 2, $\mu_2 = (-10,-10)$. In addition the ratio of the number of blocks in A_i to the total number of blocks formed from the learning samples for

class i was fixed at 1/16, so asymptotically $P(\text{unclass}|i)$
$\leq$ .066.

In this experiment 1000 class 1 pattern vectors were
drawn from a Gaussian population $N(\underline{\mu}_1,\Sigma)$, and 1000 class 2
vectors were drawn from $N(\underline{\mu}_2,\Sigma)$, where

$$\Sigma = \begin{bmatrix} 100 & 0 \\ 0 & 25 \end{bmatrix} \tag{24}$$

Using A_i^{Ni} to denote a tolerance region based on N_i learning
samples from class i, tolerance regions A_i^{Ni} were formed for
i=1,2, and N_i=15,31,63,127,255,511, using the first 15,31,
63,127,255, and 511 samples drawn from each class. These
samples constituted the learning samples, while the remaining samples from the original 1000 were used for recognition
in each case. Component 1 was used as the initial ranking
component since asymptotically the order in which the components are used for ranking does not make any difference.

Convergence of the Empirical ε-Bayes Rule

The two class average observed percent of correct, unclassifiable and ambiguous decisions are plotted in Figure 1
for both the learning samples and the recognition samples.
Since the regions A_i^{Ni} were selected to have approximately
6.6% of the probability of the underlying distribution, the
decision regions for the Bayes Rule used for comparison were
chosen to make the probability of an unclassifiable (abstention) equal 0.066.

As can be seen from Figure 1 the empirical rule does
indeed approach the Bayes Rule as the number of samples
grows large. In fact for 255 learning samples (745 recognition samples) the average probability of correct classification is between .94 and .95, the average probability of
unclassifiable is about .05 and the average probability of
ambiguous is less than .01. The agreement with predicted
values is even closer for the 511 learning sample case.

Change in the Unclassifiable Probability When Mixing Occurs

The second portion of this experiment was to investi-
gate the change that occurs in $P(\text{unclassifiable}|1)$ when
class 1 vectors are replaced by "mixed" vectors of the form

$$\underline{X}^c = \alpha\underline{X}_1 + (1-\alpha)\underline{X}_2 \tag{25}$$

where X_i is a vector from class i. The detector structures formed in the first part of the experiment were used and the observed percentages of misclassification, unclassifiable, and ambiguous were recorded for the "mixed" vectors. The results for the 511 learning sample case are presented in graph on Figure 2. Also shown in this figure are the misclassification and unclassifiable probability curves for Bayes Rules with $P(\text{unclassifiable}|i) = 0.066$ and 0.04.

By varying α we are in a sense moving a distribution through the unclassifiable region. If the shape of this region for the empirical rule is the same as the shape for the Bayes Rule then the curve as a function of α for the empirical rule should agree with the curve as a function of α for the Bayes Rule. We see that asymptotically the empirical rule approaches the Bayes Rule since the empirical curve for 511 learning samples agrees quite well with the theoretical curves based on the Bayes Rule.

Cloud Cover Data Experiments

The purpose of these experiments was to evaluate the performance of the empirical rule on a set of data produced by a practical pattern recognition system. This system, described by Mead (11), forms the diffraction pattern of a transparency. This diffraction pattern is then sampled by an array of photodiodes in the form of rings and wedges. The energy falling on a given ring is the energy in a band of spatial frequencies independent of orientation. The energy falling on a given wedge is the energy for a given direction independent of spatial frequency.

This system was used to obtain diffraction patterns for a set of 1024 cloud covered samples and another set of 1024 "not cloud covered" samples. Due to the limitation imposed by the binary ranking algorithm, we used only 7 normalized ring measurements.

Our original intent was to use this data for experiments similar to those described above. In particular, a subset of the data would be used for learning and the remainder for recognition. Then one class would be replaced by a convex combination of the two classes to simulate mixing. In the course of the experiment, however, we discovered that the 1024 non cloud samples were not a homogeneous group. In fact we detected an actual change in the underlying distribution very similar to the idealized mixing change. This was indicated by both an increase in the unclassifiable and ambiguous probability.

Table 2 illustrates how the observed average unclassifiable and ambiguous probabilities reflect changes in the underlying distribution. The number of blocks in each A_i^{511} was set at 25 to have an approximate bound of .05 on the misclassification and unclassifiable probabilities. Since in a realistic setting we can only observe the average number of unclassifiables for both classes it is important to see if this average probability increases sufficiently. Assume we set the threshold for the expected percentage of unclassifiables and ambiguous at 5% and 3% respectively. From table 2, for the learning set we would have observed unclassifiable and ambiguous percentages of 4.8 and 1.6 respectively while for the recognition set we would have observed corresponding percentages of 8.2 and 12.8. Thus, using change detection procedures on both the observed unclassifiable and ambiguous distributions we could detect this change with a reasonably high probability (e.g. (8)).

The main conclusions drawn from this experiment were: 1) The empirical procedure can be used successfully on moderate dimensional data from a completely unknown distribution. 2) Changes in the underlying distributions actually occur in data from practical systems and these changes are reflected in the observed probabilities of unclassifiable and ambiguous.

Empirical Mutual Information

In the experiments conducted by the author, a consistent relationship between the empirical mutual information and the separation of the underlying distributions was noted. The greater the mutual information, the greater the increase in the unclassifiable probability (for a fixed size unclassifiable region) when the mixing change occurs. The following quantitative result illustrates the relationship between the mutual information and the average probability of correct classification for the learning set.

In 28 cases for which $P(A_i^{Ni}) \simeq .066$, the empirical mutual information and average observed probability of correct classification for the learning samples was computed. These cases were ranked according to the value of the empirical mutual information. The rank correlation, RC, between the mutual information and the probability of correct classification was computed using

$$RC = 1 - (6\Sigma d^2)/\left(n(n^2-1)\right) \tag{26}$$

where d is the difference between the mutual information rank and the correct classification probability rank, and n is the number of samples being ranked, 28 in this case. The value obtained was .91.

When the two samples are not correlated, RC is approximately normally distributed with zero mean and variance 1/(n-1) in the large sample case (13). Thus for our case we see that the probability of obtaining a value of $RC \geq .91$ when in fact no correlation exists is essentially zero ($1-\Phi(4.73) \simeq 0$).

Conclusion

The essential conclusion we wish to draw from this work is that the empirical ε-Bayes Rule is a readily mechanizable empirical procedure with abstention actions. In many cases it is very close to the corresponding Bayes procedure. The main rationale for using abstention actions should be that the cost of misclassification is very high compared to the cost of making no decision. Once this criterion has been satisfied, one may also take advantage of the observed abstention probabilities to detect changes in the underlying distributions that might otherwise go undetected.

References

1. Cover, T. M. and P. E. Hart, (1967), IEEE Trans. on Information Theory, Vol. IT-13, pp. 21-27.
2. Fix, E. and J. L. Hodges, Jr., (1951), USAF School of Aviation Medicine, Proj. No. 21-49--004, Rep. No. 4.
3. Fu, K. S. (1969), Methodologies of Pattern Recognition, Academic Press.
4. Anderson, M. W. and R. D. Benning, (1970), IEEE Trans. on Information Theory, Vol. IT-16, No. 5, pp. 541-548.
5. Quesenberry, C.P. and M. P. Gessaman (1968), The Annals of Math. Statistics, Vol. 39, No. 2, pp. 664-673.
6. Hellman, M. E., (1970), IEEE Trans. on Systems Science and Cybernetics, Vol. SSC-6, pp. 179-185.
7. Ferguson, T. S., (1967), Mathematical Statistics: A Decision Theoretic Approach, Academic Press.
8. Kasdan, H. L., (1971), Nonparametric Pattern Recognition, Ph.D. Dissertation, UCLA.
9. Anderson, T. W., (1966), Proc. of an Intern. Symposium on Multivariate Analysis, Academic Press.

10. Kendall, M. G., (1966), Proc. of an Intern. Symposium on Multivariate Analysis, Academic Press.
11. Mead, D. C., (1971), Feature Extraction by Simultaneous Diagonalization, Ph.D. Dissertation, UCLA.
12. Fisher, R. A., (1936), Annals of Eugenics, Vol. 7, pp. 170-188.
13. Sterling, T. D. and S. V. Pollack, (1968), Introduction to Statistical Data Processing, Prentice Hall.

TABLE 1

Versicolor - Virginica Discrimination Decisions

Learning Samples	Unclass	Amb	Versicolor	Virginica
Versicolor	0	2	29	0
Virginica	1	2	1	27
Recognition Samples				
Versicolor	1	1	17	0
Virginica	0	2	1	16

TABLE 2

Cloud Data Decision Percentages for 511 Learning Samples
1024 Total Samples, $K_i = 25$

Learning Samples	Unclass	Amb	Cloud	Not Cloud
Cloud	5.8	0.3	93.7	0
Not Cloud	3.9	2.9	0.7	92.3
Average	4.8	1.6		
Recognition Samples				
Cloud	5.4	3.8	89.6	0.9
Not Cloud	11.1	21.8	18.7	48.3
Average	8.2	12.8		

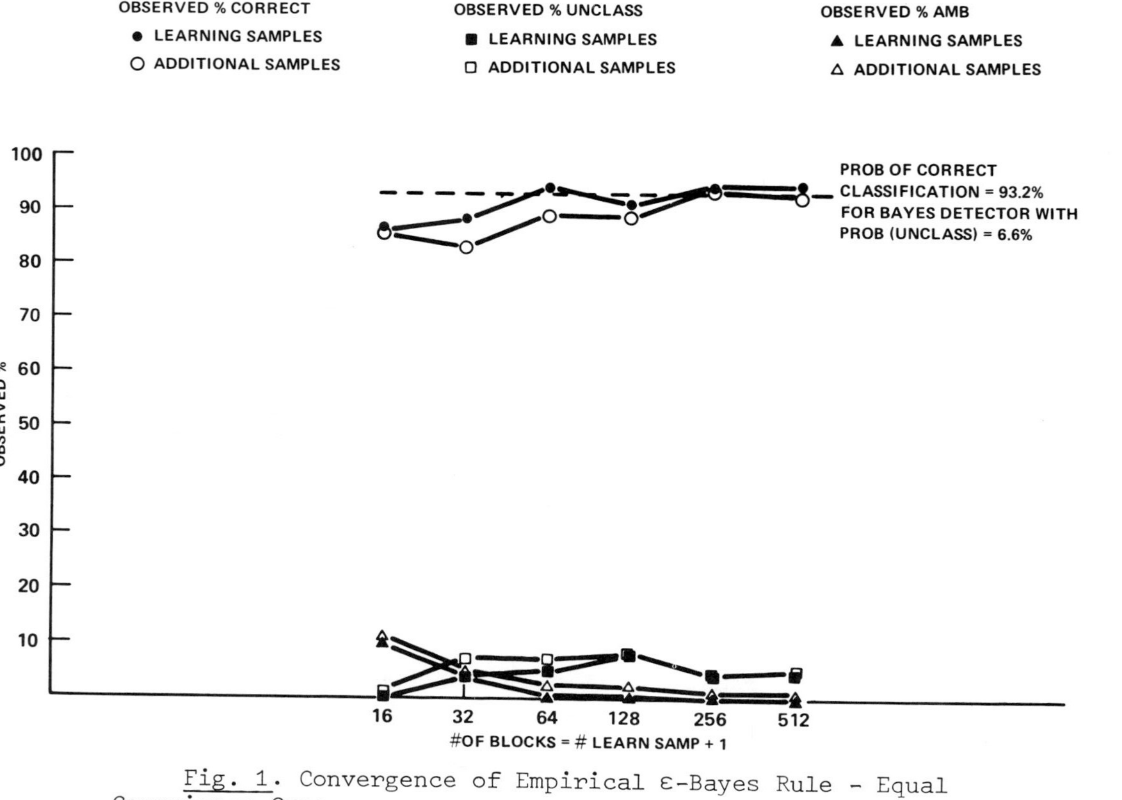

Fig. 1. Convergence of Empirical ε-Bayes Rule - Equal Covariance Case.

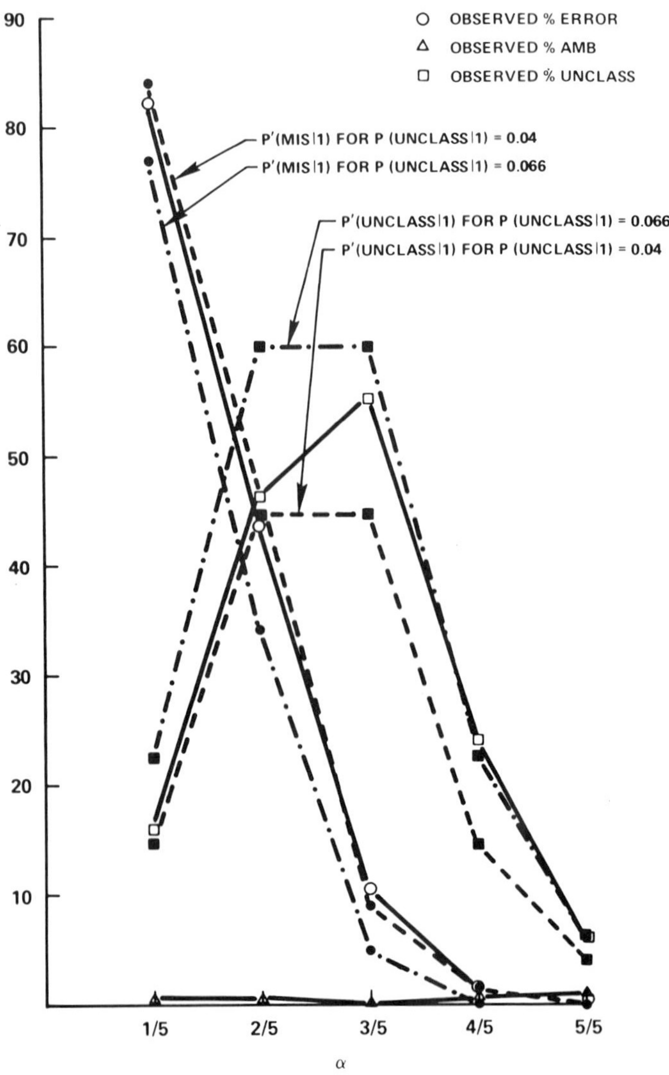

Fig. 2. Observed Probabilities for the "Mixed" Class 511 Learning Samples.

SPECTRAL ANALYSIS IN ECONOMETRICS

Phoebus J. Dhrymes*

University of Pennsylvania
Philadelphia, Pennsylvania

Introduction

Spectral analytic techniques have been employed by econometricians only recently. This is rather surprising since most economic and econometric models, which are typically formulated as difference equations, are assumed to exhibit stability; growth results only through exogenous variables (forcing functions). In addition, for the typical econometric model it has traditionally been asserted-- and found empirically quite useful--that its error component constitutes a rather low order autoregression. Thus, the typical model may be thought to constitute at least a covariance stationary process to which spectral analytic techniques are readily applicable.

The applications we shall consider, in this paper, are of three types:

i. Given an economic time series, determine its cyclical components, if any. The particular example to be examined is concerned with the efficacy of the seasonal adjustment procedure employed by the Bureau of Labor Statistics (BLS).

ii. As an adjunct to estimation of distributed lag (and other) relationships.

iii. Having estimated an economy wide model what can we say about its "business cycle" properties?

---

*The research on which this paper is based was in part supported by NSF grant GS 2289 at the University of Pennsylvania.

Seasonal Adjustment Problems

Most economic time series exhibit fairly persistent cyclical behavior. Thus, for example, if we consider the monthly series of U.S. (civilian) employment (or unemployment) one observes that it fluctuates from a low (high) point at approximately mid-winter to a high (low) point in mid-summer. Correspondingly, various output and income series exhibit analogous fluctuations. Now, such fluctuations are not precisely regular but are sufficiently so, that one may conclude that climatic conditions, the calendar of holidays and certain institutional practices may well account for this phenomenon.

Earlier on, the National Bureau of Economic Research (NBER), that studied extensively economic fluctuations, concluded that it is a convenient scheme to think of economic time series as consisting of three "components": the trend-cycle, the seasonal and the irregular component. In the particular application considered here, due to Nerlove [10], the problem is to determine whether the seasonal adjustment procedure employed by BLS removes only the seasonal component, leaving undisturbed the remaining two. Presumably, the trend-cycle component corresponds to low frequency components, the seasonal to frequencies corresponding to annual cycles and the irregular to all the rest.

The seasonal adjustment procedure employed by BLS is a rather complicated one and the question is whether it merely removes the seasonal component leaving the others undisturbed.

The actual seasonal adjustment procedure employed by BLS is as follows. Let $x(t)$ be employment at month t. First compute

$$y(t) = \frac{1}{12} \sum_{i=-5}^{6} x(t + i) \tag{1}$$

and

$$x_1(t) = \frac{x(t)}{y(t)}. \tag{2}$$

This operation is meant to remove the trend-cycle (low frequency) component. Finally,

$$x_2(t) = \frac{1}{5} \sum_{i=-2}^{2} x_1(t + 12i) \tag{3}$$

gives the seasonal component of the t-th month, at least in preliminary form. Consequently,

$$x^*(t) = \frac{x(t)}{x_2(t)}$$

gives seasonally adjusted employment for month t.

If this is written out in full it becomes

$$x^*(t) = (5/12)[x(t)/\sum_{i=-2}^{2} \{x(t+12i)/\sum_{j=-5}^{6} x(t+12i+j)\}] \qquad (4)$$

and we see that this seasonal adjustment procedure involves the use of a highly nonlinear filter. However, (4) represents only a preliminary version which is subjected to further manipulation; there is, perhaps, no need to go into this matter in more detail, the point about the complexity of the process being established by the preceding.

Due to the highly nonlinear character of the seasonal adjustment filter it is not simple to determine, analytically, its properties. Consequently, the efficacy of the seasonal adjustment process must be appraised by other means.

The approach followed by Nerlove is to compare the spectra of the various unemployment series before and after seasonal adjustment.

The results are reproduced in Figures 1-4, which are reproduced from Nerlove's article; they refer, respectively, to unemployment for males 14-19 years of age; males 20 years of age and over; females 14-19 years of age; females 20 years of age and over.

What one observes is that this procedure removes "excessive" power at all frequencies; it does not, however, introduce extraneous cyclical components. But a more useful exercise would be to consider the gain and phase angle --or time lag--implied by the filter. Such quantities can certainly be computed, numerically, from the cross spectrum of the two series (seasonally adjusted and unadjusted). The results are given in Figures 5-8, where time lag is defined as phase angle divided by the corresponding frequency.

The important findings, here, are that the filter attenuates amplitude (i.e., gain is generally between zero and one) and that it introduces appreciable phase shifts, particularly at low frequencies. Since the latter corresponds to long "periods," components of low frequency convey the bulk of information concerning the state of the economy.

Such phase shifts are not uniform and thus important lead-lag relationships are disturbed in a non-uniform manner. This may have important policy implications insofar as much policy is undertaken after consideration of seasonally adjusted information. The data on which this application is based are monthly series covering, variously, July 1947 to December 1961 or July 1948 to December 1961.

Other aspects of the problem have been investigated by Hannan [5] and Hannan, Terrell and Tuckwell [7] who are concerned with designing optimal seasonal adjustment filters.

Spectral Analytic Techniques in Estimation

Frequently, there is ample economic justification for considering the model

$$y_t = \sum_{i=0}^{\infty} w_i x_{t-i} + u_t \tag{5}$$

under a number of restrictive assumptions on the sequence of coefficients

$$\{w_i : i = 0,1,2,\ldots\}.$$

Two of the most common restrictive assumptions are

$$w_i = \alpha\lambda^i, \ i = 0,1,2,\ldots \tag{6}$$

where $\lambda \in (0,1)$, and

$$W(L) = \frac{A(L)}{B(L)} \tag{7}$$

where

$$W(L) = \sum_{i=0}^{\infty} w_i L^i, \ A(L) = \sum_{i=0}^{n} a_i L^i, \ B(L) = \sum_{j=0}^{m} b_j L^j, \tag{8}$$

and L is the lag operator defined by

$$L^k x_t = x_{t-k}. \tag{9}$$

In the engineering literature, (5) in conjunction with (7), is said to constitute a rational plant and instead of the lag operator and the associated lag (polynomial) operators one uses the z-transform.

In another paper [4] the author (in conjunction with others) has given an abbreviated dictionary of names applied to the same concepts in the literature of econometrics and control engineering.

It will be useful to reproduce this equivalent

terminology below.

Engineering	Econometrics
Identification	Specification and estimation of a model
Plant	Nonstochastic model
Rational z-transform	Rational lag distribution
Record	Sample
White (Gaussian) noise	Sequence of mutually independent, identically distributed (normal) random variables
Colored noise	Autocorrelated errors

The chief reasons why spectral analytic techniques are useful in the estimation of parameters (plant identification) of the models in (5) and (6) or (5) and (7) is that we are able to specify the stochastic process in (5) somewhat nonparametrically.

Until recently, the model

$$y_t = \alpha \sum_{}^{\infty} \lambda^i x_{t-i} + u_t \tag{10}$$

was assumed to be characterized by a white noise u-process, although much empirical evidence in several contexts would seem to indicate a colored process.

The solution of the estimation problem for the model in (10) with u-process assumed to be covariance stationary is due to Hannan [6], and the extension to the general rational distributed lag model is due to Dhrymes [3] and [2].

Let us now give an account of the estimation procedure for the case of the model in (10); in econometrics this is called the geometric distributed lag model or the Koyck model in honor of L.M. Koyck who first hypothesized such a model in his study of investment in plant and equipment.

Let us attempt to obtain estimators for the parameters α and λ by minimizing

$$S^* = (y-\alpha x^*)'K^{-1}(y-\alpha x^*) \tag{11}$$

where

$$y=(y_1, y_2, \ldots, y_T)', \quad x^*=(x_1^*, x_2^*, \ldots, x_T^*)', \quad x_t^* = \sum_{t=0}^{\infty} \lambda^i x_{t-i}. \tag{12}$$

Such estimators in econometrics are called either Aitken or minimum chi square estimators.

In (11) the matrix K is given by

$$K = [K(t-t')]t, \quad t' = 1,2,\ldots,T, \quad K(t,t') = E(u_t u_{t'}) \tag{13}$$

it being assumed that

$$E(u_t) = 0, \quad \text{all } t. \tag{14}$$

Because of covariance stationarity the <u>covariance kernel</u> $K(\cdot)$ of the u-process obeys

$$K(t \cdot t') = K(t'-t) = K(\tau), \quad \tau = |t'-\tau| \tag{15}$$

and thus we may write

$$K = \begin{bmatrix} K(0) & K(1) & \ldots & K(T-1) \\ K(1) & K(0) & \ldots & K(T-2) \\ \vdots & & & \\ K(T-1) & K(T-2) & \ldots & K(0) \end{bmatrix}. \tag{16}$$

Under certain mild conditions on the covariance kernel, we can show that there exists a unitary matrix

$$V = (v_{ts}), \quad v_{ts} = \frac{1}{\sqrt{T}} e^{i(2\pi/T)ts}, \quad t,s = 1,2,\ldots,T \tag{17}$$

such that

$$V^*KV \approx 2\pi \, \mathrm{diag}[f_{uu}(\tfrac{2\pi}{T}), \ f_{uu}(\tfrac{4\pi}{T}),\ldots, \ f_{uu}(2\pi)] = D. \tag{18}$$

This approximation can be made as exact as desired by taking T (the size of the sample, or the length of the record) large enough. To this effect, see Wahba [11].

Since

$$K^{-1} = VV^*K^{-1}VV^* \approx VD^{-1}V^* \tag{19}$$

we have

$$S^* = (y-\alpha x^*)'K^{-1}(y-\alpha x^*) \approx (y-\alpha x^*)'VD^{-1}V^*(y-\alpha x^*). \tag{20}$$

If the spectrum of the u-process is nearly constant over a band of frequencies of length, say, $\frac{\pi}{m}$, $m = [T^{\frac{1}{2}}]$, then we can effectively write the minimand as

$$S = \frac{1}{2\pi} \sum_{j=m+1}^{m} \left\{ \frac{\hat{f}_{yy}(\theta_j) - 2(1-\alpha e^{i\theta_j})^{-1}\hat{f}_{xy}(\theta_j) + \alpha^2[1-\lambda e^{-i\theta_j}]^2 f_{xx}(\theta_j)}{f_{uu}(\theta_j)} \right\} \tag{21}$$

where

$$\theta_j = \frac{\pi j}{m} \tag{22}$$

and f_{xx}, f_{yy}, f_{xy} are the estimated spectral and cross spectral densities of the x- and y-process.

In the above it is assumed, for convenience, that the

x-process is also covariance stationary and independent of the u-process. This assumption may be removed rather easily without occasioning any complications.

Minimizing (21) with respect to α and λ we obtain, after some rearrangement,

$$\alpha \sum_{j=-m+1}^{m} \left[\frac{\hat{f}_{xx}(\theta_j)}{f_{**}(\theta_j)}\right] + \lambda \sum_{j=-m+1}^{m} \left[\frac{\hat{f}_{xy}(\theta_j)}{f_{**}(\theta_j)}\right] = \sum_{j=-m+1}^{m} \left[\frac{\hat{f}_{xy}(\theta_j)}{f_{**}(\theta_j)}\right] \quad (23)$$

$$\alpha \sum_{j=-m+1}^{m} \left[\frac{f_{xx}(\theta_j)e^{-i\theta_j}}{(1-\lambda e^{-i\theta_j})f_{**}(\theta_j)}\right] + \lambda \sum_{j=-m+1}^{m} \left[\frac{f_{xy}(\theta_j)}{(1-\lambda e^{-i\theta_j})f_{**}(\theta_j)}\right]$$

$$= \sum_{j=-m+1}^{m} \left[\frac{f_{xy}(\theta_j)e^{-i\theta_j}}{(1-\lambda e^{-i\theta_j})f_{**}(\theta_j)}\right] \quad (24)$$

where $f_{**}(\cdot)$ is the spectral density of

$$u_t^* = (I-\lambda L)u_t \quad (25)$$

which is, thus, given by

$$f_{**}(\theta) = |1-\lambda e^{i\theta}|^2 f_{uu}(\theta). \quad (26)$$

It would appear that the spectrum of the u-process cannot be estimated and must be known a priori before we can obtain estimates of the parameters α and λ. This, however, is not so.

Returning to (10) we see that it implies

$$y_t = \alpha x_t + \lambda y_{t-1} + u_t - \lambda u_{t-1}. \quad (27)$$

If the x-process is independent of the u-process we can form the instrumental variables estimator implied by

$$\Sigma x_t y_t = \alpha \Sigma x_t^2 + \lambda \Sigma x_t y_{t-1}$$

$$\Sigma x_{t-1} y_t = \alpha \Sigma x_t x_{t-1} + \lambda \Sigma x_{t-1} y_{t-1}. \quad (28)$$

Denote such by $\tilde{\alpha}$ and $\tilde{\lambda}$ and from (27) obtain

$$u_t^* = y_t - \tilde{\alpha} x_t - \tilde{\lambda} y_{t-1}. \quad (29)$$

But from the residuals in (29) we can estimate the spectral ordinates $f_{**}(\theta_j)$ and, thus, render (23) and (24) quite feasible as estimating equations for the parameters of the model. Of course, these equations, even with $\tilde{f}_{**}$ replacing f_{**}, are still highly unlinear (in λ) but the computational burden can be lightened by substituting $\tilde{\lambda}$ for λ wherever the latter appears in the denominator.

It may be shown that, asymptotically, the estimators so obtained, say $\hat{\alpha}$, $\hat{\lambda}$, obey

$$\sqrt{T}\left[\begin{pmatrix}\hat{\alpha}\\\hat{\lambda}\end{pmatrix}-\begin{pmatrix}\alpha\\\lambda\end{pmatrix}\right]\sim N(0,C^{-1})\tag{30}$$

where

$$C = \frac{1}{2\pi}APA, \quad APA = \int_{\pi}^{\pi}\frac{\beta(\theta)\beta'(\theta)}{f_{**}(\theta)}\,f_{xx}(\theta)d\theta$$

$$\overline{\beta}(\theta) = (1,\ \frac{\alpha e^{i\theta}}{1-\lambda e^{i\theta}})'.\tag{31}$$

The asymptotic properties of this estimator are not altered if we iterate. Moreover, if we iterate we do not know the conditions under which the iteration converges.

Determining Cyclical Properties of Econometric Models

For almost half a century now the National Bureau of Economic Research (NBER) has carried out research in economic fluctuations in the U.S. By painstaking and often ingenious examination and juxtaposition of numerous economic time series a classification scheme has been developed for "business cycles," i.e., periods over which "the economy" expands rapidly, very slowly, contracts and expands again. This research has proceeded largely without elaborate analytical apparatus--indeed very little.

On the other hand, at least over the decade of the thirties, considerable work was done in attempting to explain movements of major components such as consumption, investment, etc., and elaborate theoretical constructs have been put forth. In recent times, notably in the decade of the sixties, and very largely because of the advent of large computer systems, many such theoretical constructs have led to the creation of models of the economy, parametrically formulated and which have actually been estimated on the basis of data provided, chiefly, by various agencies of the U.S. government. One may wonder, then, whether such models provide any information regarding the "business cycle" properties of the economy.

Despite the fact that most econometric models extant

contain, at least, a product nonlinearity* much of the theory and applications are couched in terms of linear models. The prototype econometric model is of the form

$$y_{t.} = y_{t.}B + y_{t-1.}C_0 + w_{t.}C_1 + u_t \qquad (32)$$

where $y_{t.}$ is an m-element vector of the (jointly) dependent variables, i.e., those variables which are determined by the system and $w_{t.}$ is an s-element vector of exogenous variables, i.e., those variables which are determined outside the system. The matrices B, C_0, and C_1 contain the unknown (structural) parameters of the system and $u_{t.}$ is a random vector which for $t = 1,2,\ldots$ is assumed to constitute a sequence of mutually independent identically distributed random variables, such that

$$E(u'_{t.}) = 0, \; Cov(u'_{t.}) = \Sigma. \qquad (33)$$

Using the lag operator introduced earlier we find

$$y'_{t.} = [B(L)]^{-1}(I-B')^{-1}C'_1 w'_{t.} + [B(L)]^{-1}(I-B')^{-1}u'_{t.},$$
$$B(L) = I-(I-B')C'_0 L. \qquad (34)$$

The inverse operator will be well defined as a matrix whose elements are power series in the lag operator L, provided the roots of $(I-B')^{-1}C'_0$ are less than unity in modulus. In fact, denoting by $A(L)$ the transpose of the matrix of cofactors of $B(L)$ and by $b(L)$ the determinant of $B(L)$, we can write

$$y'_{t.} = \frac{A(L)}{b(L)}(I-B')^{-1}C'_1 w'_{t.} + \frac{A(L)}{b(L)}(I-B')^{-1}u'_{t.}. \qquad (35)$$

The representation we have given in (35) suppresses all transients of the system. This is certainly permissible in view of the stability condition that the roots of $(I-B')^{-1}C'_0$ be less than unity in modulus. It implies that the system has "started up" in the indefinite past and that the effect of initial conditions has waned. It is, of course, possible to write the system in such a way that initial conditions are explicitly taken into account, and contain certain sinusoidal components of amplitude that ultimately vanishes.

---

*This is so since the economic system determines both prices (p) and quantities (q). Some behavioral equations are defined in terms of money magnitudes (pq) while others are defined in terms of "real" magnitude (q).

Now, suppose that the exogenous variables can be written as

$$w'_{t.} = m(t) + \omega'_{t.} \tag{36}$$

where $m(t)$ is some deterministic function which can be estimated from the data and that $\{\omega'_{t.} = t = 0,\pm1,\pm2,\ldots\}$ is, for simplicity, a sequence of mutually independent identically distributed random vectors with

$$E(\omega'_{t.}) = 0, \quad \mathrm{Cov}(\omega'_{t.}) = \Omega. \tag{37}$$

We may write then

$$y'_{t.} = \frac{A(L)}{b(L)}(I-B')^{-1}C_1 m(t) + \frac{A(L)}{b(L)}(I-B')^{-1}C'_1 \omega'_{t.} + \frac{A(L)}{b(L)}(I-B')^{-1}u'_{t.}. \tag{38}$$

and it is evident that the dependent variables of the system are expressed as the sum of a systematic component, that may well exhibit a trend and account for growth, and a random component. The latter consists of two terms, one contributed by the error process of the exogenous variables and the other by the structural errors of the system. Both these terms are outputs of linear time invariant filters with white noise input.

Consequently, the spectral matrix of the endogenous variables is given by

$$F(\theta) = \frac{1}{2\pi} \frac{A(\theta)}{b(\theta)}(I-B')[C'_1\Omega C_1 + \Sigma](I-B)^{-1}\frac{\overline{A}'(\theta)}{\overline{b}'(\theta)} \tag{39}$$

where $A(\theta)$, $b(\theta)$ are the quantities $A(L)$, $b(L)$ with L replaced by $e^{-i\theta}$; in (39) $\overline{A}'$ is the transpose of a matrix whose elements are the complex conjugates of corresponding elements in A; $\overline{b}(\theta)$ is simply the complex conjugate of the polynomial (in $e^{i\theta}$) $b(\theta)$.

The diagonal elements of the spectral matrix are the spectral densities of the dependent variables. If we determine that such densities exhibit significant "peaks" at certain frequencies we can interpret this to mean that the harmonic components of the period corresponding to such frequencies explain an appreciable part of the variation in the dependent variables. But this is another way of saying that such dependent variables contain a "significant cyclical component" of a certain period. In this fashion we make an identification between the business cycle constructs put forth--in a somewhat unstructured fashion--by NBER and the rather elaborately structural view of the economy portrayed in econometric models.

In addition to this cyclical aspect, the spectral matrix is useful in providing us with information regarding the lead-lag relations amongst various endogenous variables.

In particular, by viewing the coherencies and gains amongst endogenous (and exogeneous) variables we may be able to obtain information about whether certain cyclical components in one variable lead or lag similar components in other variables.

At the moment much is made in NBER and U.S. government forecasting of the behavior of so-called "leading indicators." A more systematic way of examining these issues might well be through the spectral matrix. Of course, this matrix is easily obtained once the structural parameters, the elements of B, C_0, C_1 have been estimated.

The exposition above follows Dhrymes [1] who also gives a paradigmatic example of a three equation model of the U.S. economy. A recent application to a linearized version of the somewhat larger Klein-Goldberger [9] model is given in Howrey [8]. Unfortunately, he does not consider the contribution of the exogenous variables in the spectral matrix of (39).

Application of such techniques to existing econometric models is hampered by a not sufficiently widespread knowledge of such elementary aspects of time series analysis among econometricians, and the fact that econometric models are rather nonlinear and the corresponding theory for such nonlinear systems has yet to be developed--or imported into the literature of econometrics.

References

1. Dhrymes, P. J., Econometrics: Statistical Foundations and Applications, Harper and Row, New York, 1970.

2. ______________, Distributed Lays: Problems of Estimation and Formulation, Holden-Day, San Francisco, 1971.

3. ______________, "Estimation of the General Rational Lag Structure by Spectral Techniques," Discussion Paper No. 35, University of Pennsylvania, January 1968.

4. Dhrymes, P. J., L. R. Klein and K. A. Steiglitz, "Estimation of Distributed Lags," International Economic Review, 11, 1970.

5. Hannan, E. J., "The Estimation of Seasonal Variation in Economic Time Series," Journal of the American

Statistical Association, 58, 1963.

6. __________ ,"The Estimation of Relationships Involv-
    ing Distributed Lags," Econometrica, 33, 1965.

7. Hannan, E. J., R. D. Terrell and N. E. Tuckwell, "The
    Seasonal Adjustment of Economic Times Series,"
    International Economic Review, 11, 1970.

8. Howrey, E. P., "Stochastic Properties of the Klein-
    Goldberger Model," Econometrica, 39, 1971.

9. Klein, L. R. and A. S. Goldberger, An Econometric
    Model of the United States, 1929-1952, North-
    Holland Publishing Co., Amsterdam, 1964.

10. Nerlove, M., "Spectral Analysis of Seasonal Adjustment
    Procedures," Econometrica, 32, 1964.

11. Wahba, G., "On the Distribution of Some Statistics
    Useful in the Analysis of Jointly Stationary Time
    Series," Annals of Mathematical Statistics, 39,
    1968.

APPROXIMATE REPRESENTATIONS OF
CAUSAL SYSTEMS WITH BOUNDED MEMORY

William L. Root

The University of Michigan
Ann Arbor, Michigan

Abstract

This paper is a preliminary report on one part of an attempt to develop an essentially nonrepresentational theory of identification of systems, into which ultimately identification problems framed in terms of particular representations will fit. Here, the subject is causal systems with bounded memory; most of what is done is preliminary to the identification problem. An appropriate mathematical structure including input and output function spaces and a class of mappings (called systems) from the input to output spaces is established. The relationships between these systems and the class of induced systems where only a finite observation interval is permitted are investigated, and an abstract representation of the original systems as trajectories of the induced systems is discussed. The motivation for all this is that identification is to take place in the context of the induced systems; this point is briefly mentioned. Approximations of the induced systems, such as result from an identification, leads to approximations of systems in the original class. What is done here is nonstatistical; it relates to the inversion problem that lies behind the statistical identification problem.

Introduction

The substance of this paper has to do with identification

theory*, but is really not itself identification theory. An
abstract, rather general formulation of some basic identification theory is proposed in Ref. 2. The results developed there are not immediately applicable to the interesting class of causal systems that operate for an indefinite
time period, and it is the purpose here to provide a mathematical interface between a fairly general characterization of causal, infinite-time systems and the theory of
Ref. 2. The structure developed may also be of some use
for other applications. Some conclusions regarding the
identifiability of infinite-time systems follow immediately,
but a thorough study of the identifiability, and of methods
for identification of such systems, or at least a study using the concepts of this paper remains to be done.

The basic class of systems discussed here is that of
systems which are not only causal but have bounded memory. This restriction is for technical convenience; it is
perhaps not too serious since the entire identification theory is an approximation theory. Also the systems are
taken to be noise-free, i.e., all the statistical aspects of
identification are ignored (however, see Refs 3 and 4 for a
treatment of some relevant statistical questions). This
is, of course, a serious limitation, but it allows isolating
a study of the inversion aspect of identification problems.
Finally, there is no restriction, except in special cases,
to linear or time-invariant systems.

Because of lack of space most of the proofs are omitted, although occasional ones are included if they are
short in order to give a slightly better notion of the idea
involved. A longer paper is in preparation, one part of
which will develop this material more fully and with
proofs.

---

*There is no discussion here of what is meant by the term
"identification theory", nor how the theory is used. See
Ref. 1, an excellent survey paper with an extensive bibliography.

Preliminaries

The following definitions and notations are introduced in Ref. 2. Let $\mathcal{X}, \mathcal{U}$ be metric spaces, $\mathcal{Y}$ a Banach space, and f a bounded continuous function from $\mathcal{X} \times \mathcal{U}$ (the topological product of $\mathcal{X}$ and $\mathcal{U}$) into $\mathcal{Y}$. Then $\mathcal{S} = \{\mathcal{Y}, f, \mathcal{X}, \mathcal{U}\}$ is called a class of systems, and $S = \{\mathcal{Y}, f, x, \mathcal{U}\}$, $x \in \mathcal{X}$, is called a system. $\mathcal{U}$ is the input space, $\mathcal{Y}$ the output space, and $\mathcal{X}$ the system space. For fixed x, $y = f(x, u)$, $u \in \mathcal{U}$, gives the transformation from inputs to outputs when that x is operative.

Let $\mathcal{F} = \mathcal{F}(\mathcal{U}, \mathcal{Y})$ be the class of all bounded continuous functions from $\mathcal{U}$ into $\mathcal{Y}$ made into a Banach space in the usual way. Thus, if $F \in \mathcal{F}(\mathcal{U}, \mathcal{Y})$, $\|F\| = \sup_{\mathcal{U}} \|F(u)\|$. Now $f(x, \cdot)$ defines an element, call it H_x, belonging to $\mathcal{F}(\mathcal{U}, \mathcal{Y})$. Define $g: \mathcal{F} \times \mathcal{U} \to \mathcal{Y}$ by $g(F, u) = F(u)$, $F \in \mathcal{F}$. Since g is continuous and bounded, $\mathcal{S}_{\mathcal{F}} = \{\mathcal{Y}, g, \mathcal{F}, \mathcal{U}\}$ is a class of systems; we say it is a <u>linear</u> <u>class</u> since $g(aF_1 + bF_2, u) = ag(F_1, u) + bg(F_2, u)$. If $\mathcal{H} \subset \mathcal{F}$, then $\mathcal{S} = \{\mathcal{Y}, g, \mathcal{H}, \mathcal{U}\}$ is a class of systems with the same g; we say it is <u>pre-linear</u> since the linearity equality just stated holds whenever all the terms are defined.

Now consider $\mathcal{S} = \{\mathcal{Y}, f, \mathcal{X}, \mathcal{U}\}$. The function $f(x, \cdot)$ defines an element $H_x \in \mathcal{F}(\mathcal{U}, \mathcal{Y})$; let ψ be the mapping from $\mathcal{X}$ into $\mathcal{F}(\mathcal{U}, \mathcal{Y})$ determined this way, i.e., $\psi(x) = H_x$, and put $\mathcal{H} = \psi(\mathcal{X})$. We call $\mathcal{S}_0 = \{\mathcal{Y}, g, \mathcal{H}, \mathcal{U}\}$ the <u>natural</u> <u>representation</u> of $\mathcal{S} = \{\mathcal{Y}, f, \mathcal{X}, \mathcal{U}\}$ and ψ the <u>natural</u> <u>mapping</u> from $\mathcal{X}$ into $\mathcal{F}$.

Let $\mathcal{S}_1 = \{\mathcal{Y}, f_1, \mathcal{X}_1, \mathcal{U}\}$ be another class of systems which is related to $\mathcal{S}_0$ as follows: there exists a mapping ϕ_1 from $\mathcal{H}$ into $\mathcal{X}_1$ such that $\|H - \psi_1 \circ \phi_1(H)\| \leq \epsilon$ for all $H \in \mathcal{H}$, where ψ_1 is the natural mapping from $\mathcal{X}_1$ into $\mathcal{F}$. $\mathcal{S}_1$ is said to be an <u>ϵ-representation</u> of $\mathcal{S}_0$ (and also of $\mathcal{S}$). The ϵ-representation is said to be <u>continuous</u> if ϕ_1 is continuous, and <u>linear</u> if the representing class is prelinear and ψ_1 is the restriction of a linear mapping. It is said to be <u>determined by</u> $\mathcal{U}_0$, $\mathcal{U}_0 \subset \mathcal{U}$, if the mapping ϕ depends only on the functions H restricted to $\mathcal{U}_0$, $H \in \mathcal{H}$.

A theorem quoted in Ref. 2. can be stated as follows:
Let $\mathcal{X}$ and $\mathcal{U}$ be compact metric spaces. Then, given
$\epsilon > 0$, there exists an ϵ-representation of $\mathcal{S} = (\mathcal{Y}, f, \mathcal{X}, \mathcal{U})$
which is continuous, linear, determined by a finite subset
$\mathcal{U}_\epsilon \subset \mathcal{U}$, and which has a system space $\mathcal{X}_\epsilon$ that is a subset
of a finite-dimensional Euclidean space.

This theorem guarantees the possibility of identifying a
system to within ϵ, in the norm of $\mathcal{F}(\mathcal{U}, \mathcal{Y})$, if it belongs
to a class $\mathcal{S}$ satisfying the hypotheses stated, from finitely
many measurements and observations. To identify to with-
in ϵ means (in this context) to find a continuous ϵ-repre-
sentation. The experiment to be performed for the identification consists of using as successive inputs the elements
of $\mathcal{U}_0$ and observing the corresponding outputs

This existence theorem is immediately applicable to
classes of systems for which the input and output spaces
$\mathcal{U}$ and $\mathcal{Y}$ are spaces of functions of time, providing the
entire system operation is completed in a finite time inter-
val. It cannot usefully be applied (although it is still valid)
when the time required for one operation of the system is
infinite, because then the entire output can never be available. In this paper the idea is investigated of treating a
causal, bounded memory infinite-time system as a trajectory of systems each operating in fixed, finite time. Then
under certain conditions the finite-time systems can be
identified. If in addition there is enough known a priori
about the evolutionary development of the system, the entire system can be identified.

Specifically, a function space format is first established
which is sufficiently general for most applications, and
then the relations between an infinite-time system and the
corresponding trajectory of finite-time systems is studied
in detail. This is done in the next two Sections. In the
last Section there is some brief comment about the implications for identification; however, as indicated above,
identification theory using the structure set forth is not developed.

Function Spaces for the Inputs and Outputs

Let y be a function of a real variable that is either real-valued or vector-valued with finitely many real components, and P_t denote the operator defined by

$$(P_t y)(s) = y(s) \quad , \ s \leqq t$$
$$= 0 \quad\quad , \ s > t.$$

Let $L_2(A)$ denote the L_2-space of functions $y(\cdot)$ square-integrable with respect to Lebesgue measure on A, A a Lebesgue-measurable subset of R^1. The L_2 norm of y on R^1 will be written $\|y\|$, or occasionally, to avoid confusion, $\|y\|_2$. Let $\mathcal{L}_0$ denote the linear space of functions y on R^1 that are uniformly locally L_2; i.e., $y \in \mathcal{L}_0$ iff for any $T > 0$, $T < \infty$, there is a positive number $K = K(T,y)$ such that $\|(P_{t+T} - P_t)y\| \leqq K$ for all $t \in R^1$.

We make $\mathcal{L}_0$ into a Banach space, denoted $\mathcal{L}_T$, by defining the norm:

$$\|y\|_T = \sup_t (\|(P_{t+T} - P_t)y\|_2). \tag{1}$$

The subscript T on this norm will be dropped when there is no ambiguity about what is intended.

<u>Proposition 1</u> $\mathcal{L}_T$ is a Banach space if the elements of $\mathcal{L}_T$ are interpreted to be the equivalence classes of functions $\mathcal{L}_0$ that are equal a.e. Lebesgue.

<u>Proof</u>: It is readily verifiable that $\mathcal{L}_T$ is a normed linear space. To prove it is complete we consider the particular set of intervals $(kT, (k+1)T]$, k a positive or negative integer. Since for any t, $(t, t+T] \subset (kT, (k+1)T] \cup ((k+1)T, (k+2)T]$ for some k, we have:

$$\sup \|(P_{t+T} - P_t)y\|_2 \leqq \sup_k \|(P_{(k+2)T} - P_{kT})y\|$$
$$\leqq \sup_k \|(P_{(k+1)T} - P_{kT})y\|$$

or, $\|y\|_T \leqq 2\sup_k \|(P_{(k+1)T} - P_{kT})y\| \leqq 2\|y\|_T.$

Now let $\{y_n\}$ be a sequence such that for any $\epsilon > 0$,

$\|y_n - y_m\|_T \leq \epsilon$ for $m, n \geq n_0$. Then,

$$\|(P_{(k+1)T} - P_{kT})(y_n - y_m)\| \leq \epsilon \quad \text{for } m, n \geq n_0,$$

for all k. Since $L_2(kT, (k+1)T]$ is complete, there is a limit element $y^{(k)} \epsilon L_2(kT, (k+1)T]$. Take as candidate y for the limit element for the original sequence the equivalence class of functions on R^1 that are equal a.e. on $(kT, (k+1)T]$ to any function representing $y^{(k)}$. Then $y \epsilon \mathscr{L}_T$, and

$$\|y - y_n\|_T = \sup_t \|(P_{t+T} - P_t)(y - y_n)\|_2$$

$$\leq 2 \sup_k \|(P_{(k+1)T} - P_{kT})(y - y_n)\|_2 .$$

Since,

$$\|(P_{(k+1)} - P_{kT})(y - y_n)\|_2 \leq 2\epsilon \quad \text{for } n \geq n_0, \text{ all k}$$

it follows that $\|y - y_n\|_T \leq 4\epsilon$ when $n \geq n_0$, which proves the proposition.

<u>Proposition 2</u>. The spaces $\mathscr{L}_T$ for all finite positive numbers T are topologically equivalent.

The proof is trivial and is omitted.

Henceforth we shall refer (improperly) to the elements of $\mathscr{L}_T$ or $L_2(A)$ as the functions in $\mathscr{L}_T$ or $L_2(A)$. Also, when considering a function y belonging to, say, $L_2(t, t+T]$ we shall, without comment, identify it when necessary with $\tilde{y}$ where $\tilde{y}(s) = y(s)$, $t < s \leq t + T$, $\tilde{y}(s) = 0$ otherwise, and where $\tilde{y}$ may be regarded as an element of either $L_2(-\infty, \infty)$ or $\mathscr{L}_T$. This is possible because, not only is there a 1:1 correspondence, but $\|y\|_2 = \|\tilde{y}\|_2 = \|\tilde{y}\|_T$.

Let L_c and R_c denote the operations of translation to the left or right respectively by c, i.e., $(L_c y)(t) = y(t + c)$. L_c and R_c are linear operations in $\mathscr{L}_0$; they preserve norm in any $\mathscr{L}_T$, and $L_c = R_{-c} = R_c^{-1}$. Also, for any real numbers a, b, c:

$$L_c(P_b - P_a)y = (P_{b-c} - P_{a-c})L_c y \tag{2}$$

$$R_c(P_{b-c} - P_{a-c})y = (P_b - P_a)R_c y$$

If L_c or R_c are applied to functions belonging to $L_2(t, t+T]$, say, it is intended that the correspondence with functions in $L_2(-\infty, \infty)$ or $\mathcal{L}_T$ just mentioned be made, so that the resulting expressions are meaningful.

Compactness of the input space is an essential hypothesis for the theorem quoted in the previous Section. However, in the next Section we take the input space $\mathcal{U}$ to be a subset of an $\mathcal{L}_T$ space, and compactness of $\mathcal{U}$ in $\mathcal{L}_T$ is too restrictive an assumption to permit realistic modeling of infinite-time systems. As is readily seen, a compact subset of $\mathcal{L}_T$ must consist of a set of functions whose future (and past) behavior is more and more tightly constrained as time advances (or recedes). A condition that is sufficient for the theory and is also suitable in the context of $\mathcal{L}_T$ spaces is T-compactness, defined as follows. A subset $\mathcal{C}$ of $\mathcal{L}_T$ is $\underline{\text{T-compact}}$ if $(P_{t+T} - P_t)\mathcal{C}$ regarded as a subset of $L_2(t, t + T]$ is compact for every t. One can easily verify the following assertions:

$\underline{\text{Proposition 3}}$. If $\mathcal{C}$ is a compact subset of $\mathcal{L}_T$ it is T-compact, but the converse is not necessarily true. If T_1 and T_2 are any two positive numbers and $\mathcal{B} \subset \mathcal{L}_0$ is T_1-compact (regarded as a subset of $\mathcal{L}_{T_1}$) then it is T_2-compact (regarded as a subset of $\mathcal{L}_{T_2}$).

We further characterize special subsets of $\mathcal{L}_T$ spaces in order to be able to specify satisfactory input spaces $\mathcal{U}$. A subset $\mathcal{A}$ of $\mathcal{L}_T$ is said to have $\underline{\text{property (P)}}$ if whenever $u \in \mathcal{A}$, then $P_t u$, $(I - P_t)u$ and $(P_t - P_s)u$ all belong to $\mathcal{A}$ also, for any real t and any $s \leq t$. (I is the identity operator.) A subset $\mathcal{B}$ of $\mathcal{L}_T$ is said to be $\underline{\text{shift-invariant}}$ if whenever $u \in \mathcal{B}$, then $L_t u \in \mathcal{B}$ for all real t. It is desirable to require that the input spaces for causal, bounded memory, infinite-time systems have all of these properties. That such a requirement is not self-contradictory and does not really impose a restriction on the modelling of actual physical systems is shown by the following proposition.

<u>Proposition 4</u>. Let $\mathcal{U}_0'$ be any compact set in $L_2(0,T]$. Then there exists a set $\mathcal{U} \subset \mathcal{L}_T$ with the following properties:

1) $(P_T - P_0)\mathcal{U} \supset \mathcal{U}_0'$
2) $\mathcal{U}$ is shift-invariant
3) $\mathcal{U}$ has property (P)
4) $\mathcal{U}$ is T-compact

The proof is omitted.

Continuous, Causal, Bounded Memory Systems

We now consider classes of systems $\mathcal{S} = \{\mathcal{Y}, g, \mathcal{H}, \mathcal{U}\}$ in the natural-representation form, where $\mathcal{U} \subset \mathcal{L}_{T+d}$, $0 \leq d < \infty$, $\mathcal{Y} = \mathcal{L}_T$, and $\mathcal{H} \subset \mathcal{F}(\mathcal{U}, \mathcal{L}_T)$. This is a natural and convenient setting for the study of causal systems with memory bounded by d, but some of the results to follow apply more generally; when the conditions of causality and bounded memory are required that fact will be indicated explicitly. However, it will be required in all that follows that $\mathcal{U}$ be shift-invariant, T-compact, and have property (P). Since the transformations H are considered as elements of the Banach space $\mathcal{F}(\mathcal{U}, \mathcal{L}_T)$, one has

$$\|H\| = \sup_{u \in \mathcal{U}} \|Hu\|_T = \sup_{u \in \mathcal{U}} \sup_t \|(P_{t+T} - P_t)Hu\|_2 .$$

<u>Definition</u>. $H \in \mathcal{H}(\mathcal{U}, \mathcal{L}_T)$ is <u>causal</u> if $P_t Hu = P_t H P_t u$ for all t, all $u \in \mathcal{U}$. H has <u>bounded memory</u> (d) if $(I - P_t)Hu = (I - P_t)H(I - P_{t-d})u$ for all t, all $u \in \mathcal{U}$. (Note that the right sides of these two expressions are meaningful since $\mathcal{U}$ has property (P).) The class of causal transformations in $\mathcal{H}(\mathcal{U}, \mathcal{L}_T)$ with bounded memory (d) is denoted by $\mathcal{F}_d^0(\mathcal{U}, \mathcal{L}_T)$. $\mathcal{F}_d^0(\mathcal{U}, \mathcal{L}_T)$ is a closed linear subspace of $\mathcal{F}(\mathcal{U}, \mathcal{L}_T)$.

<u>Remark</u>: Let T_1 and T_2 be any two positive numbers. It follows then from Prop. 3 that if $\mathcal{U} \subset \mathcal{L}_0$ is a T-compact, shift-invariant subset of $\mathcal{L}_{T_1+d}$ with property (P) it is also a subset of $\mathcal{L}_{T_2+d}$, and, as such, has again these properties. Further, if H is a mapping from $\mathcal{L}_0$ to $\mathcal{L}_0$ that belongs to $\mathcal{F}(\mathcal{U}, \mathcal{L}_{T_1})$ it belongs also to $\mathcal{F}(\mathcal{U}, \mathcal{L}_{T_2})$. Thus, neither the class of inputs nor the class of transformations

which may be under consideration depends on the value of T.

Let $\mathcal{U}'_{t,T} \overset{d}{=} (P_{t+T} - P_{t-d})\mathcal{U}$, and let $H'_{t,T}$ be defined in $\mathcal{U}'_{t,T}$ by

$$H'_{t,T}u' \overset{d}{=} (P_{t+T} - P_t)Hu', \quad u' \in \mathcal{U}'_{t,T} . \tag{3}$$

This equation does define a transformation $H'_{t,T}$ since u' belongs to the domain of H by property (P). Now put $\mathcal{U}_T \overset{d}{=} L_t \mathcal{U}'_{t,T}$. $\mathcal{U}_T$ can be regarded as a subset of $L_2(-\delta, T]$; $\mathcal{U}_T$ does not depend on t because of the shift-invariance of $\mathcal{U}$. Finally, define $H_{t,T}: \mathcal{U}_T \to L_2(0, T]$ by

$$\begin{aligned}
H_{t,T}z &= L_t H'_{t,T} R_t z \tag{4}\\
&= L_t(P_{t+T} - P_t)HR_t z, \quad z \in \mathcal{U}_T\\
&= (P_T - P_0)L_t HR_t z .
\end{aligned}$$

It is obvious that $H_{t,T} \in \mathcal{F}(\mathcal{U}_T, L_2(0,T])$. $H_{t,T}$ is the mapping induced by H from the contractions of the inputs on $(t-\delta, t+T]$ to the contractions of the outputs on $(t, T]$ referred back to basic input and output spaces. As t moves along the real line $H_{t,T}$ describes a trajectory in $\mathcal{F}(\mathcal{U}_T, L_2(0,T])$. We now consider such trajectories and some relationships between them and the original system transformations $H \in \mathcal{F}(\mathcal{U}, \mathcal{L}_T)$. As long as T is fixed it is convenient to write H_t for $H_{t,T}$.

Let $\pi_T: \mathcal{F}(\mathcal{U}, \mathcal{L}_T) \to \mathcal{F}(\mathcal{U}_T, L_2(0,T])$ be the mapping that carries H into H_t according to Eq. (4). Rewrite Eq. (4) in the form

$$H_t = (P_T - P_0)L_t HR_t(P_T - P_{-d}), \tag{5}$$

which, although it has an unnecessary projection on the right, is convenient to use when $\mathcal{U}_T$ is considered as a subset of $\mathcal{L}_T$.

<u>Proposition 5.</u> π_t is a linear contraction.

<u>Proof:</u> Let $H, H' \in \mathcal{F}(\mathcal{U}, \mathcal{L}_T)$ and $H_t = \pi_t H$, $H'_t = \pi_t H'$. Then,

$$\|H_t - H'_t\| = \sup_{\mathcal{U}_T} \|[P_T - P_0]L_t HR_t u - [P_T - P_0]L_t H'R_t u\|_2$$

$$= \sup_{\mathcal{U}_T} \|[P_T - P_0]L_t[HR_t u - H'R_t u]\|_2$$

$$\leq \sup_{\mathcal{U}_T} \|[H - H']R_t u\|_T = \|H - H'\|.$$

Linearity is immediately verifiable.

<u>Proposition 6.</u> $H_t = \pi_t H$, $H \in \mathcal{F}(\mathcal{U}, \mathcal{L}_T)$, is continuous in t. Further, if $\mathcal{H} \subset \mathcal{F}(\mathcal{U}, \mathcal{L}_T)$ is compact, then all the $H_t = \pi_t H$, $H \in \mathcal{H}$, are equicontinuous in t.

The proof, which is straightforward enough but is somewhat long, is omitted.

We introduce two consistency relations. The second, which is stronger than the first, can also be regarded as an interpolation formula.

$$(P_{T-\eta} - P_0)L_\eta H_t R_\eta (P_{T-\eta} - P_{-d})$$

$$= (P_{T-\eta} - P_0)H_{t+\eta}(P_{T-\eta} - P_{-d}), \quad 0 \leq \eta \leq T. \quad (6)$$

$$H_{t+\eta} = (P_{T-\eta} - P_0)L_\eta H_t R_\eta (P_{T-\eta} - P_{-d})$$

$$+ (P_T - P_{T-\eta})L_{\eta-T}H_{t+T}R_{\eta-T}(P_T - P_{T-\eta-d}),$$

$$0 \leq \eta \leq T \quad (7)$$

<u>Proposition 7.</u> i) If $H_t = \pi_t H$, $H \in \mathcal{F}(\mathcal{U}, \mathcal{L}_T)$, then H_t satisfies Eq. (6) for all t. ii) If $H \in \mathcal{F}_d^0(\mathcal{U}, \mathcal{L}_T)$, then H_t satisfies Eq. (7) for all t. iii) If H_t is any function from $\mathcal{U}_T$ into $L_2(0, T]$ that satisfies Eq. (7), then it satisfies Eq. (6).

<u>Proof:</u> The proof of i) is given by the following calculation:

$$(P_{T-\eta} - P_0)L_\eta[(P_T - P_0)L_t HR_t(P_T - P_{-d})]R_\eta(P_{T-\eta} - P_{-d})$$

$$= (P_{T-\eta} - P_0)(P_{T-\eta} - P_{-\eta})L_{\eta+t}HR_{\eta+t}(P_{T-\eta} - P_{-d-\eta})(P_{T-\eta} - P_{-d})$$

$$= (P_{T-\eta} - P_0)L_{\eta+t}HR_{\eta+t}(P_{T-\eta} - P_{-d})$$

$$= (P_{T-\eta} - P_0)[(P_T - P_0)L_{t+\eta}HR_{t+\eta}(P_T - P_{-d})](P_{T-\eta} - P_{-d})$$

$$= (P_{T-\eta} - P_0)H_{t+\eta}(P_{T-\eta} - P_{-d}).$$

To prove ii) we use i) for the first term on the right hand side of Eq. (7), and make an analogous calculation for the second term. Then the right hand side of Eq. (7) becomes

$$(P_{T-\eta} - P_0)H_{t+\eta}(P_{T-\eta} - P_{-d}) + (P_T - P_{T-\eta})H_{t+\eta}(P_T - P_{T-\eta-d}).$$

Since H causal with bounded memory (d) implies all the H_t are causal with bounded memory (d), the expression above reduces to

$$(P_{T-\eta} - P_0)H_{t+\eta} + (P_T - P_{T-\eta})H_{t+\eta} = H_{t+\eta} \; ;$$

which proves ii). The proof of iii) is an immediate verification.

We now define mappings that go in the other direction, i.e., that carry trajectories in $\mathscr{F}(\mathcal{U}_T, L_2(0,T])$ into elements of $\mathscr{F}(\mathcal{U}, \mathscr{L}_T)$. Let $\{H_n\}$, $n = \ldots -2, -1, 0, 1, 2, \cdots$, be a sequence of functions that belong to $\mathscr{F}(\mathcal{U}_T, L_2(0,T])$. We require that $\{H_n\}$ be a bounded subset of $\mathscr{F}(\mathcal{U}_T, L_2(0,T])$ and that the H_n be equicontinuous functions. Put

$$G \stackrel{d}{=} \sum_{n=-\infty}^{\infty} \Delta_{n,t} R_{t-nT} H_n L_{t-nT} \Delta^*_{n,t} \tag{8}$$

where

$$\Delta_{n,t} = P_{t-(n-1)T} - P_{t-nT}$$

and

$$\Delta^*_{n,t} = P_{t-(n-1)T} - P_{t-nT-d}$$

For any real number t define the mapping ρ_t by $\rho_t(\{H_n\}) = G$.

<u>Proposition 8.</u> ρ_t is a mapping from the bounded equicontinuous sequences as specified above into $\mathscr{F}(\mathcal{U}, \mathscr{L}_T)$. Also, if $\{H_n\}$ and $\{\tilde{H}_n\}$ are two sequences as specified and

$\|H_n - \tilde{H}_n\| \le \delta$ for all n (the norm is that of $\mathscr{F}(\mathcal{U}_T, L_2(0,T]))$ then $\|\rho_t(\{H_n\}) - \rho_t(\{\tilde{H}_n\})\| \le 2\delta$ (the norm is that of $\mathscr{F}(\mathcal{U}, \mathscr{L}_T))$.

The proof is omitted.

If by π we denote the mapping carrying $H \in \mathscr{F}(\mathcal{U}, \mathscr{L}_T)$ into the entire trajectory $\{\pi_t(H)\}$, then it will develop that π and ρ_t are more or less inverse to each other. However in general the precise relationship between these mappings is somewhat complicated and simplifies only when H is causal with bounded memory d.

<u>Proposition 9</u>. Let $\{H_t\}$, $-\infty < t < \infty$, be a family of mappings $\in \mathscr{F}(\mathcal{U}_T, L_2(0,T])$ which is bounded and in which the H_t are equicontinuous. For any fixed s consider $\{H_{s-nT}\}$, $n = \ldots, -2, -1, 0, 1, 2, \cdots$. Put

$$\tilde{H}_{s-nT+\eta} = (P_{T-\eta} - P_0) L_\eta H_{s-(n-1)T} R\eta (P_{T-\eta} - P_{-d}$$

$$+ (P_T - P_{T-\eta}) L_{\eta-T} H_{s-nT} R_{\eta-T} (P_T - P_{T-\eta-d}),$$

$$0 \le \eta \le T$$

This defines $\tilde{H}(t)$ for all t, and $\tilde{H}_{s-nT} = H_{s-nT}$. Then

$$H_t^{(1)} \overset{d}{=} \pi_t \circ \rho_s(\{H_{s-nT}\}) = \tilde{H}_t \qquad \text{for all } t.$$

Also, if $H^{(1)} \overset{d}{=} \rho_s(\{H_{s-nT}\})$ and

$$H^{(2)} \overset{d}{=} \rho_s \circ \pi(H^{(1)}) \quad \text{then } H^{(2)} = H^{(1)}.$$

The proof is omitted. Note that $H_t^{(1)}$, $H_t^{(2)}$, etc, satisfy the strong consistency condition, Eq. (7). If H_t happens to satisfy Eq. (7), then $H_t = \tilde{H}_t$ and we have $H_t^{(1)} = H_t$. In case the original H is causal with bounded memory d, this is true. In fact, a simple verification gives the stronger assertion:

<u>Proposition 10</u>. If $H \in \mathscr{F}_d^0(\mathcal{U}, \mathscr{L}_T)$ then

$$\rho_t(\{\pi_{t-nT}H\}) = H.$$

Comments on Identification

The mappings described in the last Section provide a suitable mathematical structure for the approximate identification of continuous, causal, bounded memory (d) systems for which the inputs and outputs are locally L_2. If $\{\mathcal{Y}, g, H, \mathcal{U}\}$ is such a system where $\mathcal{U}$ is shift-invariant, T-compact and has property (P), then for any fixed T, a trajectory of systems $\{L_2(0,T], g_T, H_t, \mathcal{U}_T\}_t$ is defined. If each member of the family $\{H_{nT}\}_n$ is identified to within ϵ in $\mathcal{F}(\mathcal{U}_T, L(0,T])$ then through one of the maps ρ_S H is identified to within 2ϵ in $\mathcal{F}(\mathcal{U}, \mathcal{L}_T)$. Furthermore, if H does not have bounded memory, but can be approximated by an $H \in \mathcal{F}^0{}_d(\mathcal{U}, \mathcal{L}_T)$, then it can be approximately identified if each of the family $\{H_{nT}\}$ can be. There is nothing in this structure that requires that the approximate identifications $\hat{H}$ be causal with bounded memory (d), nor is there anything that prevents that requirement being imposed as a constraint. However, the approximate identifications $\hat{H}$ will be roughly causal with bounded memory, as is obvious from the nature of the mappings ρ_S. As T becomes smaller, these $\hat{H}$ will become more nearly causal with bounded memory.

Now it is part of the ground rules that we do not consider a class $\mathcal{H}$ of systems to be identifiable unless an arbitrarily good approximation to each $H \in \mathcal{H}$ can be obtained from observations made during a finite time interval. This obviously requires some element of predictability into the future for the systems under consideration. One situation in which there is predictability is the following:

All the $H_{nT} = \pi_{nT} H, \cdots, -2, -1, 0, 1, 2, \cdots$ are the same, $H \in \mathcal{F}^0{}_d(\mathcal{U}, \mathcal{L}_T)$. Then H is either time-invariant or periodic. In either case $H_{nT} = H_0$ is a "repeatable system", and if H_0 is known to belong to a compact class $\mathcal{H}_0 \subset \mathcal{F}(\mathcal{U}_T, L_2(0,T])$ then H_0 is identifiable to within an arbitrary ϵ with a finite number of observations (i.e., total observation time $= N(\epsilon)T$) by the theorem quoted from Ref. 2. Hence H is approximately identifiable.

More generally, suppose $H \in \mathcal{H} \subset \mathcal{F}^0{}_d(\mathcal{U}, \mathcal{L}_T)$,

$$H_{nT} = \pi_{nT}(H) = \theta_n^{(T)}(H_0), \quad H_0 \in \mathcal{H}_0 \subset \mathscr{F}(\mathcal{U}_T, L_2(0,T]),$$

where $\{\theta_n^{(T)}\}_n$ is a known family of transformations. It can be shown easily that if $\mathcal{H}_0$ is compact, then $\mathcal{H}$ is compact iff the $\theta_n^{(T)}$ are equi-continuous. In this situation, where $\mathcal{H}$ is compact, one has some hope of the possibility of approximate identification with observations over a finite interval; however trivial examples show that in general this is not possible, and still further constraints are necessary. Note that the theorem quoted in the second Section does not apply directly because the condition that the observations be taken over a finite time interval is meaningless there, and because we do not want to require that $\mathcal{U}$ be compact. No reasonably complete theory has yet been worked out (as far as I know), but some positive results can be expected because of special cases where it can be shown that approximate identification is possible. For example, if $\theta_n^{(T)} = \theta^n$ where θ is a contraction, then the truncation of H which acts only on the future (which is usually what is desired) can be approximately identified in finite time.

It is desirable of course to have the identification not depend in any critical way if possible on the particular observation interval length T. The mathematical framework has been set up so that neither the input space $\mathcal{U}$ nor the class of functions $\mathscr{F}(\mathcal{U}, \mathcal{L}_T)$ depends on T. However, the nature of the trajectories can obviously change with T. For example, if $H_{nT} = \theta_n^{(T)}(H_0)$ as mentioned above, then one can always define mappings $\{\theta_n^{(T')}\}$ so that $H_{nT'} = \theta_n^{(T')}(H_0)$ if $T' > T$, but cannot in general if $T' < T$.

References

1. K. J Åström and P. Eykhoff, Automatica 7, 123 (1971).
2. W. L. Root, Proc. of the Fifth Annual Princeton Conference on Info. Sciences and Systems, 13 (1971).
3. W. L. Root, Automatica 7, 219 (1971).
4. W. L. Root, Proc. of the Ninth Annual Allerton Conference on Circuit and System Theory, (1971) to appear.

NUMERICAL METHODS

DUAL, FEASIBLE DIRECTION ALGORITHMS

O. L. Mangasarian[†]

ABSTRACT

An algorithm for the solution of nonlinearly constrained optimization problems is proposed by minimizing a sequence of convex quadratic problems with essentially nonnegativity constraints only. The quadratic subproblems are solved by principal pivoting or other fast quadratic methods. A new method for preventing jamming (or zigzagging) is proposed. Also a general convergence theorem for optimization algorithms is given using the concept of a general necessary optimality function and incorporating the antijamming feature. All algorithms work under a choice of a number of step size selection methods.

1. INTRODUCTION

We develop in this work (Section 2) an algorithm for solving nonlinear programming problems with nonlinear constraints (problem 2.1). At each step a convex quadratic problem with linear constraints (2.3a, 2.3a', or 2.3a") is solved by principal pivoting [7, 8] or any fast quadratic programming algorithm [2, 18]. Although no rates of convergence are given, superlinear order effects could be incorporated

† Computer Sciences Department, University of Wisconsin
   Madison, Wisconsin 53706.

This research was supported by U. S. Army Contract No. _DA-31-124-ARO-D-462 and NSF Grant GJ-362.

by appropriate choice of a matrix $H(x_i)$ appearing in the quadratic subproblems. For example this matrix can be taken to be the inverse of a Hessian matrix or can be updated by methods similar to those of the variable metric methods [4, 11]. Preliminary computational results are encouraging and indicate that by appropriate choice of $H(x_i)$ convergence can be increased appreciably.

We also develop a new antijamming procedure which is a modification and improvement of the Topkis-Veinott [19] procedure. It is simpler than the Zoutendijk [21] and Zangwill [20] procedures in that one fixed ε tolerance is used throughout the algorithm.

Our quadratic subproblems, are <u>transformed</u> dual problems to the feasible direction subproblems of Topkis-Veinott [19] with the modified antijamming procedure. The Topkis-Veinott subproblems have quadratic constraints and hence are not computationally tractable. Ours have a quadratic objective function and linear constraints and can be solved by fast quadratic programming algorithms.

The convergence of the algorithm is established by establishing a general theorem for the convergence of algorithms (Section 3) using the concept of a general optimality function. The antijamming procedure is included in this general theorem and so are various step size selection methods.

The proofs of the theorems of Section 2 are in the Appendix.

2.   THE ALGORITHM

We consider the constrained minimization problem

2.1 $\underset{x \in X}{\text{minimize}} \, f(x), \quad X = \{x \mid x \in R^n, \ g(x) \leq 0\}$

where R^n is the n-dimensional real Euclidean space, f is a function on R^n into the reals and g is a function on R^n into R^m . We shall need certain assumptions on the problem which we list below and impose throughout this section of the paper, Section 2.

2.2a
$$f \in C^1 .$$

2.2b Each component g_j of g has a Lipschitz continuous gradient that is

$$\|\nabla g_j(y) - \nabla g_j(x)\| \leq K\|y - x\|, \quad \text{for all } x, y \in X ,$$

$$j = 1, \ldots, m$$

for some positive number K, where $\|x\|$ denotes the Euclidean norm $(xx)^{\frac{1}{2}}$.

2.2c $\|\nabla f(x)\| \leq \alpha$, $\|\nabla g_j(x)\| \leq \alpha$ for all $x \in X$ and some positive number α .

We divide the algorithm into two parts: a direction finding part and a step size part.

2.3. <u>Direction Finding</u>. At x_i solve the quadratic programming problem (with linear constraints).

2.3a
$$-\theta(x_i, I(x_i)): = \underset{\substack{(\eta, u_{I(x_i)}) \geq 0 \\ \eta + eu_{I(x_i)} = 1}}{\text{minimum}} \{\|\eta \nabla f(x_i) +$$

$$+ u_{I(x_i)} \nabla g_{I(x_i)}(x_i)\|^2_{H(x_i)} - u_{I(x_i)} g_{I(x_i)}(x_i)\}$$

and denote its solution by (η_i, u_i) . Here $\eta \in R$, $u \in R^m$, e is a vector of ones,

$$u_I \nabla g_I(x) = \sum_{j \in I} u_j \nabla g_j(x), \quad u_I g_I(x) = \sum_{j \in I} u_j g_j(x) ,$$

$$\text{2.3b} \begin{cases} I(x) = \{ j \mid -\varepsilon \leqq g_j(x) \leqq 0 \} \\[2ex] \varepsilon = \text{any positive number, fixed throughout algorithm,} \end{cases}$$

$$\text{2.3c} \begin{cases} H(x_i) \text{ is any continuous, symmetric, uniformly positive} \\ \quad \text{definite } n \times n \text{ matrix, that is} \\[2ex] M_1 \| z \|^2 \leqq z H(x)z \leqq M_2 \| z \|^2 \text{ for all } z \in R^n, x \in X, \text{ and} \\[2ex] \quad \text{some numbers } M_2, M_1 > 0 , \end{cases}$$

$\| z \|_H^2$ denotes zHz. $u_{I(x_i)}$ denotes a vector with component indices in $I(x_i)$. Similarly $g_{I(x_i)}$ denotes a vector with component indices in $I(x_i)$. For convenience we refer to $g_{I(x)}$ as the $\underline{\varepsilon\text{-active constraints at } x}$, and the remaining constraints as the $\underline{\varepsilon\text{-inactive constraints at } x}$.

The direction p_i at x_i is defined by

$$\text{2.3d} \qquad\qquad p_i = \nu_i H(x_i)q_i$$

where

$$\text{2.3e} \qquad\qquad q_i = -2(\eta_i \nabla f(x_i) + u_i \nabla g_{I(x_i)}(x_i))$$

$$\text{2.3f} \qquad \nu_i = \max\{1, \tfrac{1}{2}, \tfrac{1}{4}, \dots \} \text{ such that } x_i + \mu p_i \in X$$

$$\text{for } 0 \leqq \mu \leqq 1 .$$

2.4. <u>Step Size</u>. If either $\nabla f(x_i)p_i = 0$ or $q_i = 0$ terminate (x_i is stationary). If neither, use any of the following step size methods to find $x_{i+1} = x_i + \lambda_i p_i$:

2.4a (Minimization along p_i) $f(x_i + \lambda_i p_i) = \min\limits_{0 \le \lambda \le 1} f(x_i + \lambda p_i)$

$$\text{or} \qquad f(x_i + \lambda_i p_i) = \min\limits_{x_i + \lambda p_i \epsilon X} f(x_i + \lambda p_i)$$

2.4b (Armijo [1]) $\lambda_i = \text{maximum} \{1, \frac{1}{2}, \frac{1}{4}, \dots \}$ such that

$$f(x_i) - f(x_i + \lambda_i p_i) \ge -\lambda_i^2 \nabla f(x_i) p_i .$$

For this step-size method f is assumed to have a Lipschitz continuous gradient. (See 2.2b)

2.4c (Goldstein [13])

$$\lambda_i = \begin{cases} 1, & \text{if } \gamma_i(1) \ge \rho \\[2em] \hat{\lambda}_i \text{ such that } \rho \le \gamma_i(\hat{\lambda}_i) \le 1 - \rho, & \text{if } \gamma_i(1) < \rho \end{cases}$$

$$\text{where} \qquad \gamma_i(\lambda) = \frac{f(x_i) - f(x_i + \lambda p_i)}{-\lambda \nabla f(x_i) p_i}$$

and ρ is any fixed number satisfying $0 < \rho \le \frac{1}{2}$.

For this step-size method f is assumed to be bounded below on X .

Before stating the convergence result for the above algorithm we define the concept of a stationary point.

2.5 <u>Stationary Point.</u> A point $\bar{x} \epsilon X$ is said to be stationary if $\theta(\bar{x}, I(\bar{x})) = 0$, where θ is defined by 2.3a.

The significance of such a point follows from the following result.

2.6. <u>Necessary and Sufficient Optimality Theorem (Necessity)</u>. If $\bar{x}$ solves the minimization problem 2.1 then $\theta(\bar{x},\ I(\bar{x})) = 0$. If in addition a constraint qualification such as the Arrow–Hurwicz–Uzawa, Karlin, or Slater constraint qualification is satisfied at $\bar{x}$ [15, pp. 102–105] then the Kuhn–Tucker conditions are also satisfied at $\bar{x}$[†] that is $\nabla f(\bar{x}) + \bar{u}\nabla g(\bar{x}) = 0$, $\bar{u}g(\bar{x}) = 0$, $g(\bar{x}) \leqq 0$, $\bar{u} \geqq 0$. (Sufficiency) If $\bar{x} \in X$, $\theta(\bar{x},\ I(\bar{x})) = 0$, $\bar{\eta} > 0$ (where $(\bar{\eta}, \bar{u})$ is the solution of 2.3a with $x_i = \bar{x}$), f is pseudoconvex at $\bar{x}$ and g_j, $j = 1,\ldots,m$, are quasiconvex at $\bar{x}$, then $\bar{x}$ is a global solution of (2.1).

For the sake of readability we shall collect all proofs of this section of the paper in the Appendix.

Theorem 2.6 indicates then that a stationary point is a desirable point to have, since it is a Kuhn–Tucker point and under additional conditions it is also a global solution. This is precisely what the algorithm 2.3–2.4 leads to, as indicated by the following convergence result.

2.7. <u>Convergence Theorem</u>. Either the sequence $\{x_i\}$ generated by the algorithm 2.3–2.4 terminates at a stationary point $x_{\bar{i}}$, that is $\theta(x_{\bar{i}},\ I(x_{\bar{i}})) = 0$, or each accumulation point $\bar{x}$ is stationary, that is $\theta(\bar{x},\ I(\bar{x})) = 0$.

[†]We remark here that the Kuhn–Tucker constraint qualification does not guarantee that the Kuhn–Tucker conditions hold at an $\hat{x}$ for which $\theta(\hat{x},\ I(\hat{x})) = 0$. For example the problem minimize $\{-x_1 \mid x_2 \geqq x_1^3,\ x_2 \leqq x_1^2\}$ has a solution at $(1,1)$, but at the origin $\theta(0,\ I(0)) = 0$, the Kuhn–Tucker constraint qualification is satisfied but not the Kuhn–Tucker conditions. Such examples are ruled out if we assume the original form of the Arrow–Hurwicz–Uzawa constraint qualification, namely, that there exists a $z \in R^n$ such that $\nabla g_i(\hat{x})z > 0$ for $i \in \{i \mid g_i(\hat{x}) = 0\}$.

We now make a few remarks about the algorithm and in particular about the quadratic minimization problem 2. 3a. We observe first that the tolerance $\varepsilon > 0$ is fixed once for all during the algorithm. If it is set to $\varepsilon = \infty$, then $I(x) = \{1, 2, \ldots, m\}$ and we have an antijamming procedure similar to that of Topkis and Veinott [19]. It is more efficient to take ε some positive number so that we exclude ε-inactive constraints at x_i from 2. 3a.

For the unconstrained case, $I(x)$ is empty and there remains no direction finding problem 2. 3a because the only remaining variable η in 2. 3a becomes $\eta = 1$, and $q_i = -2\nabla f(x_i)$. By 2. 3d and 2. 3f, the direction p_i is given by $p_i = -2H(x_i) \nabla f(x_i)$. If we take $H(x_i) = I$, we get Cauchy's method [16], and if we take $H(x_i) = \nabla^2 f(x_i)^{-1}$, which is the $n \times n$ inverse Hessian matrix of second partial derivatives, we get the damped Newton method [16] which has a quadratic rate of convergence. However any matrix $H(x_i)$ satisfying condition 2. 3c will also work.

For later use and for comparison purposes it is convenient to state a quadratic programming problem the dual of which is problem 2. 3a

$$2. 8 \qquad \varphi(x_i, I(x_i)) := \min_{(\delta, s)}$$

$$\left\{ \delta \left| \begin{array}{l} \nabla f(x_i)s + \tfrac{1}{4} s\, H(x_i)^{-1} s \leq \delta \\[2mm] g_j(x_i) + \nabla g_j(x_i)s + \tfrac{1}{4} sH(x_i)^{-1} s \leq \delta, \quad j \in I(x_i) \end{array} \right. \right\}$$

where $I(x) = \{j \,|\, -\varepsilon \leq g_j(x) \leq 0\}$, the variable s is related to the variable q of 2. 3e by

$$2. 9 \qquad\qquad\qquad s = H(x_i)q \quad .$$

We observe that 2. 8 is a quadratically constrained problem whereas 2. 3a is a linearly constrained quadratic programming problem. This makes problem 2.3a considerably easier to solve by principal pivoting methods [7,8] or any of the recent fast quadratic programming algorithms [2,18]. Problems 2.3a and 2.8 are related through the following duality theorem.

2.10. <u>Duality Theorem.</u> If (η_i, u_i) solves 2.3a then $\delta_i = \theta(x_i, I(x_i))$ and $s_i = -2H(x_i)(\eta_i \nabla f(x_i) + u_i \nabla g(x_i))$ solves 2.8. Conversely if (δ_i, s_i) solves 2.8 then some (η_i, u_i) satisfying $\eta_i \nabla f(x_i) + u_i \nabla g(x_i) = -\frac{1}{2} H(x_i)^{-1} s_i$ solves 2.3a. In both cases $\theta(x_i, I(x_i)) = \varphi(x_i, I(x_i))$.

We point out here that the problem 2.8 is related to the Topkis-Veinott [19] subproblem in the following way. If the quadratic term in s is removed from the constraints in $I(x_i)$ and if $\varepsilon = \infty$, then 2.8 becomes the Topkis-Veinott direction finding problem given in their Theorem 3 [19]. We feel that our formulation 2.3a has two advantages over their formulation. 1) We consider only a subset of the constraints, that is the ε-active set, whereas they consider all the constraints. 2) The formulation 2.8 is impractical for taking into account quadratic terms, whereas 2.3a can accommodate such effects. We also point out the difference between our fixed-ε technique and the ε-halving method of Zoutendijk [21] and Zangwill [20].

If the ε-active constraints at each x are divided into concave and noncave constraints, that is

$$I(x) = I_1(x) \cup I_2(x), \quad I_1(x) = \{j \mid -\varepsilon \le g_j(x) \le 0, \ g_j \text{ is concave }\},$$

$$I_2(x) = \{j \mid -\varepsilon \le g_j(x) \le 0, \ g_j \text{ is not concave}\}$$

then problem 2.3a can be replaced by

$$2.3a' \quad -\theta(x_i, I(x_i)) := \operatorname*{minimum}_{\substack{(\eta, \, u_{I(x_i)}) \ge 0 \\ \eta + e u_{I_2}(x_i) = 1}} \{\|\eta \nabla f(x_i) + u_{I(x_i)} \nabla g_{I(x_i)}(x_i)\|^2_{H(x_i)} - u_{I(x_i)} g_{I(x_i)}(x_i)\}$$

and problem 2.8 by

$$2.8' \qquad \varphi(x_i, I(x_i)) := \min_{(\delta, s)}$$

$$\left\{ \delta \left| \begin{array}{ll} \nabla f(x_i)s + \tfrac{1}{4} s\, H(x_i)^{-1} s \leq \delta & \\[4pt] g_j(x_i) + \nabla g_j(x_i)s + \tfrac{1}{4} s\, H(x_i)^{-1} s \leq \delta, & j \in I_2(x_i) \\[4pt] g_j(x_i) + \nabla g_j(x_i)s \leq 0, & j \in I_1(x_i) \end{array} \right. \right\} .$$

All the above theorems hold for 2.3a and 2.8 replaced by 2.3a' and 2.8' respectively provided we make the additional assumption that the $\nabla g_j(x_i)$, $j \in I_1(x_i)$, are uniformly positively linearly independent, that is $\| u_{I_1(x)} \nabla g_{I_1(x)}(x) \|^2 \geq \omega \| u_{I_1(x)} \|^2$ for some $\omega > 0$ and all $u_{I_1}(x) \geq 0$. Problem 2.3a' simplifies further if we consider the case when $I_{2(x_i)}$ is empty. Then $\eta = 1$ and we have:

$$2.3a'' \qquad -\theta(x_i, I(x_i)) := \underset{u_{I_1(x_i)} \geq 0}{\text{minimum}}$$

$$\left\{ \| \nabla f(x_i) + u_{I_1(x_i)} \nabla g_{I_1(x_i)}(x_i) \|^2_{H(x_i)} - u_{I_1(x_i)} g_{I_1(x_i)}(x_i) \right\} .$$

We remark further that if in 2.8' we had that in each constraint in which $H(x_i)^{-1}$ appears, $H(x_i)^{-1}$ is a different matrix, say twice the Hessian of each function, as would be the case if we are taking quadratic approximations of f and g, then the corresponding dual problem 2.3a' would have an $H(x_i)$ given by

$$H(x_i) = \tfrac{1}{2}(\eta \nabla^2 f(x_i) + \sum_{j \in I_2(x_i)} u_j \nabla^2 g_j(x_i))^{-1} .$$

Problem 2.3a' would then be highly impractical to solve,

except for the case when $I_2(x_i)$ is empty, in which case $\eta = 1$ and we essentially have the counterpart of the constrained damped Newton method. However knowing the form of the matrix $H(x_i)$ given above which results in a Newton method, should help in constructing updating schemes, such as the variable metric method and others [11, 4], for constrained optimization.

A preliminary code using principal pivoting has been written by Toby J. Teorey and is being tested now. On Colville's problem number 1 [6] with $H(x_i) = I$, the time was .0057 standard units, better than any reported method. On test problem number 7 [6] again with $H(x_i) = I$, the time was .0202 standard units, again better than any reported method.

3. ALGORITHM CONVERGENCE THEOREM

In this section we give a theorem which can be utilized to prove a wide class of optimization algorithms, both constrained and unconstrained. We have not striven for as broad a generality as that of the Topkis-Veinott [19], Zangwill [20] or Polak [17] convergence algorithms. However, because our result includes a new fixed-ε antijamming technique, and because it uses the novel concept of a general necessary optimality condition, we have included it here rather than appealing to more general results.

3.1. <u>Algorithm Convergence Theorem.</u> Consider the problem $\min_{x \in X} f(x)$, where $X = \{x \mid x \in R^n,\ g(x) \leq 0\}$, f is a function with continuous first derivatives from R^n into R and g is a continuous function from R^n into R^m. Define a general optimality function $\theta(x, I(x))$ on X, not necessarily the same function of 2.3a, where $I(x) = \{j \mid -\varepsilon \leq g_j(x) \leq 0\}$ for some arbitrary but fixed $\varepsilon > 0$, as follows

3.2 $\qquad x \in X$, implies that $\theta(x, I(x)) \leq 0$

3.3 $\qquad \bar{x}$ solves $\min_{x \in X} f(x)$, implies that $\theta(\bar{x}, I(\bar{x})) = 0$.

3.4 For each $x \in X$ and fixed $J \subset \{1, 2, \ldots, m\}$, $\theta(x, J)$ is continuous in x .

3.5
$$x \in X, \quad J \subset L \subset \{1, 2, \ldots, m\}$$
implies that $\theta(x, J) \leq \theta(x, L)$.

Consider the following algorithm. Start with any x_0 in X . Having x_i determine x_{i+1} as follows:

3.6 <u>Direction Finding</u>. Choose any direction $p_i \in P$ where P is some compact set in R^n such that

   a) $x_i + \mu p_i \in X$ for all μ, $0 \leq \mu \leq 1$.

   b) $- \nabla f(x_i) p_i \geq \sigma(-\theta(x_i, I(x_i)))$

   where σ [5] is any increasing continuous function mapping $(0, \infty)$ into itself and such that $\sigma(0) = 0$.

3.7 <u>Step Size</u>. If $\nabla f(x_i) p_i = 0$, terminate, otherwise choose $x_{i+1} = x_i + \lambda_i p_i \in X$ according to any rule such that if (x_{i_j}, p_{i_j}) converges to $(\hat{x}, \hat{p})$ then $\nabla f(\hat{x})\hat{p} = 0$. (The step size methods defined in 2.4 have this property, see theorem 3.10 below.)

   Either the sequence $\{x_i\}$ generated by the algorithm 3.6-3.7 terminates at a stationary point $x_{\bar{i}}$, that is $\theta(x_{\bar{i}}, I(x_{\bar{i}})) = 0$, or each accumulation point $\bar{x}$ is a stationary point, that is $\theta(\bar{x}, I(\bar{x})) = 0$.

<u>Proof.</u> If for some $\bar{i}$, $\nabla f(x_{\bar{i}}) p_{\bar{i}} = 0$, then the algorithm terminates and, by 3.6b, $\theta(x_{\bar{i}}, I(x_{\bar{i}})) = 0$. Suppose now $-\nabla f(x_i) p_i > 0$ for all i, and let $\bar{x}$ be any accumulation point of $\{x_i\}$. Take any subsequence $\{x_i\}_{L_1}$ of $\{x_i\}$ converging to $\bar{x}$. Extract a further subsequence $\{x_i\}_{L_2}$ such that $\{x_i, p_i\}_{L_2}$ converges to $(\bar{x}, \bar{p})$. By 3.7

3.8
$$\nabla f(\bar{x})\bar{p} = 0$$
and by 3.6b

3.9 $\qquad -\nabla f(x_i)p_i \geq \sigma(-\theta(x_i, I(x_i)))$ $\qquad$ for $i \in L_2$.

Since there is a finite number of constraints $g_j(x) \leq 0$, $j = 1, 2, \ldots, m$, one subset $J \subset \{1, 2, \ldots, m\}$ must occur an infinite number of times in the sequence of sets $\{I(x_i)\}_{L_2}$. Extract a further subsequence $L_3 \subset L_2$ such that only J occurs in $\{I(x_i)\}_{L_3}$. Hence 3.9 becomes $-\nabla f(x_i)p_i \geq \sigma(-\theta(x_i, J))$ for $i \in L_3$ and in the limit $-\nabla f(\bar{x})\bar{p} \geq \lim \sup_{i \in L_3} \sigma(-\theta(x_i, J))$. By using 3.8 and the fact $\sigma(-\theta(x, J))$ is continuous in x (which follows from 3.4 and 3.6b) this last inequality becomes $0 \geq \sigma(-\theta(\bar{x}, J))$ which implies by 3.6b that $\theta(\bar{x}, J) = 0$. Now if $I(\bar{x}) \subset J$ and $I(\bar{x}) \neq J$, then for $j \in J$, $j \notin I(\bar{x})$ and $i \in L_3$, we have that $-\varepsilon \leq g_j(x_i) \leq 0$ and in the limit, $-\varepsilon \leq g_j(\bar{x}) \leq 0$, and hence $j \in I(\bar{x})$ which is a contradiction. So $J \subset \bar{I}(\bar{x})$ and by 3.5, and 3.2 we have $0 = \theta(\bar{x}, J) \leq \theta(\bar{x}, I(\bar{x})) \leq 0$. Hence $\theta(\bar{x}, I(\bar{x})) = 0$ and $\bar{x}$ is stationary. $\qquad\qquad$ Q.E.D.

We note here that 3.6a although not used explicitly in the proof, is needed in all the step size methods suggested in 2.4. Also the condition $\varepsilon > 0$ can be relaxed to $\varepsilon \geq 0$ in the above theorem, however in establishing 3.6b in practical cases (such as in the proof of algorithm 2.3-2.4 here) $\varepsilon > 0$ is needed to prove that 3.6b holds.

We also note that ε can be set to $+\infty$ in which case the above theorem holds with $I(x) = \{1, 2, \ldots, m\}$ for all $x \in X$. This is essentially the anti-jamming procedure of Topkis-Veinott [19] which is less efficient than the one proposed here.

We show now that the three step sizes proposed here 2.4a, 2.4b and 2.4c have the stated property in 3.7 above. (Step-size results in more generality are given in [9].)

3.10. <u>Step Size Theorem.</u> All step size methods, 2.4a, 2.4b and 2.4c have the property that if (x_{i_j}, p_{i_j}) converges to $(\hat{x}, \hat{p})$ then $\nabla f(\hat{x})\hat{p} = 0$.

<u>Proof.</u> (<u>Proof of 2.4a</u>). For $0 \le \mu \le 1$ we have that $f(x_{i_j} + \mu p_{i_j}) \ge f(x_{i_j+1}) \ge f(x_{i_{j+1}})$ and in the limit $f(\hat{x} + \mu\hat{p}) - f(\hat{x}) \ge 0$ for $0 \le \mu \le 1$. Since f is differentiable this implies that $\nabla f(\hat{x})\hat{p} \ge 0$, and by 3.6b (See A20) $\nabla f(\hat{x})\hat{p} \le 0$. Hence $\nabla f(\hat{x})\hat{p} = 0$.

(<u>Proof of 2.4b</u>). Since f has a Lipschitz continuous gradient with constant K, we have $f(x_i) - f(x_{i+1}) \ge \lambda_i(-\nabla f(x_i)p_i - \frac{K}{2}\lambda_i \| p_i \|^2)$. Hence

3.11 $\quad \dfrac{K}{2}\lambda_i \| p_i \|^2$

$$\le -(1 - \lambda_i)\nabla f(x_i)p_i \implies f(x_i) - f(x_{i+1}) \ge -\lambda_i^2 \nabla f(x_i)p_i$$

and

3.12 $\quad \dfrac{K}{2}\lambda_i \| p_i \|^2$

$$> -(1 - \lambda_i)\nabla f(x_i)p_i \impliedby f(x_i) - f(x_{i+1}) < -\lambda_i^2 \nabla f(x_i)p_i \quad .$$

If λ_i is the largest of $\{1, \frac{1}{2}, \frac{1}{4}, \dots \}$ satisfying the right inequality of 3.11 (which is the same as given under 2.4b) then either $\lambda_i = 1$ or $2\lambda_i$ must satisfy the right inequality of 3.12 and hence the left inequality of 3.12, that is $K\lambda_i \| p_i \|^2 > -(1 - 2\lambda_i)\nabla f(x_i)p_i$, but from 2.3d, 2.3e, 2.2c, 2.3c and 2.3f we have $\| p_i \| = \| \nu_i H(x_i)q_i \| \le 2 M_2 \alpha$ hence

$$\lambda_i > \frac{-\nabla f(x_i)p_i}{2(2 M_2^2 \alpha^2 K - \nabla f(x_i)p_i)} > 0 \quad .$$

Thus

$$\lambda_i \geq \min\{1, \frac{-\nabla f(x_i)p_i}{2(M_2^2\alpha^2 K - \nabla f(x_i)p_i)}\}.$$

But by 2.4b

$$f(x_i) - f(x_{i+1}) \geq -\nabla f(x_i)p_i(\min\{1, \frac{-\nabla f(x_i)p_i}{2(M_2^2\alpha^2 K - \nabla f(x_i)p_i)}\})^2$$

$$= \min\{-\nabla f(x_i)p_i, -(\frac{-\nabla f(x_i)p_i}{2(M_2^2\alpha^2 K - \nabla f(x_i)p_i)})^2 \nabla f(x_i)p_i\} > 0.$$

Let (x_{i_j}, p_{i_j}) converge to $(\hat{x}, \hat{p})$. Then since $f(x_i) - f(x_{i+1}) > 0$

$$f(x_{i_j}) - f(x_{i_{j+1}}) > f(x_{i_j}) - f(x_{i_j+1}) \geq \min\{-\nabla f(x_{i_j})p_{i_j},$$

$$-(\frac{-\nabla f(x_{i_j})p_{i_j}}{2(M_2^2\alpha^2 K - \nabla f(x_{i_j})p_{i_j})})^2 \nabla f(x_{i_j})p_{i_j}\}.$$ Since f is bounded

below, $\{f(x_i)\}$ converges and hence $\{f(x_{i_{j+1}}) - f(x_{i_j})\}$ converges to zero which forces the last term above to converge
to zero which in turn forces $\nabla f(x_{i_j})p_{i_j}$ to converge to 0.

(Proof of 2.4c) By the differentiability of f we have that

$$\gamma_i(\lambda) = \frac{f(x_i) - f(x_i + \lambda p_i)}{-\lambda \nabla f(x_i)p_i} = 1 + \tau(x_i, \lambda p_i)\|p_i\|/\nabla f(x_i)p_i$$

where $\lim_{\lambda \to 0} \tau(x_i, \lambda p_i) = 0$. So $\lim_{\lambda \to 0} \gamma_i(\lambda) = 1$. If $\gamma_i(1) \geq \rho$
then $\lambda_i = 1$. If $\gamma_i(1) < \rho \leq \frac{1}{2}$, then by the continuity of γ_i
and the fact that $\lim_{\lambda \to 0} \gamma_i(\lambda) = 1$, there exists some $\hat{\lambda}_i \epsilon (0, 1)$
such that

3.13 $$0 < \rho \leq \gamma_i(\hat{\lambda}_i) \leq 1 - \rho < 1 \ .$$

Let $(x_{i_j}, p_{i_j}, \lambda_{i_j})$ converge to $(\hat{x}, \hat{p}, \hat{\lambda})$. Now if $\hat{\lambda} = 0$, then $\lambda_{i_j} = \hat{\lambda}_{i_j}$ (defined by 3.13) an infinite number of times. Then for a further subsequence $\hat{\lambda}_{i_j}$ converges to $\hat{\lambda} = 0$ and by

3.13, $1 + \tau(x_{i_j}, \hat{\lambda}_{i_j}, p_{i_j}) \| p_{i_j} \| / \nabla f(x_{i_j}) p_{i_j} \leq 1 - \rho < 1$ which gives a contradiction in the limit. Hence $\hat{\lambda} > 0$. Now $f(x_{i_j}) -$

$$f(x_{i_{j+1}}) \geq f(x_{i_j}) - f(x_{i_j+1}) = -\gamma_{i_j}(\lambda_{i_j})\lambda_{i_j} \nabla f(x_{i_j}) p_{i_j} \geq$$

$-\rho \lambda_{i_j} \nabla f(x_{i_j}) p_{i_j} \geq 0$. Since f is bounded below, $\{f(x_i)\}$ converges and so does $\{f(x_{i_j})\}$. Hence $\{f(x_{i_j}) - f(x_{i_{j+1}})\}$ converges to zero which forces $\{-\nabla f(x_{i_j}) p_{i_j}\}$ to converge to zero, since $\{\lambda_{i_j}\}$ converges to $\hat{\lambda} > 0$. Q. E. D.

    If we assume that X is compact and convex, and at each x_i solve the problem: $\theta(x_i, I(x_i)) = \min_{q \in X} \nabla f(x_i) q$, and take $p_i = q_i - x_i$, where q_i is the solution of $\min_{q \in X} \nabla f(x_i) q$, we can easily check that conditions 3.2 to 3.6 are satisfied. Algorithm 3.1 then becomes the conditional gradient method of Demyanov-Rubinov [10] or for the case of linear g, the Frank-Wolfe algorithm [12].

APPENDIX

PROOFS OF THEOREMS OF SECTION 2

<u>Proof of Theorem 2.6</u> (Necessity). If $\bar{x}$ solves 2.1, then by the Fritz John Theorem [15, Theorem 2, p. 99] there exists $(\hat{\eta}, \hat{u}) \geq 0$ such that

Al. $\bar{\eta} \nabla f(\bar{x}) + \bar{u}_{I(\bar{x})} \nabla g_{I(\bar{x})}(\bar{x}) = 0, \ \bar{u} g(\bar{x}) = 0, \ g(\bar{x}) \leq 0, \ \bar{\eta} + e\bar{u} = 1 \ .$

These conditions imply that $u_{I(\bar{x})}g_{I(\bar{x})}(\bar{x}) = 0$ and hence $-\theta(\bar{x}, I(\bar{x})) = 0$. If any of the mentioned constraint qualifications are satisfied, then the Arrow-Hurwicz-Uzawa constraint qualification [15, p. 102] is satisfied [15, Lemma 6, p. 103] and hence [15, Theorem 7, p. 105] the Kuhn-Tucker conditions are satisfied at $\bar{x}$. (Sufficiency) If $\theta(\bar{x}, I(\bar{x})) = 0$, then $\bar{\eta}\nabla f(\bar{x}) + \bar{u}\nabla g(\bar{x}) = 0$, $\bar{u}_{I(\bar{x})}g_{I(\bar{x})}(\bar{x}) = 0$, $g(\bar{x}) \le 0$, $\bar{u}_{I(\bar{x})} \ge 0$. If $\bar{\eta} > 0$ then x solves the minimization problem [15, Theorem 2, p. 153].

Proof of Theorem 2.7.   We invoke Theorem 3.1 to prove the convergence of the algorithm 2.3-2.4. Hence we have to show that conditions 3.2 to 3.6 are satisfied. For notational simplicity we shall drop the indices $I(x_i)$ and i, and the arguments of the functions. All functions are evaluated at x_i. All g are $g_{I(x_i)}$, u are $u_{I(x_i)}$ and all x are x_i.

By using the Kuhn-Tucker condition [15, Theorem 7, p. 105] we have that at the solution (η, u) of the quadratic problem 2.3a, (η, u) and some real number ξ must satisfy the Kuhn-Tucker conditions (where a prime denotes a transpose of matrix).

A2. $$2\eta\|\nabla f\|_H^2 + 2u\nabla gH\nabla g - \xi \ge 0$$

A3. $$2\eta\nabla gH\nabla f + 2\nabla gH\nabla g'u - g - e\xi \ge 0$$

A4. $$2\eta^2\|\nabla f\|_H^2 + 2\eta u\nabla gH\nabla f - \xi\eta = 0$$

A5. $$2\eta u\nabla gH\nabla f + 2u\nabla gH\nabla g'u - ug - eu\xi = 0$$

A6. $$\eta + eu - 1 = 0.$$

By using A4, A5, A6 and 2.3a we get that

A7. $$0 \ge \theta(x, I(x)) = \|\eta\nabla f + u\nabla g\|_H^2 - \xi.$$

Hence

A8.
$$\xi \geq \| \eta \nabla f + u \nabla g \|_H^2$$

Using A7 and A2 gives

A9.
$$-\theta(x, I(x)) \leq -(\eta \nabla f + u \nabla g) H(\eta \nabla f + \nabla g'u - 2\nabla f) \ .$$

Let

A10.
$$q = -2(\eta \nabla f + u \nabla g) \ .$$

Note that by A9 and A10, if $q = 0$, then $0 \leq -\theta(x, I(x)) \leq 0$ and hence x is stationary and the algorithm terminates. So for $q \neq 0$, set

A11.
$$\beta(x) = \text{minimum} \left\{ 1, \ \frac{M_1 \xi}{M_2^2 K \|q\|_H^2}, \ \frac{-1 + (1 + \varepsilon K/\alpha^2)^{\frac{1}{2}}}{2KM_2} \right\} \ .$$

By A8 and A10 we have that

A12.
$$\frac{M_1 \xi}{M_2^2 K \|q\|_H^2} = \frac{M_1 \xi}{4M_2^2 K \|\eta \nabla f + u \nabla g\|_H^2} \geq \frac{M_1}{4M_2^2 K} \ .$$

Hence by A11 and A12

A13.
$$\beta(x) \geq \gamma := \text{minimum} \left\{ 1, \frac{M_1}{4M_2^2 K}, \ \frac{-1 + (1 + \varepsilon K/\alpha^2)^{\frac{1}{2}}}{2KM_2} \right\} > 0 \ .$$

Define

A14.
$$t: = \beta(x)Hq \ .$$

We will now show that $x + \mu t \in X$ for $0 \leq \mu \leq 1$.

For $j \in I(x)$, and $0 \leq \mu \leq 1$ we have (dropping the argument of β)

$$g_j(x+\mu t) = g_j(x+\mu\beta Hq)$$

A15.
$$\leqq g_j + \mu\beta\nabla g_j Hq + \frac{K\mu^2\beta^2}{2}\|Hq\|^2$$

$$\text{(by Lipschitz continuity of } \nabla g_j)$$

$$= \mu\beta(g_j + \nabla g_j Hq) + (1-\mu\beta)g_j + \frac{K\mu^2\beta^2}{2}\|Hq\|^2$$

$$\leqq \mu\beta(-\xi + \frac{K\mu\beta M_2^2}{2M_1}\|q\|_H^2)$$

$$\text{(by A3, } g_j \leqq 0, \text{ and } 0 < \beta \leqq 1)$$

$$\leqq \frac{-\mu\beta\xi}{2} \quad \text{(by A11 since } \frac{\mu\beta}{2} \leqq \frac{\beta}{2} \leqq \frac{M_1\xi}{2M_2^2 K\|q\|_H^2})$$

$$\leqq 0 .$$

For $j \notin I(x)$, and $0 \leqq \mu \leqq 1$ we have that

$$g_j(x+\mu t) \leqq g_j + \mu\beta\nabla g_j Hg + \frac{K\mu^2\beta^2}{2}\|Hq\|^2 \quad \text{(by A15 above)}$$

$$< -\varepsilon + \mu\beta\nabla g_j Hq + \frac{K\mu^2\beta^2}{2}\|Hq\|^2 \quad \text{(since } j \notin I(x))$$

$$\leqq -\varepsilon + \mu\beta\alpha M_2 2\alpha + \frac{K\mu^2\beta^2}{2} M_2^2 4\alpha^2 \quad \text{(by A10, } \|q\| \leqq 2\alpha)$$

$$= [-\frac{\varepsilon}{2} + 2\alpha^2 M_2\mu\beta + 2\alpha^2 M_2^2 K\mu^2\beta^2] - \frac{\varepsilon}{2}$$

$$\leqq -\frac{\varepsilon}{2} < 0$$

where the next to the last inequality follows from the fact
that the strictly convex quadratic expression (in $\mu\beta$) in the

square bracket has a negative root and a positive root equal
to $\dfrac{-1 + (1 + \varepsilon K/\alpha^2)^{\frac{1}{2}}}{2KM_2}$ and by A11 $\mu\beta$ is less than or equal this
positive root, so the square-bracketed quadratic expression
is nonpositive.

Hence we have just shown that $x + \mu\beta(x)\,Hq \in X$ for $0 \le \mu \le 1$ and by A13

A16. $\qquad x + \mu\gamma Hq \in X$ for $0 \le \mu \le 1$ $\quad (\gamma > 0)$.

But by 2.3f and A16 we have that

A17. $x + \mu\nu Hq \in X$, $\quad \nu = \max\{1, \frac{1}{2}, \frac{1}{4}, \ldots\}$, $\quad$ for $0 \le \mu \le 1$.

So we either have that $\nu = 1$ or $2\nu > \gamma$ (otherwise by A16
we would have had $x + 2\nu Hq \in X$) . Hence

A18. $\qquad\qquad \nu \ge \min\{1, \frac{\gamma}{2}\} =: \gamma' > 0$.

Since $p = \nu Hq$, A17 implies that

A19. $\qquad\qquad x + \mu p \in X$ for $0 \le \mu \le 1$

which is condition 3.6a of Theorem 3.1. To verify condition
3.6b we note that by A9 and A10 we have $-\theta(x, I(x)) \le -\frac{1}{4}\|q\|_H^2 - q H \nabla f$. Hence

A20. $\quad -\nabla fp = -\nu\nabla fHq \ge$

$$-\nu\theta(x, I(x)) + \frac{\nu}{4}\|q\|_H^2 \ge -\nu\theta(x, I(x)) \ge -\gamma'\theta(x, I(x))$$

and since, by A18, γ' is a positive number, condition 3.6b
is established.

It only remains to verify conditions 3.2 to 3.5. Condition 3.2 follows from the fact that for x in X, $g_{I(x)}(x) \le 0$
and hence $\theta(x, I(x)) \le 0$. Condition 3.3 follows from the
fact that if $\bar{x}$ is optimal then by the Fritz John theorem [15,
Theorem 2, p. 99] there exist $(\bar{\eta}, \bar{u})$ such that

$$\bar{\eta}\nabla f(\bar{x}) + \bar{u}\nabla g(\bar{x}) = 0, \quad \bar{u}g(\bar{x}) = 0, \quad (\bar{\eta}, \bar{u}) \geq 0, \quad g(\bar{x}) \leq 0, \quad \bar{\eta} + e\bar{u} = 1.$$

Hence $\theta(\bar{x}, I(\bar{x})) = 0$. Condition 3.4 follows from [3, Maximum theorem, p. 116] since (η, u) lie in the fixed compact set $\{(\eta, u): \eta + eu = 1, \ (\eta, u) \geq 0\}$, and the objective function of 2.3a is continuous in $(\bar{x}, \eta, u)$. Condition 3.5 follows from the fact that if $J \subset L$, then for $(\eta, u_J) \geq 0$, $\eta + eu_J = 1$, (η, u_L) defined by $(\eta, u_L) = (\eta, u_J, 0_{j \notin J})$ also satisfies $(\eta, u_L) \geq 0$ and $\eta + eu_L = 1$ and gives the same value of the objective function of 2.3a, hence $-\theta(x, L) \leq -\theta(x, J)$.

Q. E. D.

<u>Proof of Theorem 2.10.</u> By direct application of the duality theorems of non-linear programming [15, Section 8.1, Theorems 4 and 6] theorem 2.10 holds for 2.8 as a primal problem and its dual as

$$\max_{s, \eta, u} \quad -\tfrac{1}{4} s H(x_i)^{-1} s + u_{I(x_i)} g_{I(x_i)}(x_i)$$

$$\text{subject to} \quad \eta\nabla f(x_i) + u_{I(x_i)}\nabla g_{I(x_i)}(x_i) + \tfrac{1}{2}H(x_i)^{-1} s = 0$$

$$\eta + eu_{I(x_i)} = 1$$

$$(\eta, u_{I(x_i)}) \geq 0 \ .$$

Substitution of the first constraint in the objective function and changing it to a minimization problem gives problem 2.3a. We note that $s = -2H(x_i)(\eta\nabla f(x_i) + u_{I(x_i)}\nabla g_{I(x_i)}(x_i)) = H(x_i)q$,

as given by 2.9.

Q. E. D.

NOTE ADDED IN PROOF

Related algorithms are also considered in: O. Pironneau and E. Polak: Rate of convergence of a class of methods of feasible directions, Memorandum No. ERL-M301, University of California, Berkeley, 26 July 1971.

BIBLIOGRAPHY

1.   .L. Armijo: Minimization of functions having Lipschitz continuous first partial derivatives, Pacific J. Math. 16, 1966, 1-3.

2.   R. H. Bartels, G. H. Golub and M. A. Saunders: Numerical techniques in mathematical programming, in "Nonlinear Programming" ed. J. B. Rosen, O. L. Mangasarian and K. Ritter. Academic Press, 1970, 123-176.

3.   C. Berge: Topological Spaces, MacMillan, New York, 1963.

4.   C. G. Broyden: A new method of solving nonlinear simultaneous equations, Comp. J. $\underline{12}$, 1969, 94-99.

5.   J. Céa: Opimisation theorie et algorithmes. Dunod, Paris, 1971.

6.   A. R. Colville: A comparative study on nonlinear programming codes, IBM New York Scientific Center Technical Report No. 320-2949, June 1968.

7.   R. W. Cottle: The principal pivoting method of quadratic programming, in Mathematics of the decision sciences, part 1, ed. G. B. Dantzig and A. F. Veinott, Amer. Math. Soc., Providence, R. I., 1968, 144-162.

8.   R. W. Cottle and G. B. Dantzig: Complementary pivot theory of mathematical programming, Linear Algebra and Appl. $\underline{1}$, 1968, 103-125.

9.   J. W. Daniel: Convergent step-sizes for gradient-like feasible direction algorithms for constrained optimization, in Nonlinear programming, ed. J. B. Rosen, O. L. Mangasarian and K. Ritter, Academic Press, 1970, 245-274.

10. V. F. Demyanov and A. M. Rubinov: The minimization of a smooth convex function on a convex set, SIAM J. Control 5, 1967, 280-294.

11. R. Fletcher and M. J. D. Powell: A rapidly convergent descent method for minimization, Comp. J. 6, 1963, 1963-168.

12. M. Frank and P. Wolfe: An algorithm for quadratic programming, Naval Res. Log. Quart. 3, 1956, 95-110.

13. A. A. Goldstein: Constructive real analysis, Harper and Row, New York, 1967.

14. E. S. Levitin and B. T. Poljak: Constrained minimization methods, USSR Comp. Math. and Math. Phys. 6, 1966, No. 5, 1-50.

15. O. L. Mangasarian: Nonlinear programming, McGraw-Hill, New York, 1969.

16. J. M. Ortega and W. C. Rheinboldt: Iterative solution of nonlinear equations of several variables, Academic Press, New York, 1970.

17. E. Polak: Computational methods in optimization, Academic Press, New York, 1971.

18. J. Stoer: On the numerical solution of constrained least-squares problems, SIAM J. Num. Anal. 8, 1971, 382-411.

19. D. M. Topkis and A. F. Veinott: On the convergence of some feasible direction algorithms for nonlinear programming, SIAM J. Control 5, 1967, 268-279.

20. W. I. Zangwill: Nonlinear programming, Prentice-Hall, Englewood Cliffs, N. J., 1969.

21. G. Zoutendijk: Methods of feasible directions, Elsevier, Amsterdam, 1960.

ON A CLASS OF NUMERICAL METHODS WITH AN ADAPTIVE
INTEGRATION SUBPROCEDURE FOR OPTIMAL CONTROL PROBLEMS

E. Polak

Department of Electrical Engineering and Computer Sciences
and the Electronics Research Laboratory,
University of California, Berkeley, California 94720

1. Introduction

Whenever an optimization algorithm is applied to an
optimal control problem, the values and gradients, of the
cost and boundary constraint functions, must be computed at
each iterations. Now, to evaluate these functions and their
gradients, one must solve the system differential equation
together with a number of linear perturbation differential
equations. Because of this, the major portion of the computing time is spent on integrating differential equations.

In Appendix A of [1], a general scheme for incorporating approximations into an algorithm was described. The
important feature of this scheme is that it is adaptive:
coarse approximations are used when far from a solution and
the approximations are refined adaptively, in response to a
test, as a solution is approached. It was shown in [2]
that this adaptive scheme can be used to control the integration step size in a gradient method for optimal control
problems, with a reduction in computing time by a factor of
4 or more.

The approximation scheme described in Appendix A of [1]
controls the precision of approximation by means of a single
parameter j which usually denotes the number of iterations
the function evaluation subprocedures have to run within a

Research sponsored by the National Aeronautics and Space
Administration Grant NGL-05-033-016, the Joint Services
Electronics Program, Contract F44620-71-C-0087, and the
National Science Foundation, Grant GK-10656X1.

given iteration of the main algorithm. It turns out that this approach is inadequate for use with algorithms of the feasible directions type. Because of this, we begin by developing a two parameter approximation scheme which we present in the form of an algorithm prototype and establish conditions for its convergence. We then illustrate the use of this scheme by showing that it can be used to construct two implementations for the Pironneau-Polak method [3] of feasible directions for the solution of constrained optimal control problems.

2. A General Approach

Consider the problem

$$\min\{f^0(u) \,|\, f^j(u) \le 0, \ j = 1,2,\ldots,m\}, \qquad 2.1$$

where $f^j\colon L_\infty^s[0,T] \to \mathbb{R}^1$ are defined as follows:[†]

$$f^j(u) = g^j(x(T,u)), \ j = 0,1,2,\ldots,m, \qquad 2.2$$

with $g^j\colon \mathbb{R}^r \to \mathbb{R}^1$, $j = 0,1,2,\ldots,m$, are continuously differentiable and $x(T,u)$ is the solution at T of the differential equation

$$d/dt \ x(t) = h(x(t),u(t)), \ t \in [0,T], \ x(0) = \xi, \qquad 2.3$$

where $h\colon \mathbb{R}^r \times \mathbb{R}^s \to \mathbb{R}^r$ is continuously differentiable.

2.4 <u>Assumptions</u>: (i) The set $C = \{x \in \mathbb{R}^r \,|\, g^\ell(x) \le 0, \ \ell = 1,2,\ldots,m\}$ is compact; (ii) for every $M > 0$ there exists a Lipshitz constant $L(M) > 0$ such that

$$\|h(x',u) - h(x'',u)\| \le L(M)\|x'-x''\| \qquad 2.5$$

for all x', x'' in $\mathbb{R}^r$ and for all $u \in \mathbb{R}^s$ such that $\|u\| \le M$; (iii) for every $u \in L_\infty^s[0,T]$, (2.3) has a unique solution. ◻

2.6 <u>Proposition</u>: Suppose that (2.4) (ii), (iii) hold. Then, for $j = 0,1,2,\ldots,M$ and any $u \in L_\infty^s[0,T]$, the Frechet derivative $df^j(u)(\cdot)$ of $f^j(\cdot)$ exists and is given by

$$df^j(u)(\delta u) = \int_0^T \langle \nabla f^j(u)(t),\delta u(t)\rangle \ dt, \qquad 2.7$$

[†]$L_\infty^s[0,T]$ is the usual space of equivalence classes of functions from [0,T] into $\mathbb{R}^s$, with norm $|u|_\infty = \text{ess sup}_{t\in[0,T]} \|u(t)\|$, and $\|\cdot\|$ denotes the euclidean norm.

where $\langle \cdot , \cdot \rangle$ denotes the euclidean scalar product and

$$\nabla f^j(u)(t) = [\partial h/\partial u \ (x(t,u),u(t))]^T p_j(t,u), t \in [0,T], \quad 2.8$$

with $x(t,u)$ defined by (2.3) and $p_j(t,u)$ defined by

$$d/dt \ p_j(t,u) = - [\partial h/\partial x \ (x(t,u),u(t))]^T p_j(t,u), t \in [0,T],$$
$$p_j(T,u) = [\partial g^j/\partial x \ (x(T,u))]^T, \quad 2.9$$

(For a proof see [2]). ¤

We now state an obvious generalization of the F. John optimality condition [4].

2.10 <u>Lemma</u>: Suppose that $\hat{u}$ is optimal for (2.1). Then there exist multipliers $\hat{\mu}^0, \hat{\mu}^1, \ldots, \hat{\mu}^m$ such that

$$\sum_{\ell=0}^{m} \hat{\mu}^\ell \nabla f^\ell(\hat{u}) = 0, \quad 2.11$$

$$\hat{\mu}^\ell f^\ell(\hat{u}) = 0, \ \ell = 1,2,\ldots,m, \quad 2.12$$

$$\sum_{\ell=0}^{m} \hat{\mu}^\ell = 1; \ \mu^\ell \geq 0, \ \ell = 0,1,2,\ldots,m. \quad ¤ \quad 2.13$$

We shall denote by Δ the set of feasible points for (2.1) satisfying (2.11)-(2.13) i.e.,

$$\Delta = \{u \in L_\infty^s[0,T] \mid f^j(u) \leq 0, \ j = 1,2,\ldots,m; \ \phi(u) = 0\}, \quad 2.14$$

where

$$\phi(u) = \max\{ \sum_{j=1}^{m} \mu^j f^j(u) - 1/2 \ \| \sum_{j=0}^{m} \mu^j \nabla f^j(u)\|^2$$
$$\mu^j \geq 0, \ j = 0,1,2,\ldots,m, \ \sum_{j=0}^{m} \mu^j = 1\}. \quad 2.15$$

Realistically, we can only hope to compute a $\hat{u} \in \Delta$.

For the sake of simplicity, we shall assume that in the execution of the algorithms to be presented, all of our integrations are to be performed using the Euler-Cauchy formula, though it is not hard to see that the general results of this paper apply to predictor-corrector and Runge-Kutta formulas as well. Thus, for $j = 1,2,3,\ldots$, let

$$v_j(t) = 1 \text{ for } t \in [0, T/2^j]$$
$$= 0 \text{ otherwise}, \quad 2.16$$

and let U_j be the set of all 2^j segment ramp functions, i.e. for $j = 1,2,3,\ldots$

$$U_j = \{u \,|\, u(t) = \sum_{\ell=0}^{2^j-1} [\omega_\ell + (t-\ell T/2^j)(2^j/T)(\omega_{\ell+1}-\omega_\ell)]v_j(t-\ell T/2^j) \,|$$
$$\omega_\ell \in \mathbb{R}^s, \ \ell = 0,1,2,\ldots,2^j; t \in [0,T]\}, \quad 2.17$$

Next, for $j = 1,2,3,\ldots,$ and $u \in U_j$, we define $x_j(\ell T/2^j,u)$, $\ell = 0,1,2,\ldots,2^j$, as the solution of the Euler-Cauchy integration formula

$$x_j((\ell+1)T/2^j,u) = x_j(\ell T/2^j,u)$$
$$+ (T/2^j)h(x(\ell T/2^j,u), \ u(\ell T/2^j)),$$
$$\ell = 0,1,2,\ldots,2^{j-1}, \qquad\qquad 2.18$$

with $x_j(0,u) = \xi$. Thus, our algorithms will only be able to operate on the approximation sets U_j and they will construct approximate trajectories $\overline{x}_j(t,u)$ defined for $u \in U_j$ and $t \in [0,T]$ by

$$\overline{x}_j(t,u) = \sum_{\ell=0}^{2^j-1} [x_j(\ell T/2^j)$$
$$+ (2^j/T)(t-\ell T/2^j)(x_j((\ell+1)T/2^j,u)$$
$$- x_j(\ell T/2^j,u))]v_j(t-\ell T/2^j). \qquad 2.19$$

Next, let

$$D = \{u \in L^s_\infty[0,T] \,|\, f^j(u) \leq 0, \ j = 1,2,\ldots,m\} \qquad 2.20$$

and for $j = 1,2,3,\ldots$ and $k = 1,2,3,\ldots,$ let

$$D_{j,k} = \{u \in U_j \,|\, g^\ell(x_j(T,u)) \leq -1/2^k, \ \ell = 1,2,\ldots,m\}.$$

To establish the properties of the sets $D_{j,k}$, we need the following well known result.

2.21 <u>Lemma</u>: Suppose that assumptions (2.4)(ii) and (2.4)(iii) are satisfied. Then for every $M > 0$ there exists a $Q > 0$ such that

$$\|x(T,u) - x_j(T,u)\| \leq Q/2^j, \quad j = 1,2,3,\ldots \qquad 2.22$$

for all $u \in \{u' \in U_j \mid |u_j|_\infty \leq M\}$ (See corollary to theorem 1 on p. 369 of [5]). ¤

Since the set C is compact by (2.4)(i), the following corollary to (2.21) follows.

2.23 <u>Corollary</u>: Suppose that assumptions (2.4)(i)-(iii) are satisfied. For any $M > 0$, let $D^M = D \cap \{u \in L_\infty^s[0,T] \mid |u|_\infty \leq M\}$ and let $D_{j,k}^M = D_{j,k} \cap \{u \in L_\infty^s[0,T] \mid |u|_\infty \leq M\}$. Then there exist integers j_0, k_0, possibly depending on M, such that

$$D^M_{j_0+\ell_1,k_0+\ell_2} \subset D^M \qquad \forall \ell_1 \geq \ell_2, \; \ell_2 = 0,1,2,\ldots \qquad 2.24$$

$$D^M_{j_0+\ell_1+1,k_0+\ell_2+1} \supset D^M_{j_0+\ell_1,k_0+\ell_2}$$

$$\forall \ell_1 \geq \ell_2, \ell_2 = 0,1,2,\ldots, \qquad 2.25$$

and in addition,

$$\lim_{\substack{j\to\infty \\ k\to\infty}} D^M_{j,k} = D^M. \qquad \text{¤.} \qquad 2.26$$

Now, for $j = 1,2,3,\ldots$, let $f_j^\ell: U_j \to \mathbb{R}^1$ be defined by

$$f_j^\ell(u) = g^\ell(x_j(T,u)), \quad \ell = 0,1,2,\ldots,m, \qquad 2.27$$

and suppose that we have a family of iteration functions $A_{j,k}: D_{j,k} \to D_{j,k}$, $j = 1,2,\ldots$, $k = 1,2,\ldots$. Then, the results in Section A1 of [1] suggest, by extension, that algorithms of the form below can be used to find points in Δ. The algorithm below is more complex than the one in Section A1 of [1].

2.28 <u>Algorithm Prototype</u>.

<u>Step 0</u>: Select integers $j_0 \geq 1$, $k_0 \geq 1$ and parameters $\varepsilon_0 > 0$, $\alpha \in (0,1)$; compute a $u_0 \in D_{j_0,k_0}$, and set $\varepsilon = \varepsilon_0$, $i = 0$, $j = j_0$, $k = k_0$.

<u>Step 1</u>: Compute $y = A_{j,k}(u_i)$.

<u>Step 2</u>: If $f_j^0(y) - f_j^0(u_i) \leq -\varepsilon$, go to step 4; else, set $j = j+1$, $\varepsilon = \alpha\varepsilon$ and go to step 3.[+]

[+]Alternatively, set $j = j+2$, $k = k+1$, $\varepsilon = \alpha\varepsilon$.

<u>Step 3</u>: If $u_i \in D_{j,k+1}$, set $k = k+1$ and go to step 1; else, compute a $\bar{u} \in D_{j+\ell_1,k+\ell_2}$, $\ell_1 \geq \ell_2 \geq 0$, set $u_i = \bar{u}$ and go to step 1.

<u>Step 4</u>: Set $u_{i+1} = y$, set $i = i+1$ and go to step 1. ¤

The properties of algorithm (2.28) are obviously determined by the properties of the approximation functions f_j^0 and of the iteration function $A_{j,k}$. For the former, the following holds because of (2.4)(i) as a corollary to (2.22).

2.29 <u>Lemma</u>: For every $M > 0$, there exists a $\bar{Q} > 0$ such that for $s = 0,1,2,\ldots,m$,

$$\left| f_j^s(u) - f^s(u) \right| \leq \bar{Q}/2^\ell \qquad \forall u \in D_{j,k}^M, \forall j \geq \ell \qquad ¤ \qquad 2.30$$

We must endow the maps $A_{j,k}$ with the following properties.

2.31 <u>Assumption</u>: There exists an $M > 0$ satisfying $D^M \cap \Delta \neq \phi$, such that for every $u \in D^M$, $u \notin \Delta$, there exists an $\varepsilon(u) > 0$, a $\delta(u) > 0$ and integers $J(u) \geq 0$, $K(u) \geq 0$ satisfying

$$f_j^0(A_{j,k}(u')) - f_j^0(u') \leq - \delta(u), \qquad 2.32$$

for all $u' \in \{u'' \in D^M \cap D_{j,k}^M | \ |u''-u|_\infty \leq \varepsilon(u)\}$ for all $j = J(u) + \ell_1$, for all $k = K(u) + \ell_2$, $\ell_1 \geq \ell_2$, $\ell_2 = 0,1,2,\ldots$ ¤.

The algorithm model (2.28) has the following properties.

2.33 <u>Lemma</u>: Suppose that the algorithm (2.28) constructs an infinite sequence $\{u_i\}_{i=0}^\infty$, then there exists an $i_0 \geq 0$ such that the test $u_i \in D_{j,k+1}$ is satisfied in step 3 for all $i \geq i_0$ (i.e. a point $\bar{u}$ will be constructed at most a finite number of times.

<u>Proof</u>: Since each time the test $u_i \in D_{j,k+1}$ fails, j-k is increased by 1 and j-k is kept constant otherwise, the lemma follows from (2.25). ¤.

2.34 <u>Lemma</u>: Suppose that assumption (2.21) is satisfied and that algorithm (2.9) jams up at a control u_s, cycling between steps 1 and 3 so that a control u_{s+1} cannot be constructed. Then $u_s \in \Delta$.

Proof: First we note that because of (2.25), the algorithm cannot jam up in a cycle which constructs an infinite sequence of points $\bar{u}$. The lemma now follows from (2.31) by analogy with the proof of theorem (A.1.26) in [1]. ¤

The above two lemmas show that the algorithm is well defined. Next we progress to take a look at its convergence properties.

2.35 <u>Theorem</u>: Suppose that assumption (2.31) is satisfied and suppose that algorithm (2.28) has constructed a bounded sequence $\{u_i\}$. If $\{u_i\}$ is finite because the algorithm has jammed up at its last element u_s, then $u_s \in \Delta$. If $\{u_i\}$ is infinite, then every accumulation point u^* of $\{u_i\}_{i=0}^{\infty}$ is in Δ.

Proof: Since $\{u_i\}$ is bounded and (2.30) holds, lemmas (A.1.7) and (A.1.17) in [1] will be satisfied. Hence, the theorem follows from these lemmas and (2.31) by analogy with the proof of theorem (A.1.26) in [1]. ¤

3. Implementations of the Pironneau-Polak Algorithm

Suppose that the set $\{u \in L_{\infty}^{s}[0,T] \mid f^j(u) < 0,\ j = 1, 2,\dots,m\}$ is not empty. Then problem (2.1) can be solved by means of the algorithm below, provided that all integrations are performed with a high degree of precision, since it does not take into account errors due to numerical integration.

3.1 <u>Algorithm (Pironneau-Polak [3])</u>.[†]

Step 0: Select parameters $\alpha \in (0,1)$, $\beta \in (0,1)$, $\lambda > 0$. Compute a $u_0 \in \{u \in L_{\infty}^{s}[0,T] \mid f^j(u) \le 0,\ j = 1,2,\dots,m\}$ and set $i = 0$.

Comment: To compute such a u_0 apply algorithm (3.1) to the problem $\min\{z^0 \mid f^j(z) \le z^0,\ j = 1,2,\dots,m\}$. To initialize (3.1) for this problem, take z_0 to be a good guess at a feasible control for (2.1) and set $z_0^0 = \max\{f^j(z_0) \mid j = 1,2,\dots,m\}$.

Step 1: Compute $\phi(u_i)$ and $\mu^1(u_i) \ge 0$, $\mu^2(u_i) \ge 0,\dots,$ $\mu^m(u_i) \ge 0$ as a solution of the quadratic program

---

[†]This version uses a somewhat different step size rule from the one in [3].

$$\phi(u_i) = \max\{ \sum_{j=1}^{m} \mu^j f^j(u_i) \qquad\qquad 3.2$$

$$- (1/2) \int_0^T \| \sum_{j=0}^{m} \mu^j \nabla f^j(u_i)(t)\|^2 dt \,|\, \mu^j \geq 0,$$

$$j = 0,1,2,\ldots,m; \quad \sum_{j=0}^{m} \mu^j = 1\},$$

where the $\nabla f^j(u_i)(\cdot)$ are defined as in (2.8).

<u>Step 2</u>: If $\phi(u_i) = 0$, stop; else, set

$$d(u_i) = - \sum_{j=0}^{m} \mu^j(u_i)\nabla f^j(u_i). \qquad\qquad 3.3$$

<u>Comment</u>: Although the $\mu^j(u_i)$ may not be uniquely determined by (3.2), the resulting $h(u_i)$ can be shown to be unique.

<u>Step 3</u>: Find the smallest integer $k(u_i) \geq 0$ such that
$$f^j(u_i+\lambda\beta^{k(u_i)} h(u_i)) \leq 0, \quad j = 1,2,\ldots,m, \text{ and}$$

$$f^0(u_i+\lambda\beta^{k(u_i)} d(u_i)) - f^0(u_i)$$

$$- \lambda\beta^{k(u_i)} \alpha \int_0^T \langle d(u_i)(t), \nabla f^0(u_i)(t) \rangle \, dt \leq 0.$$

<u>Step 4</u>: Set $u_{i+1} = u_i+\lambda\beta^{k(u_i)} d(u_i)$.

<u>Step 5</u>: Set $i = i+1$ and go to step 1. ◘

  If we set $k_0 = \infty$ and $j_0 = \infty$ and $\varepsilon_0 = 0$ in algorithm prototype (2.28), we obtain a simpler prototype without approximations for which theorem (2.35) still holds (see Sec. 1.3 in [1]). Algorithm (3.1) is of that form and satisfies the assumptions of theorem (2.35) under this simplification, hence the conclusions of theorem (2.35) hold also for algorithm (3.1). However, it so happens that one can establish a somewhat stronger result for (3.1). (See [3]).

3.4 <u>Theorem</u>: Let $\{u_i\}$ be a sequence of controls constructed by algorithm (3.1) in the process of solving problem (2.1) under the assumptions stated. If $\{u_i\}$ is finite, then its last element is in Δ. If $\{u_i\}$ is infinite and bounded and there exists a control u^* such that for some infinite subsequence $J \subset \{0,1,2,\ldots,m\}$, either

$$\lim_{\substack{i \in J \\ i \to \infty}} |u_i - u^*|_\infty = 0 \quad \text{or} \quad \lim_{\substack{i \in J \\ i \to \infty}} \int_0^T \|u_i(t) - u^*(t)\|^2 dt = 0, \quad \text{then } u^*$$

$\in \Delta$. ¤

To account for the effects of numerical integration and to avoid wasting time integrating too precisely in the early iterations, we must extend algorithm (3.1) into a form which corresponds to the prototype (2.28). For this purpose, we define gradients approximations $\nabla_j f^\ell(u)(\cdot)$ to the gradients $\nabla f^\ell(u)(\cdot)$ as follows. For any $u \in U_j$, let $x_j(\ell T/2^j, u)$, $\ell = 0,1,2,\ldots,2^j-1$ be defined by (2.18) and let $p_{qj}(\ell T/2^j, u)$, $\ell = 0,1,2,\ldots, 2^j-1$, $q = 0,1,2,\ldots,m$, be defined by

$$p_{qj}(\ell T/2^j, u) = p_{qj}((\ell+1)T/2^j, u) \qquad 3.5$$

$$+ (T/2^j)\,\partial h/\partial x(x(\ell T/2^j, u), u(\ell T/2^j))((\ell+1)T/2^j, u),$$

with

$$p_{qj}(T,u) = [\partial g^q/\partial x(x_j(T,u))]^T, \quad q = 1,2,\ldots,m. \qquad 3.6$$

Finally, for $q = 0,1,2,\ldots,m$ and $\ell = 0,1,2,\ldots,2^j-1$ let

$$\gamma_{q\ell}(u) = [\partial h/\partial u(x((\ell+1)T/2^j, u), u(\ell T/2^j))]^T \times$$

$$p_{qj}((\ell+1)T/2^j, u). \qquad 3.7$$

Then for $u \in U_j$, and for $t \in [0,T]$, we define

$$\nabla_j f^q(u)(t) = \sum_{\ell=0}^{2^j-1} [\gamma_{q\ell} + (2^j/T)(t - \ell T/2^j)(\gamma_{q(\ell+1)} - \gamma_{q\ell}) \times$$

$$v_j(t - \ell T/2^j)], \quad q = 0,1,2,\ldots,m. \qquad 3.8$$

The following result was established in [2].

3.9 <u>Lemma</u>: For every $M > 0$, for every $\gamma > 0$ and every $u \in B_M \triangleq \{u \mid |u|_\infty \leq M\}$, there exists an $\varepsilon > 0$ and an integer $N \geq 0$ such that for $\ell = 0,1,2,\ldots,m, |\nabla_j f^\ell(u') - \nabla f^\ell(u')|_\infty \leq \gamma$, for all $u' \in B_M$, $|u' - u|_\infty \leq \varepsilon$, for all

$j \geq N$. $\blacksquare$

The continuity of $d(\cdot)$, as defined by (3.3), together with (3.9) and (2.20) ensure that the algorithm below satisfies assumption (2.31) which is required for theorem (2.35) to hold.

3.10 <u>Implementation I.</u>

<u>Step 0</u>: Select integers $j_0 \geq 1$, $k_0 \geq 1$ and parameters $\varepsilon_0 > 0$, $\alpha \in (0,1)$, $\beta \in (0,1)$, $\lambda > 0$, $\lambda_{min} \in (0,\lambda)$; an initial guess $u_0 \in U_{j_0}$. Set $i = 0$, set $j = j_0$, set $k = k_0$, set $\varepsilon = \varepsilon_0$.

<u>Step 1</u>: Call algorithm (3.10) as a subroutine and go to step 1.1.

<u>Step 1.1</u>: Set ε_0, α, β, λ as in step 0; set $j_0 = j$, $k_0 = k$, $i = 0$ (set $i = 0$ in subroutine only!).

<u>Step 1.2</u>: Apply algorithm (3.10) to the problem $\min\{\bar{u}^0 \mid f^\ell(\bar{u}) - \bar{u}^0 \leq 0$, $\ell = 1,2,\ldots,m\}$, with the initial point $\bar{u}_0 = u_i$, $\bar{u}_0^0 = \max_{j_0} f^\ell(u_i) - 1/2^{k_0}$, until a $(\bar{u}_s^0,\bar{u}_s)$ is computed such that $\bar{u}_s^0 < 0$, with $\bar{u}_s \in D_{\bar{j},\bar{k}}(\bar{j} = j+\ell_1$, $\bar{k} = k+\ell_2,\ell_1 \geq \ell_2)$.

<u>Step 1.3</u>: For value of i as in main program, set $u_i = \bar{u}_s$, set $j = \bar{j}$, $k = \bar{k}$ and go to step 2 (end of subroutine).

<u>Step 2</u>: For $\ell = 0,1,2,\ldots,m$, compute $f_j^\ell(u_i)$, $\nabla_j f^\ell(u_i)$.

<u>Step 3</u>: Compute $\phi_j(u_i)$ and $\mu_j^\ell(u_i)$, $\ell = 0,1,\ldots,m$, by solving

$$\phi_j(u_i) = \max\{ \sum_{\ell=1}^m \mu^\ell f_j^\ell(u_i) - (1/2) \int_0^T \| \sum_{\ell=0}^m \mu^\ell \nabla_j f^\ell(u_i)(t) \|^2 dt \mid$$

$$\mu^\ell \geq 0, \ \ell = 0,1,2,\ldots,m, \ \sum_{\ell=0}^m \mu^\ell = 1\}. \tag{3.11}$$

<u>Step 4</u>: For $t \in [0,T]$, set

$$d_j(u_i)(t) = - \sum_{\ell=0}^m \mu_j^\ell(u_i) \nabla_j f^j(u_i)(t) \tag{3.12}$$

<u>Step 5</u>: Set $\gamma = \lambda$.

<u>Step 6</u>: Compute

$$\theta_j(u_i,\gamma) = f_j^0(u_i+\gamma d_j(u_i)) - f_j^0(u_i)$$
$$- \gamma\alpha \int_0^T \langle \nabla_j f^0(u_i)(t), d_j(u_i)(t) \rangle \, dt \qquad 3.13$$

and $f_j^\ell(u_i+\gamma d(u_i))$, $\ell = 1,2,\ldots,m$.

<u>Step 7</u>: If $\max\{\theta_j(u_i); (f_j^\ell(u_i+\gamma d(u_i)) + 1/2^k)$, $\ell = 1,2,$ $\ldots,m\} > 0$, set $\gamma = \beta\gamma$ and go to step 8; else, set $z = u_i+\gamma d_j(u_i)$ and go to step 9.

<u>Step 8</u>: If $\gamma \geq \varepsilon \, \lambda_{min}$, go to step 6; else, set $z = u_i$ and go to step 9.

<u>Step 9</u>: If $f_j^0(z) - f_j^0(u_i) \leq -\varepsilon$, go to step 12; else, set $j = j+1$ and go to step 10.

<u>Step 10</u>: Compute $f_j^\ell(u_i)$, $j = 1,2,\ldots,m$.

<u>Step 11</u>: If $f_j^\ell(u_i) \leq -1/2^{k+1}$, $\ell = 1,2,\ldots,m$, set $k = k+1$, $\varepsilon = \beta\varepsilon$, and go to step 3; else, set $j = j+1$, $k = k+1$ and go to step 1.

<u>Step 12</u>: Set $u_{i+1} = z$, set $i = i+1$ and go to step 2. ¤.

3.14 <u>Proposition</u>: Theorem (3.4) remains valid for algorithm (3.10). ¤

As we can see, algorithm (3.10) is much closer to a computer program than algorithm (3.1) since it incorporates the integration formulas in its description. It has the nice feature that it will use coarse integration for as long as possible and because of this it will take less time to solve a problem than algorithm (3.1), programmed to use a very small step size integration subprocedure in all the iterations. The only drawback to algorithm (3.10) is that if j_0 and k_0 are chosen poorly, then one may waste some time passing through step 1 several times. It is possible to avoid this difficulty by means of a penalty function type transformation as we shall now describe. Thus, consider the problem

$$\min\{f^0(u) + qy^2 \,|\, f^j(u) - \overline{q}y^2 \leq 0, \; j = 1,2,\ldots,m\} \qquad 3.15$$

where $y \in \mathbb{R}^1$, $u \in L_\infty^s[0,T]$, $q > 0$ and $\overline{q} > 0$ are given and $f^j(\cdot)$, $j = 0,1,2,\ldots,m$, are as in (2.1). The optimality condition (2.10), when applied to (3.15) implies that if $(\hat{u},\hat{y})$ is optimal for (3.15) then there exist $\hat{\mu}^0 \geq 0$, $\hat{\mu}^1 \geq 0, \ldots, \hat{\mu}^m \geq 0$ such that

$$\sum_{j=0}^{m} \hat{\mu}^j \nabla f^j(\hat{u}) = 0 \qquad \qquad 3.16$$

$$(q\hat{\mu}^0 - \bar{q} \sum_{j=1}^{m} \hat{\mu}^j)\hat{y} = 0, \qquad \qquad 3.17$$

$$\hat{\mu}^j [f^j(\hat{u}) - \bar{q}\hat{y}^2] = 0, \quad j = 1,2,\ldots,m, \qquad \qquad 3.18$$

$$\sum_{j=0}^{m} \hat{\mu}^j = 1. \qquad \qquad 3.19$$

Now, suppose that u* is optimal for (2.1), then (u*,0) obviously satisfies (3.16)-(3.19), i.e., every control which is stationary for (2.1) is also stationary for (3.15). Next, from (3.17) and (3.19), if $\hat{y} \neq 0$, then

$$\hat{\mu}^0 = \bar{q}/q+\bar{q}. \qquad \qquad 3.20$$

Now suppose that the vectors $\nabla f^j(u)$, $j = 1,2,\ldots,m$, are linearly independent for all u satisfying $|u|_\infty \leq M$, with M large and that

$$\inf\{\mu^0 | \sum_{j=0}^{m} \mu^j \nabla f^j(u) = 0, \mu^j \geq 0, j = 0,1,\ldots,m, \qquad 3.21$$
$$\sum_{j=0}^{m} \mu^j = 1, |u|_\infty \leq M\} = \eta > 0$$

Then, clearly, if $\bar{q}/q+\bar{q} < \eta$, then $(\hat{u},\hat{y})$ can satisfy (3.16)-(3.19) only if $\hat{y} = 0$, which implies that $f^j(\hat{u}) \leq 0$, $j = 1, 2,\ldots,m$. Thus, in this case, any $\hat{u}$ which is feasible and stationary for (3.15) is also feasible and stationary for (2.1). Obviously, there are many other situations when $\bar{q}/q+\bar{q}$ sufficiently small implies that (3.16)-(3.19) can be satisfied only if $\hat{y} = 0$.

The merit of the transformation (3.15) is that it makes the calculation of an initial feasible solution trivial. For let u_0 be any control, and let $y_0 = \max_{j = 1,2,\ldots,m} f^j(u_0)$, then (u_0,y_0) is feasible for (3.15). Now, since we do not know in advance how small $\bar{q}/(q+\bar{q})$ should be, we must incorporate into our algorithm an adaptive mechanism for adjusting this quantity. This leads to the following extension of algorithm (3.10).

3.22 Implementation II.

$\underline{\text{Step 0}}$: Select integers $j_0 \geq 1$, $k_0 \geq 1$ and parameters $q_0 > 0$, $\overline{q}_0 > 0$, $\varepsilon_0 > 0$, $\alpha \in (0,1)$, $\beta \in (0,1)$, $\sigma \in (0,1)$, $\lambda > 0$, $\lambda_{\min} \in (0,\lambda)$, and an initial guess $u_0 \in U_{j_0}$. Set $i = 0$, $j = j_0$, $k = k_0$, $\varepsilon = \varepsilon_0$, $q = q_0$, $\overline{q} = \overline{q}_0$.

$\underline{\text{Step 1}}$: Compute $f_j^\ell(u_i)$, $\ell = 0,1,2,\ldots,m$ and set $\overline{y}_{ji} = \max_{\ell=1,2,\ldots,m} f_j^\ell(u_i)$.

$\underline{\text{Step 2}}$: If $\overline{y}_{ji} \leq -1/2^{k+1}$, set $k = k+1$ and go to step 3; else, go to step 3.

$\underline{\text{Step 3}}$: Set $y_{ji} = |\max\{0,\overline{y}_{ji}+1/2^k\}/\overline{q}|^{1/2}$.

$\underline{\text{Step 4}}$: For $\ell = 0,1,2,\ldots,m$, compute $\nabla_j f^\ell(u_i)$.

$\underline{\text{Step 5}}$: Compute $\overline{\phi}_j(u_i)$ and $\mu_j^\ell(u_i)$, $\ell = 0,1,2,\ldots,m$, by solving

$$\overline{\phi}_i(u_i) = \max\{ \sum_{\ell=1}^m \mu^\ell [f_j^\ell(u_i)-y_{ji}^2] - (1/2)\int_0^T \|\mu^\ell \nabla f_j^\ell(u_i)(t)\|^2 dt$$

$$- 2(q\mu^0 - \overline{q}\sum_{\ell=1}^m \mu^\ell)^2 y_{ji}^2 |\mu^\ell \geq 0, \ \ell = 0,1,2,\ldots,m,$$

$$\sum_{\ell=0}^m \mu^\ell = 1\} \qquad 3.23$$

$\underline{\text{Step 6}}$: Set

$$d_j(u_i) = - \sum_{j=0}^m \mu_j^\ell(u_i)\nabla f_j^\ell(u_i), \qquad 3.24$$

$$\rho_j(u_i) = - 2q\mu_j^0(u_i) - \overline{q}\sum_{\ell=1}^m \mu_j^\ell(u_i)]y_{ji}. \qquad 3.25$$

$\underline{\text{Step 7}}$: Set $\gamma = \lambda$.

$\underline{\text{Step 8}}$: Compute

$$\overline{\theta}_j(u_i,\gamma) = f_j^0(u_i+\gamma d_j(u_i)) - f^0(u_i) + 2\gamma y_{ji}\rho_j(u_i) \qquad 3.26$$

$$+ \gamma^2\rho_j(u_i)^2 - \gamma\alpha[\int_0^T \langle \nabla_j f^0(u_i)(t),d_j(u_i)(t) \rangle dt$$

$$+ 2qy_{ji}\rho_j(u_i)]$$

and $f_j^\ell(u_i+\gamma d(u_i)) - (y_{ji}+\gamma\rho_j(u_i))^2$, $\ell = 1,2,\ldots,m$.

<u>Step 9</u>: If $\max\{\overline{\theta}_j(u_i); f_j^\ell(u_i+\gamma d(u_i)) - (y_{ji}+\gamma\rho_j(u_i))^2 + 1/2^k$, $\ell = 1,2,\ldots,m\} > 0$, set $\gamma = \beta\gamma$ and go to step 10; else, set $z = u_i + \gamma h_j(u_i)$, $z^0 = y_{ji} + \gamma\rho_j(u_i)$ and go to step 11.

<u>Step 10</u>: If $\gamma \geq \varepsilon\,\lambda_{min}$, go to step 8; else, set $z = u_i$, $z^0 = y_{ji}$ and go to step 11.

<u>Step 11</u>: If $f_j^0(z) - f_j^0(u_i) + (z^0)^2 - y^2 \leq -\varepsilon$, go to step 12; else, set $j = j+2$, set $k = k+1$, set $q = q/\sigma$, $\overline{q} = \sigma q$, $\varepsilon = \beta\varepsilon$ and go to step 1.

<u>Step 12</u>: Set $u_{i+1} = z$, set $i = i+1$ and go to step 1.   ⊓

Note that algorithm (3.22) reduces to algorithm (3.10) as soon as u_i is in $D_{j,k}$, and it avoids the time consuming step 1 of algorithm (3.10).

Conclusion

It is important to note that the algorithm prototype (2.28) is not merely a device for explaining the two specific implementations of the Pironneau–Polak method. Rather, it represents a general scheme for constructing multi-parameter algorithms. As far as the two implementations of the Pironneau–Polak method are concerned, there is still not enough experimental evidence to indicate a preference, though on the basis of heuristic reasoning the second one appears as if it might be more efficient than the first one. Finally, it should be pointed out that there exists a more efficient, but also more complex version of the Pironneau–Polak method (see [6]) which can be implemented in essentially the same way.

References

1. E. Polak, <u>Computational Methods in Optimization; A Unified Approach</u>, Academic Press, New York, 1971.
2. R. Klessig and E. Polak, "An Adaptive Algorithm for Unconstrained Optimization with an Application to Optimal Control", Univ. of Calif. ERL Memo #M297, Feb. 22, 1971.
3. O. Pironneau and E. Polak, "A Dual Method for Optimal Control Problems with Initial and Final Boundary Con-

straints", Univ. of Calif. ERL Memo #M299, March 30, 1971.

4. F. John, "Extremum Problems with Inequalities as Side Conditions", in Courant Aniversary Volume (K.). Friedriechs et.al. eds.). pp 187-204, Interscience, New York, 1948.

5. E. Isaacson and H. B. Keller, <u>Analysis of Numerical Methods</u>, Wiley, New York, 1966.

6. E. Polak, H. Mukai, O. Pironneau, "Methods of Centers and Methods of Feasible Directions for the Solution of Optimal Control Problems", Proc. IEEE Decision and Control Conference, Miami Beach, Dec. 15-17, 1971.

CONVERGENT STEP SIZES FOR CURVILINEAR-PATH
METHODS OF MINIMIZATION

James W. Daniel[*]

The University of Texas at Austin

Austin, Texas

1. Introduction and General Step-size Criteria

We consider the problem of minimizing a real-valued functional f over a subset C of the real Banach space X by means of an iterative method. Most such methods generate a sequence $\{x_n\}$ of approximate minimizers via the iteration $x_{n+1} = x_n + t_n p_n$ where t_n is chosen by means of some step-size method along the ray $x_n + t p_n$. For various reasons it has been suggested [2,10,12,15,19] for this case that the ray be replaced by a more general curvilinear path.

Therefore we shall now consider paths of the form $x_n + p_n(x_n,t)$, where for convenience we drop notationally the dependence of p_n on x_n and assume that, for each n, p_n is a continuous function of t with $x_n + p_n(t) \in C$ for $0 \le t \le T_n$, where the path is scaled so that $T_n \ge T$ for some fixed $T \in (0,+\infty]$.

We shall assume that $f(x_n + p_n(t))$ is differentiable from the right at $t = 0$, with its derivative $-\gamma_n$ being negative. More precisely, we assume:

<u>H1</u>: For each n there exists $\gamma_n \ge 0$ such that: for every $\epsilon > 0$ there exists $\delta > 0$ independent of n such that

$$\left| \frac{f(x_n + p_n(t_n)) - f(x_n) + \gamma_n t_n}{t_n} \right| < \epsilon \text{ for all } n \text{ and all } t_n \text{ satisfying}$$

$0 < t_n < \delta, \ t_n \le T_n, \ f(x_n + p_n(t_n)) \le f(x_n)$.

It will be the goal of our analysis to select step

---

*Mathematics, Computer Sciences, and the Center for Numerical Analysis at The University of Texas at Austin.

sizes t_n, with $x_{n+1} = x_n + p_n(t_n)$, such that γ_n converges to zero. To see what this convergence to zero means, we consider first some examples.

Example 1. Let us consider unconstrained minimization problems, that is, where $C = X$; we consider using linear paths, that is, $p_n(t) = tq_n$ for some vector q_n. If, for example, f is differentiable at each x and has gradient $\nabla f(x) \in X^*$, the dual space, then $-\gamma_n$ is given simply as $-\gamma_n = \langle \nabla f(x_n), q_n \rangle$. Under various hypotheses [7,13,14,16,18] we can then conclude that $\{x_n\}$ is a minimizing sequence, or that the limit points in some topology of $\{x_n\}$ are minimizers for f, or that $\{x_n\}$ converges to a unique minimizer. Many of the standard popular methods are of this type [4,7].

Example 2. Let us consider the constrained problem where C is a closed convex set, and we use again the linear path $p_n(t) = tq_n$ where q_n is a so-called feasible direction, that is, $\langle -\nabla f(x_n), q_n \rangle \geq 0$ and $x_n + tq_n \in C$ for $0 \leq t \leq 1$. Again one finds $\gamma_n = \langle -\nabla f(x_n), q_n \rangle$, and q_n must be chosen so that the convergence of γ_n to zero is useful. For example, [6,7], this condition gives the same kinds of results as in Example 1 when q_n is generated by the variable-metric gradient-projection method, by the conditional-gradient method, by Newton's method, et cetera.

Example 3. Consider minimizing f over a closed, convex subset C of a Hilbert space X, and let Π_C denote the metric projection operator onto C. One version of a gradient projection algorithm has been proposed [10,15] which leads to a curvilinear path, namely $p_n(t) = \Pi_C[x_n - t\nabla f(x_n)] - x_n$. In this case it can easily be shown [15] that $\gamma_n = \|\Pi_{B_n}[-\nabla f(x_n)]\|^2$ where B_n is the closure of the tangent cone to C at x_n. Driving γ_n to zero here is useful since $\|\Pi_{B_n}[-\nabla f(x_n)]\| = 0$ is a well-known necessary condition for x_n to minimize f over C; under some hypotheses, this condition is sufficient [7].

Example 4. Let $C = \{x; Q(x) = 0\}$, where Q maps the Banach space X into the Banach space Z, is nonlinear in general and is differentiable. Choose $y_n \in X$ such that $\|y_n\| = 1$ and $\langle \nabla f(x_n), y_n \rangle = \sup\{\langle \nabla f(x_n), y \rangle; y \in X, \|y\| = 1,$

$Q'_{x_n}y = 0\}$. It is shown in [2] that, if Q'_x maps X onto Z for each x in C, then there exists a mapping $s(t,x_n,y_n)$ such that $x_n + p_n(t) \equiv x_n - ty_n + s(t,x_n,y_n)$ lies in C for $0 \le t \le T > 0$ and has $\gamma_n = \langle \nabla f(x_n), y_n \rangle$. Here driving γ_n to zero is related to trying to satisfy the optimality criterion $\langle \nabla f(x), y \rangle = 0$ for all y satisfying $Q'_x y = 0$.

Example 5. For unconstrained or constrained problems where a deep-curving valley is present in the function's graph, it can be profitable to fit a space curve through the last few iterates to approximate the valley's shape; at least one approach to this has appeared to be successful [19].

A further example of curvilinear paths appears in [12].

Having motivated our goal of driving γ_n to zero, we now proceed with the general analysis. As in the usual approaches that have been used before [4,6,7,8,14], we shall attack the problem by attempting to show that $f(x_n) - f(x_{n+1}) \ge c(\gamma_n)$, where $c(\cdot)$ is a forcing function. We recall [7]:

Definition 1.1. A function c, defined for $t \ge 0$, is a forcing function if and only if $c(t) \ge 0$ for all t and the convergence of $c(t_n)$ to zero implies the convergence of t_n to zero.

Remarks. Some slightly more general approaches can also be used to treat this problem [4,7,17]. If one only wants to prove that norm limit points of $\{x_n\}$ are critical points, hypotheses weaker than H1 may be used as in [15].

We shall now prove a fundamental theorem to use as the main tool for the subsequent analysis. First we need:

Lemma 1.1. Assume the hypotheses H1, and define the function r via $r(0) = 0$ and

$$r(\epsilon) = \sup\{R(\epsilon'); \ 0 \le \epsilon' < \epsilon\} \qquad (1\text{-}1)$$

where $R(\epsilon') = \inf\{t; \ 0 \le t \le T_n, \ f(x_n + p_n(t)) \le f(x_n),$
$\left| \dfrac{f(x_n + p_n(t)) - f(x_n) + \gamma_n t}{t} \right| \ge \epsilon'$ for some n}. Then r is a non-decreasing left-continuous forcing function.

Remarks. The function r will play the role of the reverse modulus of continuity of ∇f in [6,7,9,16]. In reality H1 is usually proved for a larger set of points and direction paths than $\{x_n\}$ and $\{p_n\}$.

Theorem 1.1. Let $t \to c_1(t)$ and $t \to t-c_2(t)$ be forcing functions, let f be bounded below on C, and let H1 hold. (A) Suppose step-sizes $t_n \leq T_n$ satisfy $c_1(\gamma_n) \leq t_n \leq r[c_2(\gamma_n)]$; then $f(x_n+p_n(t)) \leq f(x_n)$ for $t \leq t_n$, and γ_n converges to zero. (B) If t'_n is chosen as is t_n in part (A) and x_{n+1} is chosen in C such that $f(x_n) - f(x_{n+1}) \geq \beta[f(x_n) - f(x_n+p_n(t'_n))]$ for a fixed $\beta > 0$, then γ_n converges to zero. The same conclusion follows if t'_n is chosen instead as is t_n in the first sentence of part B.

Proof: Consider part (A). Note that if $\gamma_n = 0$ then there is nothing to prove, so we suppose $\gamma_n > 0$. First we prove that $f(x_n+p_n(t)) \leq f(x_n)$ for $t \leq t_n$. We define $\alpha_n = \sup\{\alpha; \alpha \leq T_n, f(x_n+p_n(t)) < f(x_n)$ for $0 < t \leq \alpha\}$; clearly $\alpha_n > 0$ and either $\alpha_n = T_n$ or $f(x_n+p_n(\alpha_n)) = f(x_n)$. If $\alpha_n = T_n$ then $t_n \leq \alpha_n$ and hence $f(x_n+p_n(t)) \leq f(x_n)$ as asserted. On the other hand, if $f(x_n+p_n(\alpha_n)) = f(x_n)$ then

$$\left| \frac{f(x_n+p_n(t))-f(x_n)+\gamma_n t}{t} \right|$$

converges to γ_n from below as t tends to α_n. From (1-1) this implies $\alpha_n \geq r(\gamma_n)$ which in turn gives $\alpha_n \geq r(\gamma_n) \geq r(c_2(\gamma_n)) \geq t_n$, so that $f(x_n+p_n(t)) \leq f(x_n)$ as asserted. Next, since $t_n \leq r(c_2(\gamma_n))$ and $f(x_n+p_n(t_n)) \leq$

$$f(x_n), \text{we deduce from (1-1) that } \left| \frac{f(x_n+p_n(t_n))-f(x_n)+t_n\gamma_n}{t_n} \right| <$$

$c_2(\gamma_n)$, which gives $f(x_n) - f(x_{n+1}) \geq t_n[\gamma_n-c_2(\gamma_n)] \geq [\gamma_n-c_2(\gamma_n)]c_1(\gamma_n)$. Since the mapping $t \to [t-c_2(t)]c_1(t)$ is a forcing function, part (A) is proved. For part (B), the lower bound on $f(x_n) - f(x_{n+1})$ is merely multiplied by β.

Q.E.D.

Remark. The presence of β in the above theorem is to allow for sloppiness in determining t_n and to allow for acceleration methods in which after choosing t_n we further decrease the function values by some different iterative method, arriving at the point called x_{n+1}. For example t_n might be generated by exact minimization along $x_n - t\nabla f(x_n)$, the steepest descent method, and then x_{n+1} is chosen by doing several conjugate-gradient steps; this yields a global convergence proof for such "restart" conjugate-gradient methods [7,17].

2. Step Sizes Based on Minimization

As indicated in Section 1, our approach is to guarantee a sufficient decrease in the value of f from x_n to x_{n+1}; what better way theoretically than to minimize $f(x_n+p_n(t))$ for $0 \leq t \leq T_n$? Numerically of course one cannot expect to minimize the function exactly; one way of analyzing this possibility is via inverse error analysis in which we assume that we minimize exactly a "nearby" function, say $f(x_n+p_n(t)) + \alpha_n t\gamma_n$ where $\alpha_n \in [0,1]$ describes a small perturbation of the function f. The reader should note also that, were $f(x_n+p_n(t))$ assumed to be differentiable in t, this approach would be equivalent to reducing the derivative of $f(x_n+p_n(t))$ to a factor α_n of its value $-\gamma_n$ at $t = 0$.

Theorem 2.1. Let H1 and the general hypotheses of Section 1 hold, and let f be bounded below on C. For numbers $\alpha_n \in [0,\alpha]$ where $\alpha < 1$, let t_n be chosen so as to minimize $f_n(t) \equiv f(x_n+p_n(t)) + \alpha_n t\gamma_n$ over $0 \leq t \leq T_n$. Let $x_{n+1} \in$ C be any point such that $f(x_n) - f(x_{n+1}) \geq \beta[f(x_n) - f(x_n + p_n(t_n))]$ for a fixed $\beta > 0$. Then γ_n converges to zero.

Proof: Consider the algorithm based on the step size choice $t_n' = \min\left\{\frac{1}{2}T_n, \frac{1}{2}r\left(\frac{(1-\alpha)\gamma_n}{4}\right)\right\}$. By part (A) of Theorem 1.1, with $c_1(t) = \min\left\{\frac{1}{2}T, \frac{1}{2}r\left(\frac{(1-\alpha)t}{4}\right)\right\}$ and $c_2(t) = \frac{(1-\alpha)t}{4}$, this step-size choice yields a convergent method. We claim that $t_n \geq t_n'$. If not, then $2t_n \leq r\left(\frac{(1-\alpha)\gamma_n}{4}\right)$ and we can use part (A) of Theorem 1.1 and its proof to deduce $\Big|f_n(2t_n) -$

$$f_n(t_n) + (1-\alpha_n)\gamma_n t_n\Big| = \Bigg|\left[f_n(0)-(1-\alpha_n)\gamma_n 2t_n +\left(\frac{f(x_n+p_n(2t_n))}{2t_n}-\right.\right.$$
$$\left.\left.\frac{f(x_n)+\gamma_n 2t_n}{2t_n}\right)\cdot 2t_n\right] - \left[f_n(0)-(1-\alpha_n)\gamma_n t_n +\left(\frac{f(x_n+p_n(t_n))-f(x_n)}{t_n}+\right.\right.$$
$$\left.\left.\frac{\gamma_n t_n}{t_n}\right)\cdot t_n\right]+ [(1-\alpha_n)\gamma_n t_n]\Bigg| \leq 2t_n\left[\frac{(1-\alpha)\gamma_n}{4}\right]+ t_n\left[\frac{(1-\alpha)\gamma_n}{4}\right]<$$

$(1-\alpha)\gamma_n t_n$. In other words, $f_n(2t_n) < f_n(t_n) - (1-\alpha_n)\gamma_n t_n + (1-\alpha)\gamma_n t_n \leq f_n(t_n)$, contradicting the global minimizing property of t_n. Therefore $t_n \geq t_n'$. Thus we have $\frac{1}{\beta}[f(x_n) - f(x_{n+1})] \geq f(x_n) - f(x_n + p_n(t_n)) = f_n(0) - f_n(t_n) + \alpha_n t_n \gamma_n \geq f_n(0) - f_n(t_n') + \alpha_n t_n \gamma_n = f(x_n) - f(x_n + p_n(t_n')) + \alpha_n(t_n - t_n')\gamma_n \geq f(x_n) - f(x_n + p_n(t_n'))$. Since we have already shown t_n' to yield a convergent method, part (B) of Theorem 1.1 completes our proof. Q.E.D.

Searching for a global minimum of f or f_n presents formidable difficulties; it is slightly simpler to consider searching for the first local minimum or the first zero of the derivative (assuming differentiability) of f or f_n along $x_n + p_n(t)$. Unfortunately we cannot prove the convergence of these choices without assuming more regularity than required by H1. In fact, we assume:

__H1'__: For each n, the derivatives $\frac{d}{dt}f(x_n + p_n(t))$ exist on the set of t values in the intersection of $[0, T_n]$ with the convex hull of $\{t; f(x_n + p_n(t)) \leq f(x_n)\}$ and form an equicontinuous family there.

Under this hypothesis just as in Lemma 1.1 we can define the non-decreasing left-continuous forcing function

$$\tilde{r}(\epsilon) = \sup\{\inf\{|t_1 - t_2|; \ 0 \leq t_1 \leq T_n, \ 0 \leq t_2 \leq T_n, \qquad (2\text{-}1)$$

$$|\frac{d}{dt}f(x_n + p_n(t_1)) - \frac{d}{dt}f(x_n + p_n(t_2))| \geq \epsilon'$$

$$\text{for some n and } x_n\}; \ 0 \leq \epsilon' < \epsilon\}, \ \tilde{r}(0) = 0.$$

It is simple to show that H1' implies H1 and that $\tilde{r}(\epsilon) \leq r(\epsilon)$ for all ϵ. Therefore we have:

__Theorem 2.2.__ The Theorem 1.1 remains valid with H1 replaced by H1' and r replaced by $\tilde{r}$.

We can now analyze other minimization-oriented step-size methods. Because of space limitations in this volume, we provide few details. The arguments follow those of [6,7] with modifications as for Theorem 1.1 and Theorem 2.1. Complete details can be found in [8], where our general hypotheses are made more specific for the examples of Section 1.

__Theorem 2.3.__ Let H1' and the general hypotheses of Section 1 hold, and let f be bounded below on C. For numbers

$\alpha_n \in [0,\alpha]$ with $\alpha < 1$, define $f_n(t) \equiv f(x_n + p_n(t)) + \alpha_n \gamma_n t$. Let t_n be chosen as any number satisfying (i) $0 \le t \le T_n$, and (ii) $f_n(t) \ge f_n(t_n)$ for $0 \le t \le t_n$, and <u>either</u> (iii) $0 = \frac{d}{dt} f_n(t_n) = \frac{d}{dt} f(x_n + p_n(t_n)) + \alpha_n \gamma_n$, <u>or</u> (iv) $t_n = T_n$. Let $x_{n+1} \in C$ be any point such that $f(x_n) - f(x_{n+1}) \ge \beta [f(x_n) - f(x_n + p_n(t_n))]$ for for some fixed $\beta > 0$. Then γ_n converges to zero.

The above result proves convergence for several well-known choices of step-size.

<u>Theorem 2.4.</u> Let f and α_n be as in Theorem 2.3 above. Let t_n be chosen in any of the following ways: (i) as any point minimizing $f_n(t) \equiv f(x_n + p_n(t)) + t \gamma_n \alpha_n$ over $[0, T_n]$, (ii) as the smallest positive number providing a local minimum for $f_n(t)$ over $[0, T_n]$, (iii) as the first positive root of the equation $\frac{d}{dt} f_n(t) = 0$ in $[0, T_n]$, if such exists, and as T_n otherwise, (iv) as $t_n = \sup\{t; \frac{d}{dt} f_n(t) \le 0$ for $0 \le t \le T$ and $0 \le t \le T_n\}$. Let $x_{n+1} \in C$ be a point such that $f(x_n) - f(x_{n+1}) \ge \beta [f(x_n) - f(x_n + p_n(t_n))]$ for a fixed $\beta > 0$. Then γ_n converges to zero.

<u>Remarks.</u> We remind the reader that the introduction of α_n is essentially an artifice so that our analysis will cover inexact computations. The condition $\frac{d}{dt} f_n(t_n) = 0$ for example is really just $|\frac{d}{dt} f(x_n + p_n(t_n))| \le \alpha \gamma_n$, which says that the derivative need not be zero, just smaller relative to its value at zero.

We wish to indicate another way in which our analysis can cover inexact computation. This result will be important when we consider certain computationally simple search methods for computing the step size.

<u>Theorem 2.5.</u> Let the general hypotheses of Section 1 hold, and let f be bounded below on C. Let $\alpha_n \in [0, \alpha]$, $\alpha < 1$, and let $f_n(t) \equiv f(x_n + p_n(t)) + \gamma_n \alpha_n t$. Let t_n be chosen as any number satisfying $f_n(t_n) \le f_n(t)$ for $0 \le t \le T_n$ if only H1 holds, or as any number satisfying (i) and (ii) and either (iii) or (iv) of Theorem 2.3 if H1' holds. In addition, suppose that (v) $f_n(t)$ is non-increasing for $0 \le t \le t_n$. Let $\hat{t}_n = \lambda_n t_n$ where λ_n is a relaxation factor satisfying

$d(\gamma_n) \leq \lambda_n \leq 1$ for some forcing function d. Let $x_{n+1} \in C$ be any point such that $f(x_n) - f(x_{n+1}) \geq \beta[f(x_n) - f(x_n + p_n(\hat{t}_n))]$ for a fixed $\beta > 0$. Then γ_n converges to zero.

3. Step Sizes Based on Sufficient Decrease

With the possible exception of the method (iii) of Theorem 2.4, none of the above algorithms is simple computationally; we wish now to consider methods that can be implemented more simply in practice.

First we consider the method of [1,2], in more generality. We make use of the <u>range function</u> [10,11]

$$g_n(t) \equiv \frac{(f(x_n) - f(x_n + p_n(t)))}{t\gamma_n} \text{ for } t > 0, \; g(0) = 1 \qquad (3\text{-}1)$$

which is continuous on $[0, T_n]$.

We need a stronger hypothesis at this point than H1 or H1'; in essence we need $g_n(t)$ to be uniformly Lipschitz continuous near $t = 0$. We assume

<u>H2</u>: There exists a constant L independent of n such that $\gamma_n \leq L$ and, for $0 < t \leq T_n$ with $f(x_n + p_n(t)) \leq f(x_n)$, we have $\left| \dfrac{f(x_n + p_n(t)) - f(x_n) + \gamma_n t}{t^2} \right| \leq L.$

<u>Theorem 3.1.</u> Let H2 and the general hypotheses of Section 1 hold, and let f be bounded below on C. Let c be a forcing function such that $\dfrac{c(t)}{t} \leq c_0 < 1$, let $q \in (0,1)$, and let x_0 and τ_{-1} be given; determine x_{n+1} and τ_n for $n \geq 0$ as follows. If $\tau_{n-1}\gamma_n \leq T_n$ and $g_n(\tau_{n-1}\gamma_n) \geq \dfrac{c(\gamma_n)}{\gamma_n}$ then set $\tau_n = \tau_{n-1}$; otherwise choose $\tau_n \in (0, q\tau_{n-1}]$ such that

$g_n(\tau_n\gamma_n) \geq \dfrac{c\gamma_n}{\gamma_n}$ and $\tau_n\gamma_n \leq T_n$. Let $t_n = \tau_n\gamma_n$ and $x_{n+1} \in C$ be any point such that $f(x_n) - f(x_{n+1}) \geq \beta[f(x_n) - f(x_n + p_n))]$ for a fixed $\beta > 0$. Then γ_n converges to zero and there exists k such that $\tau_n = \tau_k$ for all $n \geq k$.

<u>Proof:</u> By the choice of τ_n, we have $\dfrac{1}{\beta}[f(x_n) - f(x_{n+1})]$ $\geq \tau_n\gamma_n c(\gamma_n)$ which must therefore tend to zero. If our

theorem is false, then there is a subsequence of $\{\tau_n\}$ converging to zero; since $\{\tau_n\}$ is monotonic, it follows that τ_n and $\tau_{n-1}\gamma_n$ both converge to zero. But then for large n, $\tau_{n-1}\gamma_n \leq T_n \geq T > 0$ and

$$|g_n(\tau_{n-1}\gamma_n)-1| = \left|\frac{f(x_n)-f(x_n+p_n(\tau_{n-1}\gamma_n))}{t_{n-1}^2\gamma_n} - 1\right|$$

$$= \left|\frac{f(x_n)-f(x_n+p_n(\tau_{n-1}\gamma_n))-(\tau_{n-1}\gamma_n)\gamma_n}{(\tau_{n-1}\gamma_n)^2}\right| \tau_{n-1} \leq L\tau_{n-1},$$

which yields $g_n(\tau_{n-1}\gamma_n) \geq 1 - L\tau_{n-1} \geq c_0 \geq \dfrac{c(\gamma_n)}{\gamma_n}$. Therefore, $\tau_n = \tau_{n-1}$ for all large n, a contradiction. Q.E.D.

   <u>Remarks.</u> Note that τ_n satisfying $g_n(\tau_n\gamma_n) \geq \dfrac{c(\gamma_n)}{\gamma_n}$, $\tau_n \leq q\tau_{n-1}$, $\tau_n\gamma_n \leq T_n$ always exists; a suitable τ_n could be found for example by considering $q^i\tau_{n-1}$ for values of i until the requirements were met.

   A somewhat similar computational procedure which requires fewer hypotheses has also been proposed [10,11] in terms of the range function $g_n(t)$. We consider it next in a general form [6,7,9,16].

   <u>Theorem 3.2.</u> Let H1 and the general hypotheses of Section 1 hold, and let f be bounded below on C. Let c be a forcing function such that $\dfrac{c(t)}{t} \leq c_0 \leq \dfrac{1}{2}$, let t_n^e be numbers such that $0 < T \leq t_n^e \leq T_n$, and let x_0 be given. We determine x_n for $n \geq 1$ as follows. If

$$g_n(t) \geq \frac{c(\gamma_n)}{\gamma_n} \tag{3-2}$$

for $t = t_n^e$, then set $t_n = t_n^e$; otherwise find t_n in $(0,T_n)$ satisfying both (3-2) and

$$|g_n(t_n)-1| \geq \frac{c(\gamma_n)}{\gamma_n}. \tag{3-3}$$

Let $x_{n+1} \in C$ be any point such that $f(x_n) - f(x_{n+1}) \geq$

$\beta[f(x_n)-f(x_n+p_n(t_n))]$ for a fixed $\beta > 0$. Then γ_n converges to zero.

<u>Remark.</u> Computationally one might well take $c(t) = c_0 t$ as originally proposed [10,11].

Although the above method has proved very important in practice, it has the drawback that it may be difficult computationally to locate a t_n satisfying the two-sided inequalities of (3-2) and (3-3) together. In the case of linear paths, an algorithm is known which circumvents this drawback very simply [3]; the method works more generally.

<u>Theorem 3.3.</u> Let f, C, c(t), c_0, t_n^e, and x_0 be as in Theorem 3.2. Let $q \in (0,1)$. Determine x_n for $n \geq 1$ as follows. Choose t_n as the first number from the sequence $q^0 t_n^e$, $q^1 t_n^e$, $q^2 t_n^e$, ... such that (3-2) is satisfied, and let $x_{n+1} \in C$ be any point such that $f(x_n) - f(x_{n+1}) \geq \beta[f(x_n) - f(x_n+p_n(t_n))]$ for a fixed $\beta > 0$. Then γ_n converges to zero.

4. Step Sizes Based on Direct Search

For computational purposes it is unpleasant to contemplate such a procedure as minimizing $f(x_n+p_n(t))$ over all values of t on $[0,T_n]$. It is more reasonable to consider minimizing $f(x_n+p_n(t))$ for t ranging over some finite point set, just as in [4,6,7] for linear paths. For the purpose of this analysis, we assume

<u>H3</u>: $f_n(t) \equiv f(x_n+p_n(t))$ has a unique local minimizer on $[0,T_n]$ for each n.

If f satisfies H3, then minimizers can easily be located since, given three t-values $0 \leq t_1 < t_2 < t_3 \leq T_n$ such that $f_n(t_2) < f_n(t_1)$ and $f_n(t_2) < f_n(t_3)$, it follows that f_n has its local minimum in (t_1,t_3). We can base a search method on this simple fact.

<u>Theorem 4.1.</u> Let H1, H3, and the general hypotheses of Section 1 hold, and let f be bounded below on C. Suppose that for each n there are values $t_{n,1} < t_{n,2} < \cdots < t_{n,k_n+1}$ such that (i) $t_{n,k_n+1} \leq T_n$, and (ii) $\dfrac{t_{n,k_n-1}}{t_{n,k_n+1}} \geq q > 0$ for

some fixed q, and (iii) $f_n(0) > f_n(t_{n,1}) > \ldots > f_n(t_{n},k_n)$, and <u>either</u> (iv) $f_n(t_{n,k_n}) \leq f_n(t_{n,k_n+1})$ <u>or</u> (v) $t_{n,k_n+1} = T_n$. Let $t_n \in [t_{n,k_n-1}, t_{n,k_n}]$ and let $x_{n+1} \in C$ be any point such that $f(x_n) - f(x_{n+1}) \geq \beta[f(x_n) - f(x_n+p_n(t_n))]$ for a fixed $\beta > 0$. Then γ_n converges to zero.

   <u>Corollary 4.1.</u> Under all the hypotheses of Theorem 4.1, if in addition $t_{n,i+1} - t_{n,i} = h_n$ for all i and n, then $k_n \geq 2$ is sufficient to guarantee that choosing $t_n \in [(k_n-1)h_n, k_n h_n]$ will force γ_n to converge to zero.

   A simple search algorithm that implements the above corollary has been developed for linear paths [4,6,7]; it is simple to see that the resulting procedure, called <u>Search routine</u> in [6,7,8], works as well in the present more general setting.

References

1. Altman, M., "Generalized gradient methods of minimizing a functional," <u>Bull. Acad. Polon. Sci.</u>, vol. 14 (1966), 313-318.
2. Altman, M., "A generalized gradient method for the conditional minimum of a functional," <u>Bull. Acad. Polon. Sci.</u>, vol. 14 (1966), 445-451.
3. Armijo, L., "Minimization of functions having Lipschitz continuous first partial derivatives," <u>Pacific J. Math.</u>, vol. 16 (1966), 1-3.
4. Cea, J., <u>Optimisation théorie et algorithmes</u>, Dunod (1971), Paris. [French]
5. Curry, H. B., "The method of steepest descent for nonlinear minimization problems," <u>Quart. Appl. Math.</u>, vol. 2 (1944), 258-263.
6. Daniel, J. W., "Convergent step-sizes for gradient-like feasible direction algorithms for constrained optimization," in <u>Nonlinear Programming</u>, Rosen, <u>et al.</u> (eds.), Academic Press (1970), 245-274.
7. Daniel, J. W., <u>The Approximate Minimization of Functionals</u>, Prentice-Hall (1971).
8. Daniel, J. W., "Convergent step sizes for curvilinear-path methods of minimization," CNA-29 (1971), Center for Numerical Analysis, University of Texas at Austin.
9. Elkin, R. M., "Convergence theorems for Gauss-Seidel and other minimization algorithms," Comput. Sci. Rep.

#68-59, Univ. of Maryland (1968).

10. Goldstein, A. A., "Convex programming in Hilbert space," *Bull. AMS*, vol. 70 (1964), 709-710.

11. Goldstein, A. A., "On steepest descent," *SIAM J. Control*, vol. 3 (1965), 447-451.

12. Kelley, H. J., I. J. Johnson, "Curvilinear projection," IFIP 4th Colloq. on Optimization, this volume.

13. Levitin, E., B. T. Poljak, "Convergence of minimizing sequences in conditional extremum problems," *Sov. Math. Dokl.*, vol. 7 (1966), 764-767.

14. Levitin, E., B. T. Poljak, "Constrained minimization methods," *USSR Comp. Math. Math. Phys.*, vol. 6 (1968), 1-50.

15. McCormick, G., R. Tapia, "The gradient projection method under mild differentiability conditions," to appear in *SIAM J. Control* (1971).

16. Ortega, J., W. Rheinboldt, *Iterative Solution of Nonlinear Equations in Several Variables*, Academic Press (1970).

17. Ortega, J., W. Rheinboldt, "A general convergence result for unconstrained minimization methods," to appear in *SIAM J. Control* (1971).

18. Poljak, B. T., "Existence theorems and convergence of minimizing sequences in extremum problems with restrictions," *Sov. Math. Dokl.*, vol. 7 (1966), 72-75.

19. Witte, B., W. Holst, "Two new direct minimum search procedures for functions of several variables," presented to Spring Joint Computer Conference, Washington, D. C. (1964).

AN ALGORITHM FOR MINIMIZING A DIFFERENTIABLE FUNCTION THAT USES ONLY FUNCTION VALUES

Jane Cullum

IBM Watson Research Center
Yorktown Heights, New York

Introduction

Let f denote a real-valued function of n real-variables x. The problem is to determine a point x^* such that $f(x^*) \leq f(x)$, $x \in E^n$, under the assumptions that (A-1) x^* exists, (A-2) f is twice continuously differentiable, (A-3) there is a neighborhood of x^* in which f can be accurately approximated by a quadratic function, and (A-4) the derivatives of f are not available for use in the minimization scheme. (In practice, of course, one can determine only a local minimal point of f.) The critical assumption is (A-4). The algorithm can utilize only function values.

Several algorithms have been proposed for this problem; the best ones are (1) Powell's no derivative scheme [1] and (2) Stewart's modification of the Davidon-Fletcher-Powell scheme [2].

Powell's scheme, if used as stated in [1], may generate, even for a quadratic function [3], a nonminimal point. Powell suggested a modification [3] that corrects this problem but the resulting scheme may degenerate into slow successive coordinate searches. Zangwill [3] suggested an alternative modification. He recommended inserting searches along coordinate lines at the beginning of each cycle. The coordinate lines are considered cyclically, and the searches continue until a line of descent is obtained. Clearly, if no coordinate line is a line of descent the algorithm has converged to a minimal point of f since f is continuously differentiable. This idea of allowing searches along coordinate lines 'periodically' is used in the new algorithm.

Stewart's algorithm is quite different from Powell's. In Powell's algorithm no apparent approximations to the gradient of f are made. Stewart on the other hand, makes explicit approximations to the gradient of f which are then used in the Davidon-Fletcher Powell (DFP) [4] recursive formula to obtain approximations, H_i, to the inverse of the Hessian of f at the iterates $x_i, i=1, 2, \ldots$ At x_i a search is made along the ray through x_i in the direction $- H_i g_i^a$, where g_i^a denotes the gradient approximation. In particular, the jth component of g_i^a is given by $(g_i^a)^j = [f(x_i + h_i^j e_j) - f(x_i)] / h_i^j$ where the function differences are computed along the coordinate lines $x = x_i + a\, e_j, 1 \leq j \leq n$, through x_i. Stewart has a clever formula for computing h_i^j. The matrices $H_i, i=1, 2, \ldots$ are obtained from the following formula,

$$H_{i+1} = H_i - (H_i \Delta g_i^a > <H_i \Delta g_i^a) / <\Delta g_i^a, H_i \Delta g_i^a >$$
$$+ (\Delta x_i > <\Delta x_i) / <\Delta x_i, \Delta g_i^a >$$

where $\Delta g_i^a \equiv g_{i+1}^a - g_i^a$, and $\Delta x_i = x_{i+1} - x_i$.

If f is a positive definite quadratic function, $f = <Qx, x> / 2 + <c, x>$, then the DFP scheme converges to the minimal point of f in n iterations. This finite convergence depends upon three relations hips. (g_i denotes the gradient of f). (R-1) The directions of search d_i are $a\, H_i g_i$. (R-2) The linear searches are accurate, $<g_{i+1}, d_i> = 0$. (R-3) $(\Delta g)_i = Q(\Delta x)_i$. Clearly, if $g_i^a \neq g_i$, then (R-1) is not satisfied. Moreover, if h_i varies with i, (R-3) is not satisfied either. Hence, if the gradient approximations are inaccurate Stewart's scheme may converge very slowly or even terminate prematurely since the inner product $<g_i, H_i g_i^a>$ may vanish even if $g_i \neq 0$. Furthermore, unless $g_i^a \to g_i$ as $i \to \infty$, the terminal convergence of Stewart's scheme is weak even when g_i^a is an 'accurate' approximation. However, for the cases tested in [2] Stewart's scheme required 1/2 as many function evaluations to obtain a specified reduction in $f^* \equiv f - f(\min)$ (where $f(\min)$ = minimum value of f) as Powell's no derivative scheme did, so it is a scheme that is worthy of further consideration. Actually, one is lead one to consider the rank 1 scheme [5],

$$H_{i+1} = H_i - (w_i > < w_i)/ < \Delta g_i, w_i >$$

where $w_i \equiv H_i(\Delta g_i) - \Delta x_i$. The finite convergence properties of this scheme depend only upon the following two conditions (R-3) and (R-4) H_i is invertible for each i. The directions of search can be arbitrary as long as they are independent, and minimizing linear searches are not required. (R-3) is satisfied whenever h_i^j is constant over i (not j). Hence, searches along coordinate lines can be inserted without introducing errors into the Hessian approximation, and perhaps a further reduction in the number of function evaluations required can be obtained by using a nonminimizing linear search.

Hence, the Rank 1 recursion formula and the coordinate searches were combined with several other new ideas to obtain a new algorithm, MR1. The differences between this algorithm and Stewart's algorithm other than the recursion formula used for computing the matrices $H_i, i = 1, 2, \ldots$ are given in Table 1.

Table 1 : Differences

MR1	Stewart
Approximation to the jth component of the gradient	
$g^j = (\Delta f)^j / h^j - h^j a^{jj} / 2$	$g^j = (\Delta f)^j / h^j$
Formula for h_i^j	
$h_i^j = 2 \| f(x_i) \eta / a_i^{jj} \|^{\frac{1}{2}}$ <u>or</u> it is fixed apriori for all i	Depends on several factors. Basically there are two formulas for h_i^j and a central differencing option too

Table 1 (continued)

MR1	Stewart
Directions of search, d_i	
$\pm\, H_i g_i^a$, and the coordinate lines through x_i	$-H_i g_i^a$
Initial step size in the linear searches	
$\min\{2, <g_i^a, d_i>/<G_i d_i, d_i>\}$	$\min\{1, -2(f(x_i)-f_I)/<g_i^a, d_i>\}$
Restarting rules	
The number of function evaluations per linear search exceeds a given value for several successive searches, and $(n+1)$ iterations have been completed since the latest restart	$a_i^{jj} < 0$ for some j <u>OR</u> $<g_i^a, H_i g_i^a> < 0$ <u>OR</u> $+H_i g_i^a$ is a direction of descent at x_i
Restart directions at x_i	
the coordinate lines through x_i	$-g_i^a$
$\pm\, H_i g_i^a$ are not directions of descent at x_i	
searches along the coordinate lines through x_i	algorithm terminates

Table 1 (continued)

MR1	Stewart
Stopping rule	
Each coordinate line through x_i is not a line of descent	$\|d_i^j\| < \epsilon$ and $\|\Delta x_i\| < \epsilon$ for some ϵ specified apriori

Table 1 requires some explanatory comments. The subscripts i refer to the iteration number, super-scripts j refer to the component of the vector under consideration. In some cases the iteration indicator has been omitted for simplicity. Moreover, the subscript j has been used on the coordinate directions e_j.

Since directions of search other than $\pm H_i g_i^a$ are used in MR1, the algorithm must keep track of H_i and its inverse, the approximation to the Hessian of f, G_i. a^{jj} in the approximation to the jth component of the gradient of f at x_i is the jth diagonal element of G_i. This second order correction does more than just im-prove the terminal convergence of the algorithm. In fact if f is a quadratic function and $G_i = Q$ then g_i^a in MR1 equals the gradient of f, g_i, for any size h^j. In MR1 lines of search, not rays are used. As stated earlier, the rank one scheme allows arbitrary directions of search. The coordinate lines (any appropriate set of independent directions would work just as well) prevent premature convergence of the algorithm.

Two linear search strategies were used. One scheme, Search Crude (C) is a nonminimizing scheme that attempts only to determine a point x on the line of search through x_i such that $f(x) < f(x_i)$. The other scheme, Search 'Accurate' (A) is a minimizing scheme that attempts to bracket the minimal point of f on the line of search and then fits a parabola to data on this line. In both searches, the initial step is that given in Table 1. Usually, this step is 1 unit in size and takes one to the stationary point of the current approximating quadratic,

$$q_i(x) = <G_i x, x>/2 + <g_i^a - G_i x_i, x>.$$

Stewart used the F-P search scheme [5] which requires an apriori lower bound f_L on the function f. Such a bound may not be easy to determine.

Stewart's restarting strategy parallels the DFP strategy, checking the positive definiteness of the matrices H_i. The strategy in MR1 is totally different. Since the initial step size in each of the linear searches utilizes the current approximations to the gradient and the Hessian of f, it seems reasonable to connect the restarting procedure to the number of function evaluations required to perform the linear searches. Hence, if at each of q successive iterations more than r function evaluations are required (where q and r depend upon the search strategy used) and n+1 iterations have been completed since the latest restart then it is assumed that the current approximation to the Hessian is inaccurate and the algorithm is restarted. When the algorithm is restarted, $G_i = H_i = I$ and a coordinate line of descent is used. (The coordinate lines are considered cyclically).

The stopping rule in MR1, that a successive check of the coordinate lines through x_i yields no lines of descent prevents premature convergence but does not provide sharp termination if g_i^a is not an accurate approximation to g_i. The algorithm was programmed and run on an IBM 360-91 computer with the standard test functions Rosenbrock, Powell, Fletcher-Powell, Chebyquad, and a 7-dimensional quadratic function. Stewart's paper [2] contains results only for cases where g_i^a is an accurate approximation to the gradient of the function being minimized. His results were compared with results obtained using MR1 with Search A, or Search C, MDFP

with Search A or Search C and MDFP1 with Search A or
Search C. MDFP denotes the algorithm obtained from
MR1 if the DFP scheme is used in the updating of the
approximation to the Hessian; and MDFP1 denotes the
algorithm obtained from MDFP if the gradient approximation are replaced by those used by Stewart. Hence,
MDFP1 is a rank 2 algorithm with a Stewart type gradient
approximation but with the MR1 rules with respect to the
linear search schemes, restarting, directions of search,
formula for h^j_i and stopping rule. For those tests with
inaccurate gradient approximations MR1 was compared
with MDFP and MDFP1.

In those cases with g^a_i an accurate approximation
to the gradient, and $g^a_i \rightarrow g_i$ as $i \rightarrow \infty$, Stewart's scheme,
and MR1, MDFP, and MDFP1 all with Search A behaved
very similarly except on the one test function whose
Hessian is singular at the minimal point, Powell's function of four variables. For this function MR1 with Search
A or C outperformed Stewart's scheme and the other
rank 2 schemes. When g^a_i was not an 'accurate' approx-
imation to the gradient of f, $g^a_i \nrightarrow g_i$ as $i \rightarrow \infty$; MR1
with either Search A or Search C produced significantly
better reductions in $f^* \equiv f\text{-}f(\min)$ than either rank 2
scheme before it degenerated into pure coordinate
searches.

The question, can the number of function evaluations required to reduce f^* to a specified value be reduced by use of a nonminimizing search must be an-
swered, not usually. Although in certain cases significant reductions were obtained. Algorithm MR1 with
Search C generally required at each iteration 0 or at
most 1 extra function evaluation (other than the values
$f(x_i)$ and $f(x_{i+1})$) to perform a crude linear search.)
(The first n iterations after a start or restart each generally required 2 or 3 function evaluations). However,
this decrease was balanced by a corresponding increase
in the number of iterations. Table 2 demonstrates this
balancing effect. The test function was the Chebyquad
function with n=8. The results are compared with

Powell's no derivative scheme because Stewart [2] did not test this function. In Table 1, g_i^a is an accurate approximation to the gradient of f; the simplest version of Stewart's formula was used to compute each h_i^j. Observe that $h_i \nrightarrow 0$ as $i \rightarrow \infty$ because f(min)=.0035... > 0. $f^* \equiv f(x_i)-f(min)$, FE denotes the number of function evaluation and ITN denotes the iteration number. R denotes the number of times the algorithm restarted, SW denotes the number of times + $H_i g_i^a$ was the direction of descent, CD denotes the number of coordinate lines used, and ND denotes the number of times the line $x=x_i+a H_i g_i^a$ was not a line of descent at x_i.

In Table 3, f=Powell's function of 4 variables, and the simplest version of Stewart's formula was used to compute h_i. Observe that $h_i \rightarrow 0$ as $i \rightarrow \infty$ and h_i is small everywhere. Observe that MR1 with Search A outperformed MR1 with Search C and Stewart's scheme.

In Table 4, again f=Powell's function of 4 variables. Here, h_i is fixed apriori; $h_i^j = .05$, i=1, 2,... j=1,...,n. Four algorithms are compared, MR1 with Search A, MR1 with Search C, MDFP with Search A, and MDFP1 with Search A. Clearly, MDFP1 which uses the Stewart gradient approximation is out of the running. The other three algorithms performed very similarly (the basis of comparison is the number of function evaluations required to obtain a specified reduction in f^*) down to $f^* = 3 \times 10^{-3}$. After which MR1 with Search A and MR1 with Search C both outperformed MDFP with Search A.

One unsolved problem is the question of termination of the algorithm when g_i^a is not necessarily an accurate gradient approximation. Each of the algorithms considered degenerates into ordinary coordinate searches after some unspecified number of iterations, for example MR1 with Search C in Table 4 degenerated into coordinate searches at step 42.

The last table, Table 5, considers the Fletcher-Powell function with $h_i^j = .05$, j=1,...,n, i=1, 2,... . In this case both MDFP1 and MDFP with Search A performed very badly. MDFP1 degenerated into coordinate searches at iteration 46 and the decrease to 6×10^{-2} was due to coordinate searches. MDFP died at iteration 42 and the decrease to 9×10^{-4} was also due to coordinate searches. MR1 with Search C died near iteration 40 but the decrease to $f = 7 \times 10^{-5}$ was legitimate and not from the coordinate searches. MR1 with Search A died at iteration 26 so again a reduction to $f = 6 \times 10^{-5}$ was attained before the algorithm degenerated. The numerical results clearly demonstrate that the rank 1 scheme MR1 will outperform a rank 2 scheme with Stewart's gradient approximation when the change in x allowed to compute the gradient approximations cannot be made arbitrarily small. However, they also demonstrate that the number of function evaluations required by MR1 with Search A or C to achieve a specified reduction in f^* depends critically upon the degree of accuracy of the gradient approximation. In fact, for example see Tables 3, 4 with $f =$ Powell's function of 4 variables. MR1 with Search A and accurate gradient approximations required 71 function evaluations to reduce f to 10^{-4}. Powell's no derivative scheme required 138 function evaluations to make the same reduction. A ratio of almost 2 to 1. However, with $h_i^j \equiv .05$ MR1 with Search A required 160 function evaluations to make the same reduction; and the balance tipped in favor of Powell's no derivative scheme. The question of which algorithm to use when h_i^j is restricted is open. If however, this restriction is really not very restrictive, for example $h_i^j = 10^{-4}$ for $f = $ PF4V then the numerical results indicate that MR1 with either Search A or Search C should be used because the terminal convergence is superior to that of the rank 2 schemes and the number of function evaluations required is about $1/2$ that required by Powell's no derivative scheme.

More detailed computational results and a few theoretical results are available in reference [6].

References

1. M. J. D. Powell, <u>Comp. J.</u> 7, 155, (1965).

2. G.W. Stewart, J. Assoc. Comp. Mach. 14, 72 (1967).

3. W.L. Zangwill, Comp. J. 11, 293 (1967).

4. R. Fletcher and M.J.D. Powell, Comp. J. 6, 163 (1963).

5. M.J.D. Powell, Integer Programming, ed J. Abadie, North-Holland, 139 (1970).

6. J. Cullum, IBM J. Res. Dev., (1972).

TABLE 2

$$f = \text{Chebyquad } (n=8), \quad h_i^j = 2\left| f(x_i) \, 10^{-8} / G_i^{jj} \right|^{\frac{1}{2}}$$

Itn	MR1-C		MR1-A		Powell	ND
	f^*	FE	f^*	FE	f^*	FE
0	4×10^{-2}	1	4×10^{-2}	1	4×10^{-2}	1
4	3×10^{-2}	46	1×10^{-2}	60	7×10^{-3}	91
8	5×10^{-3}	85	4×10^{-3}	107	2×10^{-3}	194
16	2×10^{-3}	161	3×10^{-4}	201	2×10^{-5}	385
24	4×10^{-6}	238	2×10^{-12}	294	6×10^{-13}	5 37
30	3×10^{-13}	29 2				
R	0		0		-	
SW	2		5		-	
CD	0		1		-	
ND	0		1		-	

TABLE 3

$$f = \text{Powell's function } (n=4), \quad h_i^j = 2\left| f(x_i) \, 10^{-8} / G_i^{jj} \right|^{\frac{1}{2}}$$

Itn	MR1 - C		MR1-A		Stewart	
	f^*	FE	f^*	FE	f^*	FE
0	2×10^{2}	1	2×10^{2}	1	2×10^{2}	1
4	5×10^{1}	29	2×10^{-1}	31	6×10^{-2}	37
8	1×10^{0}	49	9×10^{-4}	63	3×10^{-3}	69
12	1×10^{-2}	69	8×10^{-8}	94	3×10^{-5}	104
16	1×10^{-4}	89	2×10^{-10}	1 27	1×10^{-8}	139
18	2×10^{-5}	99	8×10^{-11}	148	9×10^{-9}	15 8
24	3×10^{-8}	1 29	5×10^{-12}	212		
R	1		1			
SW	2		3			
CD	1		1			
ND	0		0			

TABLE 4

f = Powell's function (n= 4) , $h_1^j = .05$

Itn	MR1-A		MR1-C		MDFP-A		MDFP1-A	
	f	FE	f	FE	f	FE	f	FE
0	2×10^2	1	2×10^2	1	2×10^2	1	2×10^2	1
4	6×10^{-1}	30	5×10^0	21	6×10^{-1}	33	6×10^{-1}	29
8	1×10^{-1}	68	1×10^0	41	1×10^{-1}	71	5×10^{-1}	57
15	5×10^{-3}	137	8×10^{-2}	76	3×10^{-3}	140	4×10^{-1}	107
25	2×10^{-6}	245	3×10^{-3}	126	3×10^{-4}	222	8×10^{-2}	199
30	3×10^{-7}	305	9×10^{-4}	151	2×10^{-4}	270	1×10^{-2}	246
	$\downarrow$		$\downarrow$		$\downarrow$		$\downarrow$	
R	2		1		5		1	
SW	1		2		0		0	
CD	7		6		7		1	
ND	5		5		2		0	

TABLE 5

f = Fletcher - Powell function (n = 3), $h_1^j = .05$

Itn	MR1-A		MR1-C		MDFP-A		MDFP1-A	
	f	FE	f	FE	f	FE	f	FE
0	2×10^3	1	2×10^3	1	2×10^3	1	2×10^3	1
4	3×10^1	26	1×10^1	17	3×10^1	26	2×10^1	26
12	4×10^0	77	5×10^0	55	7×10^0	89	9×10^0	81
20	2×10^{-2}	138	2×10^0	95	4×10^0	150	6×10^0	140
28	3×10^{-6}	218	2×10^{-2}	143	2×10^0	217	2×10^0	203
36	3×10^{-7}	319	1×10^{-3}	192	2×10^{-1}	296	5×10^{-1}	272
40	$\downarrow$		7×10^{-5}	229	1×10^{-1}	332	3×10^{-1}	322
			$\downarrow$		$\downarrow$			
R	1		1		4		3	
SW	4		4		0		0	
CD	9		7		8		5	
ND	7		4		4		2	

AN APPROXIMATION ALGORITHM FOR n-th DERIVATIVES
AND ITS APPLICATIONS TO ANALYSIS
AND TO STOCHASTIC INTEGRALS

L. C. Young

The University of Wisconsin
Madison, Wisconsin

1.   We write the geometric series identity in the
form expressing as a convex combination of powers P^j the
ratio

$$Q = \frac{1}{n} \frac{P^n - 1}{P - 1} = \frac{1}{n} \sum_{j=0}^{n-1} P_j$$

We shall need a simple consequence. Let $N=2M$ be an even
positive integer, and z a complex number $\neq 0, 1$. We choose

$$P = \frac{1}{N} \frac{(z^N-1) \, z^{-N/2}}{(z - 1) \, z^{-\frac{1}{2}}} = \frac{1}{2} \frac{1}{M} \sum_{m=0}^{M-1} (z^{m+\frac{1}{2}} + 2^{-m-\frac{1}{2}})$$

Since products of convex combinations are convex
combinations of products, the powers of P are convex combinations of positive and negative half-powers of z, and
the product $(2P-2)Q$ is a convex combination of the products
$(z^{m+\frac{1}{2}} + z^{-m-\frac{1}{2}}-2)P^j$ and therefore of the expressions

$$(z^{m+\frac{1}{2}} + z^{-m-\frac{1}{2}} - 2) \, z^{v/2}$$

where $0 \leq m \leq M-1$, $0 \leq |v| \leq (n-1)(N-1)$. Hence the
quantity

$$d(z) = N^{-n} ((z^N-1) \, z^{-N/2})^n - ((z-1) \, z^{-\frac{1}{2}})^n ,$$

which may be written

$$z^{-\frac{1}{2}n} (z-1)^n (P^n-1) = \frac{n}{2} z^{-\frac{1}{2}n} (z-1)^n (2P-2)Q \, ,$$

is a convex combination

$$d(z) = \sum_{m,v} \alpha_{mv} \, d_{mv}(z) \tag{1}$$

of the expressions

$$d_{mv}(z) = \frac{n}{2} ((z-1) z^{-\frac{1}{2}})^n (z^{m+\frac{1}{2}} + z^{-m-\frac{1}{2}})z^{\frac{1}{2}v} \, .$$

From (1) we deduce a finite difference identity for a vector-valued function F of a real t. For this we write

$$\Delta_h F(t) = F(t+h) - F(t), \quad \Delta^m_k{}^* F(t)=(\Delta_h)^m F(t-\tfrac{1}{2}mh) \ (m=1,2,--),$$

and for fixed n, N, h,

$$\Lambda F(t) = ((Nh)^{-n} \Delta^n_{Nh}{}^* - h^{-n} \Delta^n_h{}^*)F(t)$$

We shall prove that $\Lambda F(t)$ is the convex combination

$$\Lambda F(t) = \sum_{m,v} \alpha_{mv} \frac{n}{2} h^{-n} \Lambda_{mv} F(t) \tag{2}$$

of the expressions $\frac{n}{2} h^{-n} \Lambda_{mv} F(t).$[*)]

To this effect we set $z = e^{ih}$, $z^{\frac{1}{2}} = e^{ih/2}$, and we first take for F(t) the function $f(t) = e^{it}$, so that $f(t+vh/2) = z^{v/2} f(t)$, and hence

$$\frac{n}{2} \Lambda_{mv} f(t) = d_{mv} (z)f(t), \quad \Lambda f(t) = h^{-n} d(z) f(t).$$

which shows that (2) holds when F=f; evidently it holds also when F is the constant unity. By magnification it holds when $F(t) = e^{ict}$, and so by addition and passage to the limit when F is a continuous periodic function with values in some Euclidean space. To extend it to an

____

[*)] The identities (1),(2) are due partly to Mr. D.B. Liu.

arbitrary F, we keep t, h, n, N fixed and we alter the function outside the finite set of arguments which actually occur, so as to make it continuous and periodic.

2.   We now suppose the values of the function F in a normed space, and that F satisfies

$$|\Delta_h^{n*} \Delta_k^{2*} F(t)| \leq \phi(h) \, \psi(k) \, , \qquad (2.1)$$

where $\phi(u)$, $\psi(u)$ are continuous increasing for $u \geq 0$, while $u^{-n} \phi(u)$, $u^{-2} \psi(u)$ are decreasing and $\phi(0) = \psi(0) = 0$. (The conditions on ϕ, ψ can be relaxed). In addition we suppose that

$$S_n(h) = \int_0^h u^{-n} \phi(u) \, d\psi(u) \text{ converges.} \qquad (2.2)$$

By (2) we find

$$|(Nh)^{-n} \Delta_{Nh}^{n*} F(t) - h^{-n} \Delta_h^{n*} F(t)| \leq \frac{n}{2} \phi(h) \, \psi(Nh)/h^n. \qquad (2.3)$$

This weak form of a Cauchy convergence condition, coupled with (2.2) suffices to ensure the existence of a limit, and so the existence of a symmetric n-th derivative $F^{(n)*}(t)$ and the validity of the inequality

$$|h^{-n} \Delta_h^{n*} F(t) - F^{(n)*}(t)| \leq KS_n(h) \, , \qquad (2.4)$$

where K is a constant depending on n.[*)]

3.   <u>First application</u>. Let $n=2\lambda$, $F=f*g$, where f is periodic and L^2, and where $g(t)$ is the conjugate of $f(-t)$. From (2.4) for $t=0$ we derive a similar estimate for the square of the L^2-norm of

$$h^{-\lambda} \Delta_h^{\lambda} f - f^{(\lambda)}$$

[*)]

The argument is similar to that of [4] & 4.

provided the square of the L^2-norm of $\Delta^\lambda_h \Delta_k f$ is $< \phi(h)\,\psi(k)$. Here $f^{(\lambda)}$ is the λ-th derivative in L^2 of f. This is related to results of Stein and Zygmund [2], which are best possible of their kind.

4. <u>Second application</u>. Let n=1, F=f*g, where f,g are periodic and L^2, and where the L^2-norms of $\Delta_h f$, $\Delta_k g$ are respectively $\leq \phi(h)$, $\psi(k)$. Then (2.1) is satisfied, and the Stieltjes integral $\int f(t)\,dg(-t)$ exists in an appropriate sense and can be approximated with an error $\leq KS_1(h)$ by the finite difference ratio of the convolution F. This extends earlier results; see also recent work announced by Orlicz and Les'niewicz [1].

5. <u>Third application</u>. With again n=1, let f be in the integrated class Lip(1,ψ), and let g(t) be a stochastic process (a function with values in an L^2-space); f(t) can be deterministic -- see [4], or stochastic -- see [5]. By modifying the above we get weak conditions for the existence of the stochastic integral $\int f dg$. The main condition on g(t) is of the following type: in absolute value, its covariance for two non-overlapping t-intervals of lengths h,k is $\leq \phi_1(h)\,\phi_2(k)$. Stochastic integrals for processes with orthogonal increments or for martingales are the special case $\phi_1 = \phi_2 = 0$.

6. <u>Fourth application</u>. (A possible argument for the Riemann hypothesis RH without going into the critical strip; another is given by Turan [3]). We consider as function of $x \leq 0$ for fixed $s=\sigma+it$, $t \geq 1$, $\sigma > n + \tfrac{1}{2}$,

$$F(x) = \sum_{m=1}^\infty p_m^{-s}\, e^{xp_m} .$$

It differs trivially from $\log \zeta(s)$ when x=0; if it satisfies (2.1) for appropriate ϕ, ψ, then RH is true.

7. <u>Fifth application</u>. We cut the plane of $s=\sigma+it$ by half-lines parallel to the negative real axis from the

zeros of $\zeta(s)$, and consider for a constant σ in $\frac{1}{2} < \sigma < 1$, and on an interval T of $t > 1$ at whose ends $\sigma + it$ is not on a cut, the function $f(t) = \overline{\log} \zeta(\sigma + it)$. We use our first application with $\lambda = 1$, i.e. $n = 2$, but without the periodicity, which can be trivially arranged for. If, for each such σ, T, there exist appropriate ϕ, ψ so that on T the square of the L^2 norm of $\Delta_h \Delta_k f(t)$ is $\leq \phi(h) \psi(k)$, then RH is true. Conversely, RH implies these hypotheses, with trivial $\phi(u)$, $\psi(u)$ which are constant multiples of u^2.

REFERENCES

1. Orlicz and Les'niewicz, paper presented at Conference on Modular and Function Spaces, September 30 – October 5, 1971, Poznan, Poland.

2. Stein and Zygmund, On the Differentiability of Functions, Studia Math 23 (1964) 247-283.

3. Turan, On Some Approximating Dirichlet Polynomials in the Theory of the Zeta Function of Riemann, D Kgl Vidensk Selskab Mat fys Medd XXIV (1948) 17.

4. Young, L. C., Some New Stochastic Integrals and Stieltjes Integrals, Part I, Analogues of Hardy-Littlewood Classes, Advances in Probability 2 (1970) 161-240.

5. Young, L. C., Some New Stochastic Integrals and Stieltjes Integrals, Part II, Stochastic Integrals for Nigh-martingales, Advances in Probability 2 (1972) to appear (preliminary version available as Math Res. Center Report 937, University of Wisconsin, Madison, Wisconsin).

DISTRIBUTED SYSTEMS

ON THE OPTIMAL CONTROL OF DISTRIBUTED PARAMETER SYSTEMS

J.L. Lions

University Paris VI

and

Institut de Recherche d'Informatique et d'Automatique

Introduction

We study in this paper the optimal control of systems governed by Partial Differential Equations, on some of the many
situations arising in practical problems.

In Section 1, we study, following Yvon [1] , a problem of
optimal heating, with a non linear model.

In Section 2, we consider problems where the control variable is a geometrical element, actually a part of the boundary
of the domain. Cf. other examples in Lions [3] .

In Section 3, we study functionals related to free surface
problems, cf. Murat-Tartar [1] .

We present briefly some numerical algorithms in Section 4,
following D.Leroy [1] , J.P. Yvon [1] . Cf. also J.P. Kernevez
[2] , R. Glowinski [1] , J. Cea, R. Glowinski [1] and the biblio-
graphies of these papers, (cf. also the survey papers Lions [3]
[9]).

Without any aim at a complete bibliography, we wish to
mention here the works of Butkowski [1] , G. Duff [1][2] ,
S.K. Mitter [1] , D.R. Russel [1][5] , P.K.C.Wang [1] , and for
stochastic systems, Bensoussan [1][2][3] , Balakrishnan [1][2]
and the bibliography of these works. One can also see the survey of Robinson [1] . For numerical algorithms, cf. also Barnes
[1][2] , Bensoussan-Bossavit-Nedelec [1] , Bensoussan-Lions-
Temam [1] , Bosarge-Johnson [1] , Bosarge-Johnson-Mcknight-
Timlake [1] , Bosarge-Johnson-Smith [1] , Cornick-Michel [1] ,
Daniel [1] , El Fattah [1] .

1.   A non linear distributed system .

1.1 Physical problem. Mathematical model.

The problem we are going to describe arises in the optimal control of the heating of a furnace. A detailed report is given in J.P. Yvon [1] . We present here the mathematical model of the system.

Let Ω_1 and Ω_2 be two open sets of $\mathbb{R}^3$ (shaded regions on Fig. 1) ; we denote by Γ_1 the boundary of Ω_1 and by Γ_2 and S the two components of the boundary of Ω_2.

The state $y = \{y_1, y_2\}$ is given by functions $y_1 = y_1(x,t)$ (*) and $y_2 = y_2(x,t)$ which are respectively defined in

$$Q_1 = \Omega_1 \times]0,T[\ , \quad Q_2 = \Omega_2 \times]0,T[\ ,$$

where T is given (>0).

After proper choice of units, y_1 and y_2 satisfy the following system of non linear P.D.E. :

(1.1)
$$\frac{\partial y_1}{\partial t} + A_1 y_1 = f_1 \quad \text{in} \quad Q_1$$

(1.2)
$$\frac{\partial y_2}{\partial t} + A_2 y_2 = f_2 \quad \text{in} \quad Q_2$$

where A_1 and A_2 are linear elliptic operators defined in Ω_1 and Ω_2 ; to simplify the exposition (but this is by no means essential), we shall assume that :

(1.3) $A_1 = - k_1 \Delta \ , \quad A_2 = - k_2 \Delta \ ,$ where k_1 and k_2 are >0

constants ; in the right hand sides of (1.1)(1.2) f_1 and f_2 are given functions of x and t.

The equations (1.1)(1.2) are linear but the problem is non linear because of the boundary conditions on Γ_1 and Γ_2 .

We impose that :

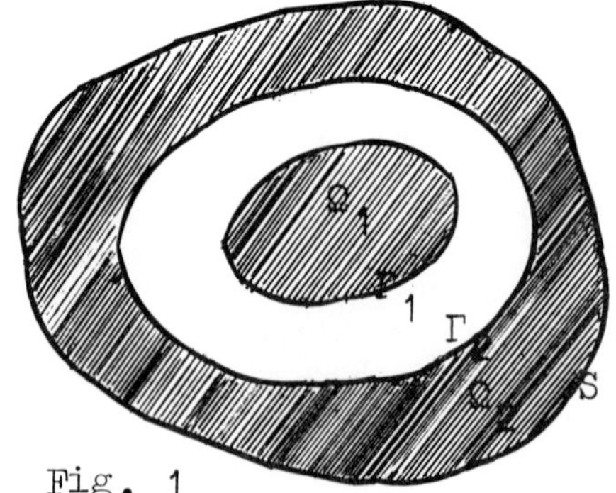

Fig. 1

(*) $x = (x_1, x_2, x_3) \in \mathbb{R}^3$.

(1.4) y_1 (resp y_2) does not depend on x on Γ_1 (resp Γ_2)
in other words :

(1.5) $y_1(x,t) = \alpha_1(t)$, $x \in \Gamma_1$; $y_2(x,t) = \alpha_2(t)$, $x \in \Gamma_2$

where α_1 and α_2 are <u>not</u> given ; we want next y_1 and y_2 to satisfy the non linear conditions (*)

$$(1.6)\begin{cases} \displaystyle\int_{\Gamma_1} \frac{\partial y_1}{\partial n}\, d\Gamma_1 + \beta_{11}\, y_1^4 + \beta_{12}\, y_2^4 = \gamma_1\, v(t) \\[2em] \displaystyle\int_{\Gamma_2} \frac{\partial y_2}{\partial n}\, d\Gamma_2 + \beta_{21}\, y_1^4 + \beta_{22}\, y_2^4 = \gamma_2\, v(t) \end{cases}$$

where β_{ij}, γ_i are given constants, $\gamma_i \geqslant 0$ and where $v = v(t)$ denotes the control at our disposal (to some extent !) (**); summing up, <u>the boundary conditions on Γ_1 and Γ_2 are given</u> by (1.5)(1.6). Of course, one has <u>a boundary condition on S</u>, for instance

(1.7) $y_2 = 0$ on S.

The <u>initial conditions</u> are :

(1.8) $y_1(x,o) = y_{o1}(x)$ in Ω_1 , $y_2(x,o) = y_{o2}(x)$ in Ω_2,

y_{o1} and y_{o2} being given functions.

Under suitable conditions on the constants β_{ij} we briefly indicate in Section 1.2 below how one can show that the problem (1.1)...(1.6), admits a solution. Therefore, for any given $v(t)$ ($\geqslant 0$), y_1 and y_2 are defined as functionals of v :
$$y_1 = y_i(x,t) = y_i(x,t\ ;\ v)\quad (***).$$
<u>The cost function</u>   is given by

$$(1.9)\qquad J(v) = \int_{Q_1} |y_1(x,t;v) - z_d(x,t)|^2 dx\ dt + \nu \int_0^T v(t)^2\ dt$$

where ν is given ($\geqslant 0$) and where z_d is given in $L^2(Q_1)$.

(*) $\frac{\partial}{\partial n}$ denotes the normal derivation on Γ_i directed towards the exterior of Ω_i.
(**) $v(t)$ represents the input of combustible at time t.
(***)   the uniqueness of the solution seems very likely but it is not proven. Therefore $y_i(x,t;v)$ denotes <u>a</u> solution of the state equations.

The set $\mathcal{U}_{ad}$ of <u>admissible controls</u> is given by

$$(1.10) \qquad \mathcal{U}_{ad} = \{ v \mid 0 \leqslant v(t) \leqslant M \}$$

<u>The problem is now to minimize</u> $J(v)$ <u>on</u> $\mathcal{U}_{ad}$.

We indicate in Section 1.3 how one can show that this problem admits a solution, necessary conditions are given in Section 1.4 ; numerical algorithms are given in Section 5.

1.2 <u>Solution of the state equations</u> :

Some notations are indispensable. We recall the definition of the <u>Sobolev space</u> $H^1(\mathcal{O})$ (see Sobolev [1] , Lions-Magenes[1]) defined on an open set $\mathcal{O}$ of $\mathbb{R}^3$:

$$H^1(\mathcal{O}) = \{ \varphi \mid \varphi, \frac{\partial \varphi}{\partial x_i} \in L^2(\mathcal{O}) \} \; ;$$

it is a Hilbert space when provided with the norm $^{(*)}$

$$\left(\int_\Omega [\varphi^2 + \sum_{i=1}^{3} (\frac{\partial \varphi}{\partial x_i})^2] \, dx \right)^{\frac{1}{2}} .$$

we define

$$(1.11) \qquad V = \{ \varphi \mid \varphi = \{ \varphi_1 , \varphi_2 \} , \varphi_1 \in H^1(\Omega_1), \varphi_2 \in H^1(\Omega_2),$$

$$\varphi_1 = \text{constant on} \Gamma_1 , \varphi_2 = \text{constant on } \Gamma_2 \; ^{(**)} ,$$

$$\varphi_2' = 0 \text{ on } S \}$$

It is easily checked that V is a <u>closed subspace of</u> $H^1(\Omega_1) \times H^1(\Omega_2)$. We shall set :

$$(1.12) \qquad \begin{cases} a_1(\varphi_1, \Psi_1) = k_1 \sum_{i=1}^{3} \int_{\Omega_1} \frac{\partial \varphi_1}{\partial x_i} \frac{\partial \Psi_1}{\partial x_i} \, dx \\ a_2(\varphi_2, \Psi_2) = k_2 \sum_{i=1}^{3} \int_{\Omega_2} \frac{\partial \varphi_2}{\partial x_i} \frac{\partial \Psi_2}{\partial x_i} \, dx , \end{cases}$$

$$(1.13) \qquad (\varphi_1, \Psi_1) = \int_{\Omega_1} \varphi_1 \Psi_1 \, dx, \quad (\varphi_2, \Psi_2) = \int_{\Omega_2} \varphi_2 \Psi_2 \, dx ,$$

$$(1.14) \qquad < \varphi_1 > = \int_{\Gamma_1} \varphi_1 \, d\Gamma_1 , \quad < \varphi_2 > = \int_{\Gamma_2} \varphi_2 \, d\Gamma_2 .$$

(*) We only consider <u>real valued</u> functions
(**)These constants <u>depend</u> on φ_1 and on φ_2 !

We shall write y_1' , y_2' instead of $\dfrac{\partial y_1}{\partial t}, \dfrac{\partial y_2}{\partial t}$.

We are going to show that, if we define b_{ij} and c_i by

$$(1.15) \quad \begin{cases} b_{11}(\text{meas } \Gamma_1)^5 = \beta_{11} \ , \ b_{11}(\text{meas } \Gamma_1)^4(\text{meas } \Gamma_2) = \beta_{12}, \\ c_1 \ (\text{meas } \Gamma_1) = \gamma_1 \ \text{and similar formula by exchanging} \end{cases}$$

indices,

then the problem of Section 1.1 is equivalent to the problem
of finding y_1, y_2 which satisfy :

$$(1.16) \quad (y_1', \varphi_1)+ a_1(y_1, \varphi_1)+(b_{11}<y_1>^4+ b_{12}<y_2>^4)<\varphi_1> =$$

$$= c_1 v(t)< \varphi_1 > + (f_1(t), \varphi_1) \ ,$$

$$(1.17) \quad (y_2', \varphi_2)+ a_2(y_2, \varphi_2)+(b_{21}<y_1>^4+ b_{22}<y_2>^4)<\varphi_2> =$$

$$= c_2 v(t) <\varphi_2 > + (f_2(t), \varphi_2)$$

$$\forall \ \varphi = \{ \varphi_1 \ , \ \varphi_2\} \in V \ ,$$

$$(1.18) \quad y(t) = \{ y_1(t) \ , \ y_2(t)\} \in V \qquad (*)$$

with the initial conditions (1.8).

It remains now to solve the problem (1.16)(1.17)(1.18)(1.8).
We proceed in two steps. We first introduce the modified system :

$$(1.19) \quad \begin{cases} (z_1', \varphi_1)+ a_1(z_1, \varphi_1)+(b_{11}<z_1>^4+ b_{12}<z_2>^4) \ \ldots \\ \qquad \ldots (\text{sign}<z_1>) <\varphi_1> = \\ \qquad = c_1 \ v(t)<\varphi_1> + (f_1(t), \varphi_1), \end{cases}$$

$$(1.20) \quad \begin{cases} (z_2', \dot{\varphi}_2)+ a_2(z_2, \varphi_2)+(b_{21}<z_1>^4+ b_{22}<z_2>^4) \ \ldots \\ \qquad \ldots (\text{sign}<z_2>) <\varphi_2> = \\ \qquad = c_2 \ v(t) <\varphi_2> + (f_2(t), \varphi_2), \end{cases}$$

the other conditions being unchanged. If, in (1.19)(1.20), we
take $\varphi_1 = z_1$, $\varphi_2 = z_2$, we obtain, adding up the two equations :

$(*) \qquad y_i(t)$ denotes the function $\longrightarrow y_i(x,t)$.

$$(1.21) \quad (z_1', z_1) + (z_2', z_2) + a_1(z_1, z_1) + a_2(z_2, z_2) + X =$$

$$= (c_1 <z_1> + c_2 <z_2>) \, v(t) + (f_1, z_1) + (f_2, z_2)$$

where

$$X = b_{11} |<z_1>|^5 + b_{12} <z_2>^4 |<z_1>| + b_{21} <z_1>^4 |<z_2>| + b_{22} |<z_2>|^5 .$$

Let us assume that

$$(1.22) \qquad\qquad b_{11} > 0 \;, \quad b_{22} > 0 \;, \quad \text{and that}$$

$$(1.23) \qquad\qquad b_{12} \geqslant 0 \;, \quad b_{21} \geqslant 0$$

<u>or</u> that $b_{12} < 0$, $b_{21} < 0$ and that there exist constants λ and μ such that

$$\begin{cases} b_{11} - (\dfrac{1}{p'\lambda^{p'}} \, |b_{12}| + \mu^p \, |b_{21}|) > 0, \\[2ex] b_{22} - (\dfrac{\lambda^p}{p} \, |b_{12}| + \dfrac{1}{p'\mu^{p'}} \, |b_{21}|) > 0, \quad p = 5/4, \; p' = 5. \end{cases}$$

Then one checks that

$$X \geqslant \beta \, (|<z_1>|^5 + |<z_2>|^5), \quad \beta > 0$$

and it is easy to get from (1.21) the a priori estimate [*]

$$|z_1(t)|^2 + \int_0^t \|z_1\|_{H^1(\Omega_1)}^2 \, d\sigma + \int_0^t |<z_1>|^5 \, d\sigma \leqslant$$

$$\leqslant c_3 \int_0^t |v(\sigma)|^2 \, d\sigma + c_3 \int_0^t |f_1(\sigma)|^2 \, d\sigma + c_3 \int_0^t |f_2(\sigma)|^2 \, d\sigma$$

and a similar one for z_2.

From these estimates and known methods (using compactness arguments) (cf. Lions [2]) we obtain <u>the existence</u> of z_1, z_2 <u>solutions of the modified problem</u>, with

$$z = \{ z_1, z_2 \} \quad \text{satisfying :}$$

____

[*] i.e. $t \longrightarrow z(t)$ is measurable from $(0,T) \longrightarrow V$ and $\int_0^T \|z(t)\|_V^2 \, dt < \infty$.

$$(1.24) \qquad \begin{cases} z \in L^2(0,T\,;\,V)\,, \qquad (*) \\ <z_i> \,\in\, L^5(0,T). \end{cases}$$

But

$$(1.25) \qquad v,\ f_1,\ f_2,\ y_{o1}\ \underline{\text{and}}\ y_{o2}\ \underline{\text{are}}\ \geq\ 0$$

and if (1.22) (1.23) hold true, <u>then one proves that any solut-</u> <u>ion</u> z <u>of the modified problem satisfies</u>

$$(1.26) \qquad z_i \geq 0\,, \qquad i = 1,\ 2.$$

Then in particular $<z_i>\ \geq 0$ and z is actually <u>a solution</u> of the <u>original problem</u>. Hence we have proved that, <u>under the</u> <u>hypothesis</u> (1.22) (1.23) <u>and</u> (1.25), <u>there exists a solution</u>

$$y = \{\ y_1,\ y_2\ \}$$

<u>of the original boundary value problem, satisfying</u>

$$y \in L^2(0,T\,;\,V),\quad <y_i>\,\in\, L^5(0,T).$$

[as we already noticed, the question of uniqueness is open.].

<u>1.3</u> <u>Existence of an optimal control</u> :

<u>We assume that</u> (1.22) (1.23) (1.25) <u>hold true. Then there</u> <u>exists</u>

$$u \in U_{ad}\ (\text{defined in } (1.10))\ \underline{\text{which satisfies}}$$

$$J(u) \leq J(v) \quad \forall\ v\ \in U_{ad}.$$

<u>Remark 1.1</u>

If the uniqueness of $y(v)$ solution of the state equat-tions seems likely, the uniqueness of the optimal control u seems doubtful but no counter example is known.

(*) where we set :

$$|\,z_1(t)\,|^2 = \int_{\Omega_1} z_1(x,t)^2\ dx$$

1.4. Necessary conditions :

The considerations of this section are partially _formal_. We assume that the _uniqueness_ is known. Then we show that $v \longrightarrow J(v)$ is _differentiable_ and _we compute_ its _derivative_. Necessary conditions are then standard. The main question is to show that the mapping $v \longrightarrow y(v)$ is differentiable from $\mathcal{U}_{ad} \longrightarrow L^2(0,T;V)$; if we admit this for a moment and if we set :

$$(1.27) \qquad y = \frac{d}{d\lambda} \left. y(u + \lambda v) \right|_{\lambda = 0} \qquad , \text{ then :}$$

$$(1.28) \quad \frac{1}{2} \frac{d}{d\lambda} \left. J(u+\lambda v) \right|_{\lambda=0} = \frac{1}{2}(J'(u),v) =$$

$$= \int_{Q_1} \hat{y}(y_1(u)- z_d) \, dx \, dt + \nu \int_0^T uv \, dt.$$

By using the definition (1.27) one can prove that $\hat{y}$ is the solution of the linearized problem

$$(1.29) \quad \begin{cases} (\hat{y}_1',\varphi_1)+a_1(\hat{y}_1,\varphi_1)+4(b_{11}<y_1>^3<\hat{y}_1>+b_{12}<y_2>^3<\hat{y}_2>)<\varphi_1> \\ \qquad\qquad = c_1 \, v(t)<\varphi_1> \, , \\ (\hat{y}_2',\varphi_2)+a_2(\hat{y}_2,\varphi_2)+4(b_{21}<y_1>^3<\hat{y}_1>+b_{22}<y_2>^3<\hat{y}_2>)<\varphi_2> \\ \qquad\qquad = c_2 \, v(t)<\varphi_2> \qquad\qquad \forall \varphi \in V, \\ \qquad\qquad \hat{y}(t) \in V \\ \qquad\qquad \hat{y}(o) = 0 \, ; \end{cases}$$

in (1.29) $\{ y_1, y_2 \} = y$ denotes the state for $v = u$.

2. Optimization of geometrical elements.

2.1 Design of the boundary. Hyperbolic operators.

Let $\mathcal{O}$ be a bounded open set in $\mathbb{R}^n$ ($n = 2$ or 3 in the applications) ; Γ_o is a fixed part of its boundary and let Ω_λ be a family of open sets satisfying, for $\lambda \in [o,1]$:

(2.1)
$$\Omega_\lambda \subset \mathcal{O}$$
$$\Omega_\lambda = \Gamma_o \cup S_\lambda .$$

We suppose that the S_λ are "smooth", $(n-1)$ dimensional varieties which depend "continuously" on λ , in the following sense :

(2.2)   if $\lambda \longrightarrow \lambda_o$, then distance $(S_\lambda , S_{\lambda_o}) \longrightarrow 0.$

Let us set :

(2.3) $Q = \mathcal{O} \times \,]0,T[\,$, $Q_\lambda = \Omega_\lambda \times \,]0,T[\,$.

<u>The state of the system</u> is given by the solution of :

(2.4) $\dfrac{\partial^2 y}{\partial t^2} - \Delta y = f$ in Q_λ ,   where   f   is given in

$L^2(Q)$ $^{(*)}$,   with the boundary conditions :

(2.5) $y = 0$ on $\partial \Omega_\lambda$, and **the initial conditions**

(2.6) $y(x,o) = y_o(x),\ \dfrac{\partial y}{\partial t}(x,o) = y_1(x),\ y_o, y_1$ given on $\mathcal{O}$.

This set of equations uniquely defines $y(x,t) = y(x,t;\lambda).^{(**)}$
<u>The cost function</u> is given by

(2.7) $J(\lambda) = \displaystyle\int_{Q_\lambda} |y(x,t;\lambda) - z_d(x,t)|^2 \ dx\ dt$

where z_d is given in $L^2(Q)$. <u>The problem is to minimize</u> $J(\lambda)$
<u>over</u> $[0,1]$. One can show that :

(2.8)   <u>there exists</u> λ_o <u>such that</u> $J(\lambda_o) \leqslant J(\lambda)\ \forall\ \lambda \in [0,1].$

2.2   Other cases.

One can consider similar functions for <u>parabolic systems</u>, with similar results (cf. Lions [3]). A physical situation leading to such a problem was indicated to us by T.L. Johnson[1].

(*)   so that in the right hand side of (2.4) f denotes actually the restriction of f to Q_λ.
(**)   so that we consider the restrictions of y_o and y_1 to Ω_λ.

One could consider as well similar problems with <u>non linear</u> state equations.

Let us consider another type of problem $(*)$. The notations are those of § 2.1. The state y is the solution of the <u>elliptic problem</u>

$$(2.9) \qquad\qquad -\Delta y = f \quad \text{in } \Omega_\lambda$$

where f is given in $L^2(\mathcal{O})$ (so that in the right hand side of (2.9), f denotes the restriction of f to Ω_λ), with the boundary conditions

$$(2.10) \qquad y = 0 \quad \text{on } \Gamma_o, \quad \frac{\partial y}{\partial n} = 0 \quad \text{on } S_\lambda .$$

Equations (2.9)(2.10) uniquely define $y = y(\lambda) \in H^1(\Omega_\lambda)$. <u>The cost function</u> is given by

$$(2.11) \qquad J(\lambda) = \int_{S_\lambda} |y(\lambda) - z_d|^2 \, dS_\lambda$$

where z_d is a given smooth function in $\mathcal{O}$ (so that in (2.11), z_d actually denotes the restriction of z_d to S_λ). In this case hypothesis (2.2) does not seem to be sufficient to imply the existence of an optimal λ . We shall make the following hypothesis :

$$(2.12) \quad \begin{bmatrix} \text{one can find a finite covering of the surfaces } S_\lambda \text{ by} \\ \text{open sets } G_1, \ldots , G_M \text{ and a family of maps } g_i \text{ which} \\ \text{are } C^1 \text{ from } G_i \longrightarrow [0,1]^n \text{ together with } g_i^{-1} \text{ and} \\ \text{which map } G_i \cap S_\lambda \text{ into }]0,1[^n \cap \{x_n = \lambda\}(**). \end{bmatrix}$$

One can show that <u>under hypothesis</u> (2.12), <u>there exists an</u> optimal λ .

3. <u>Optimization of functionals related to free surfaces.</u>

Let Ω be a bounded open set of $\mathbb{R}^n$. We define

$$(3.1) \qquad K = \{ \varphi \mid \varphi \in H^1(\Omega), \ \varphi = 0 \text{ on } \Gamma, \ \varphi \geqslant 0 \text{ a.e in } \Omega \}$$

$(*)$ a problem of this type (for a fourth **order elliptic** equation) was mentionned to me by I. Babuska $[1]$.

$(**)$ or $x_n = \lambda x_1$ in the neighborhood of $S_\lambda \cap \Gamma_o$.

and we set

$$(3.2) \qquad a(\varphi, \Psi) = \sum_{i=1}^{n} \int_{\Omega} \frac{\partial \varphi}{\partial x_i} \frac{\partial \Psi}{\partial x_i} \, dx.$$

Let f be given in $L^2(\Omega)$ and let us suppose that $v \in \mathcal{U}_{ad}$ where

$$(3.3) \qquad \mathcal{U}_{ad} = \underline{\text{bounded}} \text{ closed convex set of } L^2(\Omega).$$

The _state_ y is defined by

$$(3.4) \qquad \begin{cases} a(y, \varphi - y) \geqslant (f+v, \varphi - y) \ \forall \varphi \in K, \\ y \in K \end{cases}$$

which admits a unique solution $y = y(v)^{(*)}$.
The variational inequality (3.4) is equivalent to

$$(3.5) \qquad \begin{cases} - \Delta y - (f+v) \geqslant 0 \ , \quad y \geqslant 0 \ , \\ [-\Delta y - (f+v)]y = 0 \ \text{ a.e in } \Omega \ , \quad \text{and} \end{cases}$$

$$(3.6) \qquad y = 0 \ \text{ on } \Gamma.$$

It follows that there is a region $\mathcal{O}(v) \subset \Omega$ defined$^{(**)}$ by

$$(3.7) \qquad \mathcal{O}(v) = \{ x \mid y(x;v) = 0 \}$$

and such that $y(x;v) > 0$ a.e in $\mathcal{O}(v)$; the boundary of $\mathcal{O}(v)$ is the "free surface" (cf. C. Baiocchi [1]). We consider the cost function :

$$(3.8) \qquad J(v) = \text{meas} . \mathcal{O}(v)$$

and we are looking for

$$(3.9) \qquad \sup J(v), \quad v \in \mathcal{U}_{ad}.$$

(*)   cf. Lions-Stampacchia [1] for variational inequalities of type (3.4) with $a(\varphi, \Psi)$ <u>not</u> symmetric. In the present situation the existence and uniqueness of y, solution of (3.4), is trivial since (3.4) amounts to minimizing $\frac{1}{2} a(\varphi, \Psi) - (f+v, \varphi)$ over K.

(**)   up to a set of measure $\mathcal{O}$.

One has the following result (Murat-Tartar [1]) :

(3.10) $\qquad \left\{ \begin{array}{l} \underline{\text{under the hypothesis}}\ (3.3),\ \underline{\text{there exists}}\ u \in \mathcal{U}_{ad} \\ \underline{\text{such that}}\quad J(u) \geqslant J(v) \qquad \forall\ v \in \mathcal{U}_{ad}. \end{array} \right.$

<u>Remark 3.1</u> :

    We do not know if there exists $u \in \mathcal{U}_{ad}$ <u>minimizing</u> J over $\mathcal{U}_{ad}$.

<u>Remark 3.2</u> :

    There are many problems of optimal control where the state is given by the solution of a variational inequality. Cf. Duvaut-Lions [1][2] , J.P. Kernevez [3] , R. Glowinski [1] . For the study of variational inequalities, see H. Brézis [1][2], J.L. Lions [1] [2] [8] . For the solution of a free boundary problem (arising in dam's theory), cf. C. Baiocchi [1] .

4.   Some numerical algorithms.

<u>4.1</u>   <u>Study of a model problem. Numerical comparisons.</u>

    We briefly present in this section, the results of D. Leroy [1] . A <u>model</u> problem is considered. The state is given by the solution of

(4.1) $\qquad \left\{ \begin{array}{l} \dfrac{\partial y}{\partial t} - \dfrac{\partial^2 y}{\partial x^2} = f \ \text{ in } \ Q = \Omega \times\ ]\,0,T[\ ,\ \Omega = \ ]\,0,1[\ , \\[2ex] -\dfrac{\partial y}{\partial x}(o,t) = v(t),\quad \dfrac{\partial y}{\partial x}(1,t) = 0\ , \\[2ex] y(x,o) = y_o(x) \end{array} \right.$

and the cost function is given by

(4.2) $\qquad J(v) = \displaystyle\int_Q |y(v) - z_d|^2\ dx\ dt + \nu\int_0^T v^2\ dt,$

and we consider the "no constraint" problem i.e v spans the whole space $L^2(0,T)$.

Three numerical methods have been tested on this problem.

<u>Method 1. Penalty method.</u>

One can think of the state equations as constraints, and

then "penalize the equations" (cf. Lions [1] , Balakrishnan [3]
[4] for this idea). Numerical computations using this method
are reported in Yvon [1] (cf. also below).

Remark 4.1.

Of course, - as always in numerical analysis - there are
many choices : once it is decided to work with the penalty method, one has :

  i) the choice of the discretization method for the P.D.E
      (such as the choice between ordinary difference me-
      thods or finite element methods) ;
  ii) the choice of the algorithm for minimizing the dis-
      cretized cost function ; see Yvon [1] .

Method 2. Galerkin method. (*)

The idea is obvious : one approximates the space of con-
trols $\mathcal{U}$ (in this case $\mathcal{U} = L^2(0,T)$) by a finite dimensional sub
space generated by functions $w_1 \ldots w_m$; therefore we take

$$(4.3) \qquad v = \sum_{j=1}^{m} \mu_j w_j \ , \quad \mu_j \in \mathbb{R} \ ;$$

then $y(v) = \{$ linear form in $\mu = \{ \mu_1, \ldots, \mu_m \} \}$ + constant, and
the problem reduces to that of minimizing a quadratic function
in μ ; for this final step a method of conjugate gradient was
chosen.

Remark 4.2.

There are many possible reasonable choices for the w_j's :
step functions, piecewise polynomials functions, etc.. Cf.
numerical results below.

Method 3. Direct gradient method.

One computes $J'(v)$ - which introduces the adjoint state -
and one uses the algorithm

$$(4.4) \qquad u^{n+1} = u^n - \rho J'(u^n)$$

Numerical results. Results are reported on Fig. 2 and 3.
On this example, the gradient method is simultaneously
the most precise and the fastest.

______

(*) Or "Ritz-Galerkin method", or "Parametrization method".

4.2 Numerical results on problem of Section 1 :

We report here on the results of J.P. Yvon [1] . Computations were made by an adaptation of the Newton method (reported on Fig. 4 by Linearization + R. Galerkin or Linearization + Gradient) and by the direct gradient method (where the gradient is computed by the formula of Section 1).
The case "without constraints" was considered (taking into account (1.10) is, in any case, very simple).
The parameters ρ of the gradient method were chosen by the following (admittedly not very sophisticated) method : take ρ as large as possible and iterate ; when J starts to grow (instead of decreasing) replace ρ by $\rho/2$.
Here again the gradient method seems to be the most efficient.

Remark 4.3
The advantage of the gradient method seems to be more and more definite as ν becomes smaller.

Remark 4.4
Computations made by J.P. Kernevez [1] lead to similar conclusions. Fig. 5 shows the value of the optimal state on Γ_1 as ν gets smaller.

Remark 4.5
It seems quite clear on Fig. 5 that

$$\underset{v \in L^2(0,T)}{\text{Inf}} \int_{Q_1} \left| y_1(x,t;v) - z_d(x,t) \right|^2 dx\, dt$$

is not zero - i.e. that the set spanned by $y_1(x,t;v)$ (at least by the trace of y_1 on Γ_1) is not dense in $L^2(Q_1)$ (or, at least, in $L^2(\Gamma_1 \times]0,T[)$). Problems of this kind (in simpler non linear models) are studied by Bardos and Tartar [1] .

u(t)
1
GRADIENT computing time : 9 s.
RITZ-GALERKIN (3 diff. bases)
steps
piecewise linear
polynomials up to degree
Nbr. of elements
Computing Time s.
20   30
20   30
10.   11   12
EXACT SOLUTION
PENALIZATION
(computing time ≈5 - 10mn)
CONTROL
time
0
1

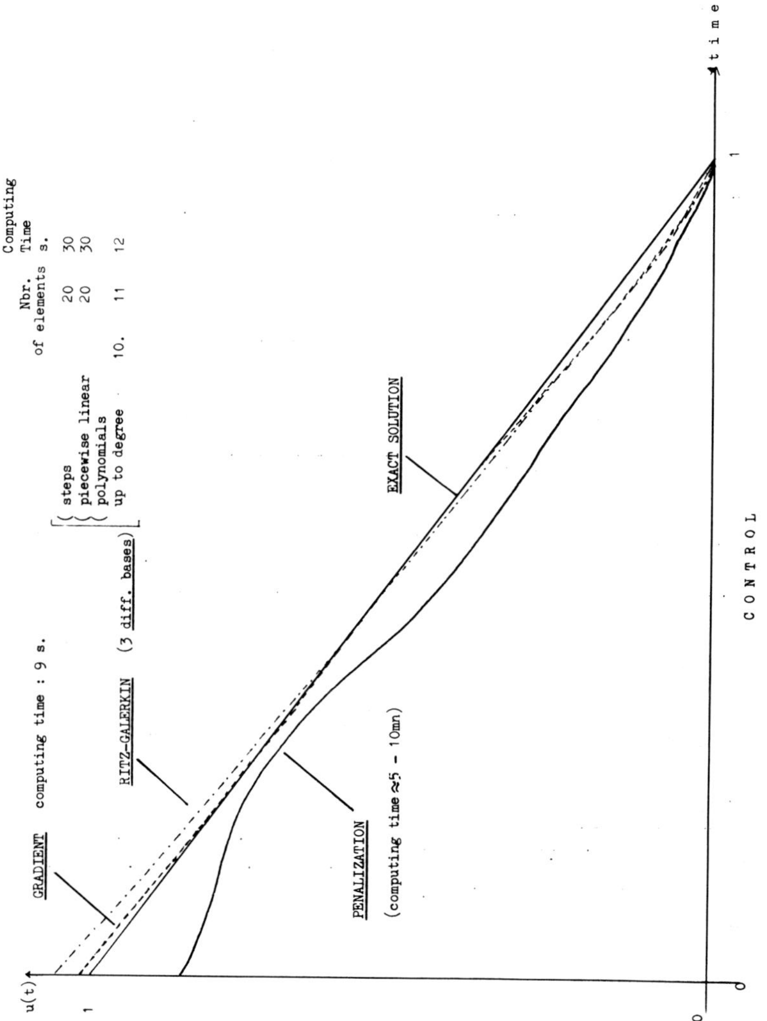

Figure 2

Number of discretization points :
in time : 40
in space : 20
Notice that the optimal control equals 0 for t = T = 1
Rel. ERROR on CONTROL %
10
5
0
PENALIZATION
GRADIENT
RITZ-GALERKIN
time
1
0
ERROR ON THE CONTROL

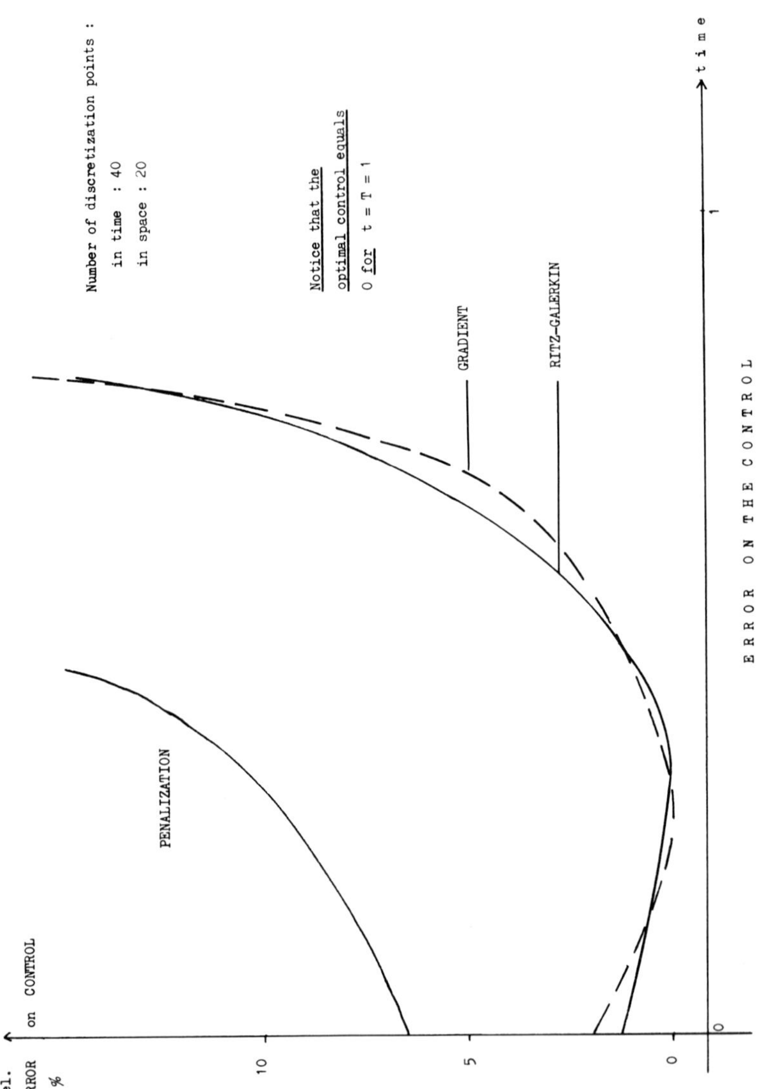

Figure 3

LINEARIZATION + R.- GALERKIN
LINEARIZATION + GRADIENT
GRADIENT
Cost value
5.09
4.94
4.99
u = optimal control
(computed for = 1)
time
COST FUNCTION :
$$J(v) = \int_0^T \int_{\Omega_1} |y_1(x,t;v) - z_d(x,t)|^2 \, dx \, dt + \nu \int_0^T |v(t)|^2 \, dt$$

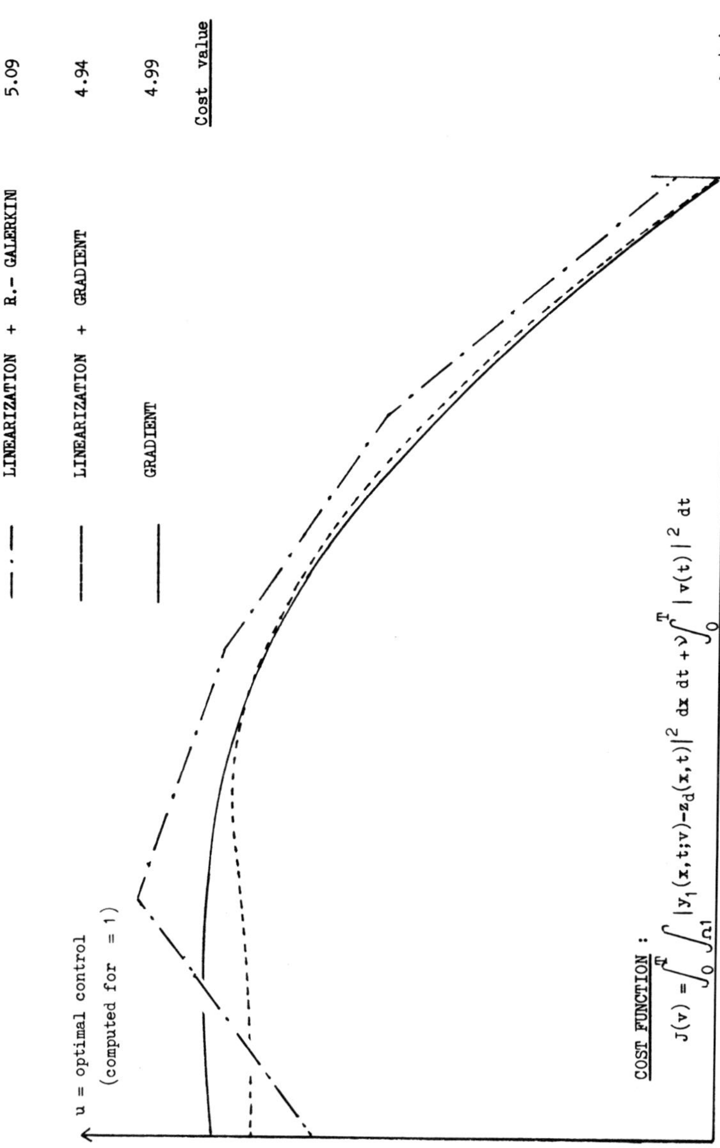

Figure 4

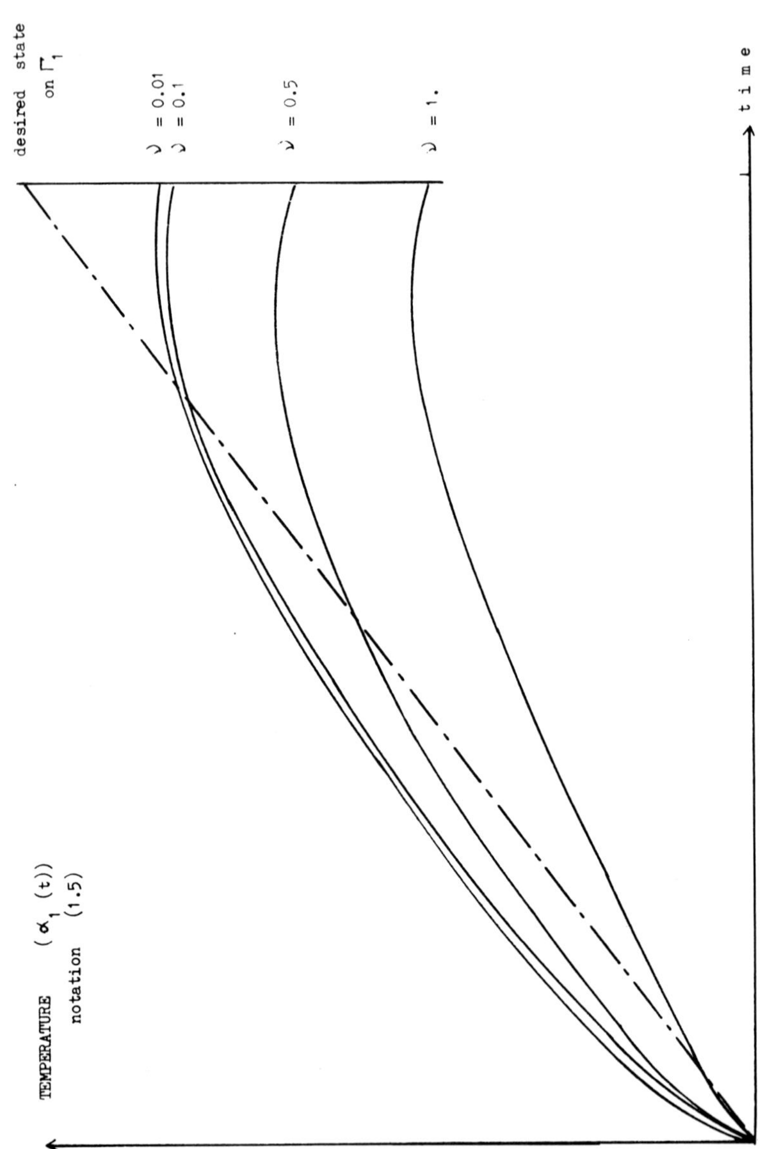

OPTIMAL STATE ON $\underline{\Gamma_1}$ (Computed by gradient method)

Figure 5

References

I. Babuska [1] Personnal communication. U.M.B.C. Aug. 1971.

C. Baiocchi [1] To appear.

A.V. Balakrishnan [1] Stochastic control : a function space approach. Siam Journal on Control, to appear.
[2] Stochastic Differential Systems. Acad.Press to appear.
[3] A new computing technique in system identification. Journal of Computer and System Science. 2 (1968), 102-116.
[4] On a new computing technique in Optimal Control. Siam Journal on Control, 6 (1968), 149-173.

C. Bardos & L. Tartar [1] To appear.

E.R. Barnes [1] Necessary and sufficient optimality conditions for a class of distributed parameter control systems. Siam Journal on Control, 9,(1), (1971), 62-82.
[2] Computing optimal controls in systems with distributed parameters. IFAC Symposium, Banff June 1971.

A. Bensoussan [1] Identification et Filtrage. Cahier de l'IRIA N° 1 (1969) p. 1-233.
[2] Filtrage optimal des systèmes linéaires. Paris, Dunod (1971).
[3] On the separation principle for distributed parameter systems. IFAC Symposium on the Control of Distributed Parameter Systems. Banff. Canada, June 1971.

A. Bensoussan, A. Bossavit & J.C. Nedelec [1] Approximation des problèmes de contrôle optimal. Cahier de l'IRIA N° 2, May 1970, 107-176.

A. Bensoussan, J.L. Lions & R. Temam [1] to appear.

W.E. Bosarge Jr. & O.G. Johnson [1] Error bounds of high order accuracy for the state regulator problem via piecewise polynomial approximation. Siam J. on Control, 9 (1971), 15-28.

W.E. Bosarge Jr., O.G. Johnson, R.S. McKnight & W.P. Timlake[1] The Ritz-Galerkin procedure for non linear control problems. I.B.M. Houston Scientific Center. May 1971.

W.E. Bosarge Jr., O.G. Johnson & C.L. Smith [1] A direct method approximation to the linear parabolic regulator problem over multivariate spline basis.

Houston Scientific Center. Dec. 1970.

A.G. Butkovskii [1] Theory of optimal control of distributed parameter systems. Moscow 1965. (English transla-
tion : American Elsevier, 1969).

H. Brezis [1] Equations et inéquations non linéaires dans
les espaces vectoriels en dualité. Annales
Inst. Fourier, 18 (1968) 115-175.
[2] Inéquations variationnelles. Journal de Mathématiques Pures et Appliquées. 1972.

J. Cea & R. Glowinski [1] Méthodes numériques pour l'écoule-
ment laminaire d'un fluide rigide viscoplas-
tique incompressible. To appear.

D.E. Cornick & A.N. Michel [1] Numerical optimization of distri-
buted parameter systems by gradient methods.
I.F.A.C. Symposium on Distributed Parameter
Systems. Banff. June 1971.

J.W. Daniel [1] Approximate minimization of functionals by
discretization : numerical methods in optimal
control. Center for Numerical Analysis. The
University of Texas, Austin, September 1970.

G.ED. Duff [1] A note on hydrodynamic optimization for tidal
barriers. To appear.
[2] Tidäl resonance and tidal barriers in the Bay
of Fundy system. J. Fish res. Bd. Canada. 27
(1970), 1701-1728.

J. Duvaut & J.L. Lions [1] Sur les Inéquations en Mécanique et
en Physique. Dunod, Paris (1971).
[2] To appear.

Y.M. El-Fattah [1] A direct method for optimization of distributed parameter systems with boundary control.
I.F.A.C. Symposium. Banff, Canada. June 1971.

R. Glowinski [1] These proceedings.
TL. Johnson [1] Personal communication. U.M.B.C., August 1971
J.P. Kernevez [1] Thesis, Paris (1971).
[2] These proceedings.
[3] To appear.
J.P. Kernevez & Thomas [1] Book. To appear. Dunod, Paris.
D. Leroy [1] Thesis 3d Cycle. Paris, 1972.
[2] IRIA Publication, 1971.
J.L. Lions [1] Contrôle optimal de systèmes gouvernés par
des équations aux dérivées partielles. Paris,
Dunod, Gauthier Villars, (1968). (English
Translation by S.K. Mitter, Grundlehren,
Springer 170, 1971).

J.L. Lions [2] <u>Quelques méthodes de résolution des problèmes aux limites non linéaires</u>. Paris, Dunod, Gauthier Villars, (1969).(English Translation by Le Van ; Holt, Rinehart, Winston, (1971).

[3] <u>Some aspects of the optimal control of distributed systems</u>. Regional Conference Séries in Applied Math. Siam Publication N° 6, (1972).

[4] Optimisation pour certaines classes d'équations d'évolution non linéaires. Annali Matematica Pura e Applic. LXXII (1966), 275-294.

[5] Sur quelques problèmes d'optimisation dans les équations d'évolution linéaires de type parabolique. In applications of Functional Analysis to Optimisation.

[6] On some non linear partial differential equations related to optimal control theory. Proc. Symposium Pure Mathematics. XVIII, Part 1. Chicago (1968). A.M.S. Publication (1970), 169-181.

[7] <u>Equations Différentielles Opérationnelles et Problèmes aux Limites</u>. Springer. (1961).

[8] On partial differential inequalities. Ouspechi Mat. Nauk. 26 : 2 (1971), (in Russian).

[9] Optimal control of Deterministic Distributed Parameter Systems. Banff. I.F.A.C. Symposium on the control of distributed parameter systems. To appear in Automatica.

J.L. Lions & E. Magenes [1] <u>Problèmes aux limites non homogènes et applications</u>. Paris; Dunod, Vol. 1,2,3 (1968, 1970). (English translation by P.Kenneth Springer, 1971,1972).

J.L. Lions & G. Stampacchia [1] Variational Inequalities. Comm. Pure applied Math. XX (1967), 493-519 .

S.K. Mitter [1] Optimal control of ditributed parameter systems. Control of Distributed Parameter Systems (1969), J.A.C.C. Boulder, Colorado (1969), 13-48.

F. Murat & L. Tartar. [1] To appear.

A.C. Robinson [1] A survey of optimal control of distributed parameter systems. Aerospace Research Laboratory. 69-0171, November (1969).

D.L. Russel [1] Optimal regulation of Linear Symmetric Hyperbolic Systems with Finite Dimensional Controls. Journal Siam 4. (1969), 194-276.

D.L. Russel [2] On boundary value control of linear symmetric hyperbolic Systems, in Mathematical Theory of Control, edited by Balakrishnan and Neustadt, Acad. Press,(1967), 312-321.

[3] Linear stabilization of the linear oscillator in Hilbert space. Journal of Mathematical Analysis Applic. 3 (1969), 663-675.

[4] Boundary value control of the higher dimensional wave equation. Siam Journal on Control, 9 (1971), 29-42.

[5] Control theory of hyperbolic equations related to certain questions in Harmonic Analysis and Spectral Theory. To appear.

S.L. Sobolev [1] Applications of Functional Analysis to Mathematical Physics. Leningrad (1950).

P.K.C. Wang [1] Optimal control of a class of Linear Symmetric hyperbolic systems with applications to plasma confinement. J. of Math. Analysis and Applications, 28 (1969), 594-608.

J.P. Yvon [1] IRIA Publication (1971).

[2] Application de la pénalisation à la résolution d'un problème de contrôle optimal. Cahier IRIA May (1970), 4-40.

[3] Applications des méthodes duales au contrôle optimal. To appear.

[4] Contrôle optimal de systèmes distribués à multi-critères. To appear.

-- § --

DUAL NUMERICAL TECHNIQUES FOR SOME VARIATIONAL
PROBLEMS INVOLVING BIHARMONIC OPERATOR
APPLICATION TO AN OPTIMAL CONTROL PROBLEM IN
THIN PLATES THEORY.

Dominique Bégis[*] and Roland Glowinski[**]

Institut de Recherche d'Informatique et d'Automatique (Paris)

Introduction

This paper is mainly devoted to numerical methods for solving
some fourth order elliptic variational problems coming from
thin plates theory.

In Part I, by using **duality** methods, via saddle-points, we
get efficient algorithms for solving various fourth order elliptic variational problems including the very classical biharmonic
problem :

$$(P_1) \quad \begin{cases} \Delta^2 u = f \text{ on } \Omega \\ u\big|_\Gamma = g_1 \\ \dfrac{\partial u}{\partial n}\big|_\Gamma = g_2 \end{cases}$$

where Δ^2 is the iterated Laplace operator (i.e $\Delta^2 = \Delta\Delta$ with

$$\Delta = \frac{\partial^2}{\partial x^2} + \frac{\partial^2}{\partial y^2}$$). Some numerical results

will be given.

In Part II, we apply Part I material to an optimal control
problem related to thin plates with unilateral condition on the
boundary ; the main difficulty is that the usual state equation
is replaced by a variational problem, of Part I type, so we have
to solve a hierarchical two levels optimization problem. Numerical
results will also be given.

On account of pages number limitation, most of the results will
be given without proofs ; for the proofs, see Bégis-Glowinski[1]
and Glowinski-Lions-Trémolières [1] , chap. 4.

* IRIA, 78 Rocquencourt, France.
** University of Paris-VI and IRIA, France.

I. Dual methods for some fourth order variational problems :

I.1 : Problems formulation.

Let Ω be an open, connected, bounded subset of R^N ($N=2$ in physical applications) and $\Gamma = \partial \Omega$ = boundary of Ω assumed to be smooth enough.

We shall denote by $\dfrac{\partial u}{\partial n}\Big|_{\Gamma}$ the outward normal derivative of the function u on Γ .

The following Hilbert spaces, of Sobolev type, are a convenient functional basis for the variational problems we shall consider :

$$H_o^1(\Omega) = \Big[v \mid v \in L^2(\Omega), \ \frac{\partial v}{\partial x_i} \in L^2(\Omega), \ i=1,N \ , \ v\big|_{\Gamma} = 0\Big]$$

$$H^2(\Omega) = \Big[v \mid v, \ \frac{\partial v}{\partial x_i}, \ \frac{\partial^2 v}{\partial x_i \partial x_j} \in L^2(\Omega), \ i,j = 1,N\Big]$$

$$H_o^2(\Omega) = \Big[v \mid v \in H^2(\Omega), \qquad v\big|_{\Gamma} = 0, \ \frac{\partial v}{\partial n}\big|_{\Gamma} = 0\Big]$$

$$V = H^2(\Omega) \cap H_o^1(\Omega).$$

We recall, Ω being bounded, that $v \longrightarrow \|\Delta v\|_{L^2(\Omega)} : V \longrightarrow R^+$ defines on V a norm equivalent to the standard norm induced by $H^2(\Omega)$. Let us consider now the following problems :

1) (P_1) definition :

(P_1) is the problem considered in the introduction with the additional assumption that $g_1 = g_2 = 0$ on Γ (*), so :

$$(1.1) \qquad (P_1) \qquad \begin{cases} \Delta^2 u = f \ , \quad f \in L^2(\Omega) \\[1mm] u\big|_{\Gamma} = 0 \\[1mm] \dfrac{\partial u}{\partial n}\big|_{\Gamma} = 0 \ . \end{cases}$$

Using Green formula, it is quite easy to prove that any solution of (P_1) is also solution of the optimization problem :

$$(1.2) \qquad \underset{v \in H_o^2(\Omega)}{\text{Min}} J_o(v) \qquad \text{and conversely.}$$

______

(*) The cases g_1 and/or $g_2 \neq 0$ are in practice no more difficult than the case $g_1 = g_2 = 0$.

In (1.2), $J_o : V \to R$ is defined by :

$$(1.3) \qquad J_o(v) = \int_\Omega |\Delta v|^2 \, dx - 2\int_\Omega fv \, dx$$

Problems are also equivalent to the variational equation :

$$(1.4) \qquad \begin{cases} \int_\Omega \Delta u \ \Delta v \, dx = \int_\Omega fv \, dx \qquad \forall \ v \in H_o^2(\Omega) \\ u \in H_o^2(\Omega) \end{cases}$$

2) (P_2) definition :

Being given a <u>positive</u> constant g and $f \in L^2(\Omega)$, $(\mathbf{P_2})$ is defined by :

$$(1.5) \qquad \begin{cases} \underset{v \in K_2}{\text{Min}} \left[J_o(v) + 2g \int_\Gamma \frac{\partial v}{\partial n} \, d\Gamma \right] \\ K_2 = \left[v \mid v \in V, \ \frac{\partial v}{\partial n} \geqslant 0 \quad \text{a.e on} \Gamma \right] \end{cases}$$

3) (P_3) definition :

The constant g and f being given as in (P_2) we define (P_3) by :

$$(1.6) \qquad \underset{v \in V}{\text{Min}} \left[J_o(v) + 2 \, g\int_\Gamma \left| \frac{\partial v}{\partial n} \right| \, d\Gamma \right]$$

For all the (P_i), $i = 1, 2, 3$, we have the following :

<u>Theorem 1.1</u> : (Existence and Uniqueness)
    All the (P_i) admit <u>one and only one</u> solution, denoted by u_i.

<u>Remark 1.1</u> : For more details about the physical and mathematical aspects of variational problems in thin plates theory, see Lions-Duvaut $[1]$, chap. IV.

I.2 : <u>Duality results</u>.

We shall use the min-max approach, as, for instance, in Glowinski-Lions-Trémolières $[1]$, chap. II, N° 4.

I.2.1. <u>Inf-sup equivalent formulation of the</u> (P_i).

Let us define the following subsets of $L^2(\Gamma)$:

$$
\left\{
\begin{array}{l}
\Lambda_1 = L^2(\Gamma) \\
\Lambda_2 = L^2(\Gamma) = [\mu/\,\mu \in L^2(\Gamma),\ \mu \leqslant 0 \text{ a.e on}\Gamma] \\
\Lambda_3 = \Lambda = [\mu/\,\mu \in L^2(\Gamma),\quad |\mu| \leqslant 1 \text{ a.e on}\Gamma]
\end{array}
\right.
$$

and we have the :

<u>Proposition 1.1</u> : The Λ_i $(i = 1,2,3)$ are closed, convex sets of $L^2(\Gamma)$. Let us now **introduce** the Lagrangians $\mathcal{L}_i$ by :

$$(1.7) \qquad \mathcal{L}_1(v,\mu) = J_0(v) + 2\int_\Gamma \mu \frac{\partial v}{\partial n}\, d\Gamma$$

$$(1.8) \qquad \mathcal{L}_2(v,\mu) = J_0(v) + 2g\int_\Gamma \frac{\partial v}{\partial n}\, d\Gamma + 2\int \mu\frac{\partial v}{\partial n}\, d\Gamma$$

$$(1.9) \qquad \mathcal{L}_3(v,\mu) = J_0(v) + 2g\int_\Gamma \mu\frac{\partial v}{\partial n}\, d\Gamma$$

The usefulness of the Λ_i and $\mathcal{L}_i$ appears in the :

<u>Proposition 1.2</u> : $\forall\, i = 1, 2, 3$, (P_i) is equivalent to :

$$
\underset{v \in V}{\text{Inf}}\ \ \underset{\mu \in \Lambda_i}{\sup}\ \mathcal{L}_i(v,\mu)
$$

<u>Proof</u> : This comes directly from the following results :

$$(1.10) \qquad \underset{\mu \in \Lambda_1}{\sup}\ \int_\Gamma \mu\frac{\partial v}{\partial n}\, d\Gamma =
\begin{cases}
0 & \text{if } \frac{\partial v}{\partial n} = 0 \ \ \mathbf{a.e} \\
+\infty & \text{if not.}
\end{cases}$$

$$(1.11) \qquad \underset{\mu \in \Lambda_2}{\sup}\ \int_\Gamma \mu\frac{\partial v}{\partial n}\, d\Gamma =
\begin{cases}
0 & \text{if } \frac{\partial v}{\partial n} \geqslant 0 \ \ \mathbf{a.e} \\
+\infty & \text{if not}
\end{cases}$$

$$(1.12) \qquad \underset{\mu \in \Lambda_3}{\sup}\ \int_\Gamma \mu\frac{\partial v}{\partial n}\, d\Gamma = \int_\Gamma \left|\frac{\partial v}{\partial n}\right|\, d\Gamma$$

I.2.2. <u>Saddle-points results.</u>

The key theoretical result of Part I is the :

<u>Theorem 1.2</u> :

$\forall\, i = 1, 2, 3,\, \mathcal{L}_i$ admits on $V \times \Lambda_i$ <u>one and only one</u> saddle-point, say (u_i,λ_i) where u_i is the solution of (P_i) and:

$$(1.13) \qquad \lambda_1 = -\Delta u_1 \big|_\Gamma$$

$$(1.14) \qquad \lambda_2 = - (\Delta u_2 + g) \text{ on } \Gamma$$

$$(1.15) \qquad \lambda_3 = - \frac{1}{g} \Delta u_3 \big|_\Gamma$$

<u>Remark 1.3</u> : The proof of Theorem 1.2 needs for u_i more regularity than we have usually in $H^2(\Omega)$; for that point and a proof of Theorem 1.2, see Bégis-Glowinski [1] and Glowinski-Lions-Trémolières [1] , chap. 4.

<u>I.2.3. An useful characterization of</u> (u_i, λ_i).

It is quite easy, and basic for computation purposes, to prove the following :

<u>Theorem 1.3</u> :
The (u_i, λ_i), $i = 1, 2, 3$, admit the following characterization :

$$(1.16) \qquad \begin{cases} \Delta^2 u_1 = f \\[4pt] u_1 \big|_\Gamma = 0 \\[4pt] \dfrac{\partial u_1}{\partial n} \Big|_\Gamma = 0 \\[4pt] \Delta u_1 \big|_\Gamma = -\lambda_1 \end{cases}$$

$$(1.17) \qquad \begin{cases} \Delta^2 u_2 = f \\[4pt] u_2 \big|_\Gamma = 0 \\[4pt] \Delta u_2 \big|_\Gamma = - (g + \lambda_2) \\[4pt] \displaystyle\int_\Gamma (\mu - \lambda_2) \frac{\partial u_2}{\partial n} \, d\Gamma \leqslant 0 \qquad \forall \, \mu \in L^2(\Gamma) \\[4pt] \lambda_2 \in L^2(\Gamma) \end{cases}$$

$$(1.18) \qquad \begin{cases} \Delta^2 u_3 = f \\[4pt] u_3 \big|_\Gamma = 0 \\[4pt] \Delta u_3 \big|_\Gamma = - g \, \lambda_3 \\[4pt] \displaystyle\int_\Gamma (\mu - \lambda_3) \frac{\partial u_3}{\partial n} \, d\Gamma \leqslant 0 \qquad \forall \mu \in \Lambda \\[4pt] \lambda_3 \in \Lambda \end{cases}$$

<u>Proposition 1.3</u> : The third solution (1.16) and the two last relations (1.17)(1.18) are, resp. equivalent to :

$$(1.19) \qquad \lambda_i = P_{\Lambda_i} \left(\lambda_i + \rho \frac{\partial u_i}{\partial n} \right) \quad \forall \rho \geqslant 0 \;, \text{ where :}$$

$$(1.20) \qquad P_{\Lambda_i} = \text{projection operator} : L^2(\Gamma) \to \Lambda_i$$

in the $L^2(\Gamma)$ norm so that :

$$(1.21) \qquad \begin{cases} P_{\Lambda_1}(\mu) = \mu & \forall \mu \in L^2(\Gamma) \\[2ex] P_{\Lambda_2}(\mu) = \inf (o,\mu) = -\mu^- & \forall \mu \in L^2(\Gamma) \\[2ex] P_{\Lambda_3}(\mu) = \dfrac{\mu}{\sup(1,|\mu|)} & \forall \mu \in L^2(\Gamma) \end{cases}$$

<u>I.3 : Dual algorithms.</u>

<u>I.3.1 : Algorithms formulation</u> :

Let **us** define for (P_i), $i = 1, 2, 3$, the following algorithms of <u>Uzawa</u> type (see Uzawa [1]) :

$$(1.22) \qquad \begin{cases} \lambda_1^o \text{ arbitrary given, } \lambda_1^n \text{ given,} \\ \quad \text{define } u_1^{n+1}, \lambda_1^{n+1} \text{ by :} \\[1ex] \Delta^2 u_1^{n+1} = f \text{ on } \Omega \\[1ex] u_1^{n+1}\big|_\Gamma = 0 \\[1ex] \Delta u_1^{n+1}\big|_\Gamma = -\lambda_1^n \\[1ex] \lambda_1^{n+1} = \lambda_1^n + \rho \dfrac{\partial u^{n+1}}{\partial n} \;, \quad \rho > 0 \end{cases}$$

and :

$$(1.23) \qquad \begin{cases} \lambda_2^o \text{ arbitrary given} \\[1ex] \Delta^2 u_2^{n+1} = f \text{ on } \Omega \\[1ex] u_2^{n+1}\big|_\Gamma = 0 \\[1ex] \Delta u_2^{n+1}\big|_\Gamma = -(g + \lambda_2^n) \\[1ex] \lambda_2^{n+1} = P_{\Lambda_2}\left(\lambda_2^n + \rho \dfrac{\partial u_2^{n+1}}{\partial n} \right), \rho > 0 \end{cases}$$

and at last :

$$(1.24) \quad \begin{cases} \lambda_3^0 \quad \text{arbitrary given} \\ \Delta^2 u_3^{n+1} = f \quad \text{on } \Omega \\ u_3^{n+1}\big|_\Gamma = 0 \\ \Delta u_3^{n+1}\big|_\Gamma = - g\,\lambda_3^n \\ \lambda_3^{n+1} = P_{\Lambda_3}(\lambda_3^n + \frac{\rho}{g}\,\frac{\partial u_3^{n+1}}{\partial n}), \quad \rho > 0. \end{cases}$$

I.3.2 : <u>Algorithms convergence.</u>

About the convergence of $(1.22),(1.23),(1.24)$, we have
the :

<u>Theorem 1.4</u> :
  $\forall\ i = 1,\ 2,\ 3$, as $n \longrightarrow + \infty$, we have :

$$(1.25) \qquad u_i^n \longrightarrow u_i \qquad \underline{\text{strongly}} \text{ in } V \text{ , when :}$$

$$(1.26) \qquad 0 < \rho < \frac{2}{C^2}$$

where C is the <u>norm</u> of the mapping $v \longrightarrow \dfrac{\partial v}{\partial u} : V \longrightarrow L^2(\Gamma)$.

For the proof, see Bégis–Glowinski, Glowinski–Lions–Trémolières,
loc. cit.

I.3.3 : <u>Some remarks</u> :

<u>Remark 1.3</u> : The algorithms $(1.22),(1.23),(1.24)$ are no more
than the gradient algorithm plus projection on Λ_i $(i=1,2,3)$,
applied to the dual problems

$$(1.27) \qquad \underset{\mu \in \Lambda_i}{\text{Max}} \ \underset{v \in V}{\text{Min}} \ \mathcal{L}_i(v,\mu) \qquad (i=1,2,3)$$

and this fact explains the terminology.
<u>Remark 1.4</u> : At each iteration of algorithms $(1.22),(1.23),$
(1.24), we have to solve a fourth order elliptic problem <u>facto-</u>
<u>rizable</u> in two <u>second order</u> Dirichlet problems as :

$$(1.28) \quad \begin{cases} \Delta^2 y = f \\ y_\Gamma = 0 \\ \Delta y|_\Gamma = \gamma \end{cases} \text{ is equivalent to } \begin{cases} \Delta z = f \\ z|_\Gamma = \gamma \\ \Delta y = z \\ y|_\Gamma = 0 \end{cases} .$$

<u>Remark 1.5</u> : The dual method related to (P_1) seems to be a new approach (more natural we think) to the Smith method to solve (P_1) ; for the Smith method, ref. Smith [1] , and also Bossavit [1] , Erlich [1] .

<u>Remark 1.6</u> : For $\Omega =]0,1[\times]0,1[$, we proved in Bégis-Glowinski that for the C constant involved in Theorem 1.4, relation (1.26) we have :

$$(1.29) \qquad\qquad C \leqslant \sqrt{\frac{2}{3}}$$

so $0 < \rho < 3$ implies the convergence for the algorithms (1.22), (1.23),(1.29).

<u>I.4</u>   :   <u>Behaviour of the (P_3) solution as a function of   g</u> .

If we consider in (P_3), g as a parameter, the (P_3) solution appears as a function of g ; we shall note u_g that corresponding solution. We have the :

<u>Theorem 1.5</u> :

   i) The mapping $g \longrightarrow u_g$ is Lipschitzian from $R^+ \longrightarrow V$.

   ii) When N = 2, it exists g_c, function of f , such that for every $g \geqslant g_c$, we have u_g = solution of (P_0).

<u>Remark 1.7</u> : We have, exactly :

$$(1.30) \qquad\qquad g_c = \underset{x \in \Gamma}{\text{Max}} \; |\Delta u_c(x)|$$

if we denote by u_c the solution of (P_0).

<u>Remark 1.8</u> : In Glowinski-Lions-Trémolières, loc. cit., in addition to Theorem 1.5 results, it is proved that the mappings :

$$g \rightarrow \|\Delta u_g\|^2_{L^2(\Omega)}$$

$$g \rightarrow \int_\Gamma \left|\frac{\partial u_g}{\partial u}\right| d\Gamma \qquad \text{are convex from } R^+ \longrightarrow R^+,$$

the second being decreasing.

I.5   Examples – Numerical results.

The variational approach of the previous sections, suggests for the (P_i) approximation, a finite elements technique $(*)$, mainly to overcome the difficulty of the normal derivative approximation ; actually and just to have a quick idea of the computational interest of the dual method, we shall limit ourselves to the very single case $\Omega =]\,0,1[\ \times\ ]\,0,1[$ and to (P_1), using finite differences approximation.

The following part of Section I.5 summarizes the numerical experiences.

Domain : $\Omega\ =\ ]\,0,1\,[\ \times\ ]\,0,1[$

Approximation :   •   finite differences method using a constant step size $\underline{\ h\ }$.

                 • Δ discretization by using the classical five points formula (see fig. 1.1)

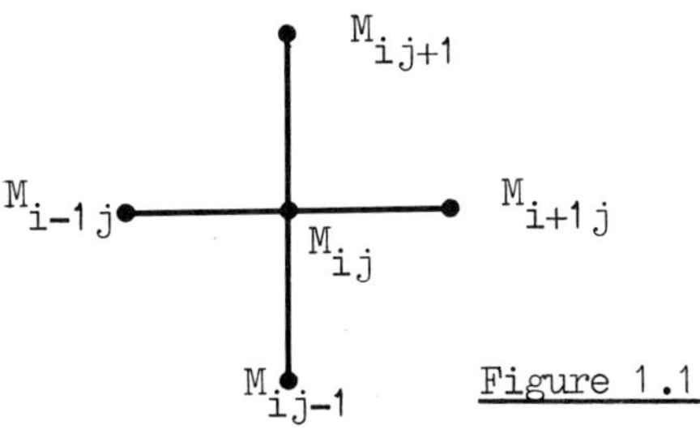

Figure 1.1

      •   normal derivative on Γ approximated by :

$$(1.31) \qquad \left(\frac{\partial u}{\partial n}\right)_M \simeq \frac{1}{2h} \left(\, 3u_M - 4u_N + u_P\right) \qquad \text{(see fig. 1.2)}$$

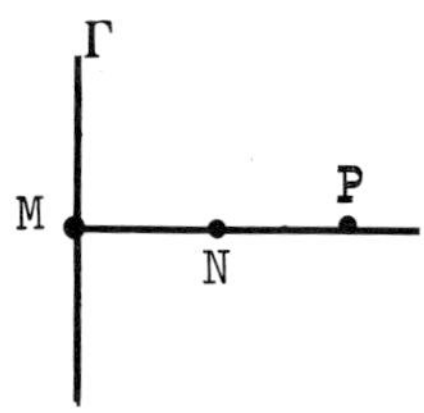

Figure 1.2

$(*)$ A such method is in study at the I.R.I.A.

<u>Test problem</u> : Let us define $u_o \in H_o^2(\Omega)$ by :

$$(1.32) \qquad u_o(x,y) = (\sin \pi x \, \sin \pi y)^2$$

$$(1.33) \qquad f = \Delta^2 u_o .$$

Then the problem will be :

$$(P_1) \qquad \begin{cases} \Delta^2 u = f \\ u|_\Gamma = 0 \\ \dfrac{\partial u}{\partial n}\Big|_\Gamma = 0 \end{cases}$$

with f defined by (1.33) so $u = u_o$.

<u>Discretization stepsize</u> : $\qquad h = \dfrac{1}{20}$

<u>Stopping rule</u> : $\qquad \sum_{i,j} |u_{ij}^{n+1} - u_{ij}^n| \leqslant 2.10^{-5}$

<u>Initial conditions</u> : $\qquad \lambda^o = 0$

<u>Results analysis</u> :

• $\underline{\rho \text{ influence}}$: Figure 1.3 shows the iterations number (NIT) in terms of ρ , for the previous stopping test ; the optimal value of ρ seems to be very close to 2.5 and, in practice, the algorithm diverges for $\rho \geqslant 4$ (all this proves the good quality of the estimate (1.29)).

... Figure 1.3 ....

• <u>computing time</u> : When solving, at each external iteration, the two Dirichlet problems by <u>point S O R</u> using the optimal value of ω the processing time is about 2" on IBM 360/91 ; the use, for solving internal problems, of a more sophisticated iterative method like ADI, or of <u>direct</u> elimination methods does not improve the processing time and needs much more storage : the main reason is as the iterations are going, it appears that, very quickly, very few iterations are needed to solve, at each external step, the two internal Dirichlet problems ; in addition the accuracy in solving internal problems is not critical and only iterative methods can take account of these facts.

<u>comparison with other methods</u> : On a square domain,

and for (P_1), the comparison with the Dorr, Golub algorithms, and others, as tested in the Golub, Dorr, Walker report $[1]$ seems to be very good.
The main other advantage of the dual approach seems to be the flexibility connected with the variational approach, the very easy programming and the possibility of solving <u>inequalities problems</u> such that (P_2), (P_3).

<u>Remark 1.9</u> : For more details on the <u>approximation</u> aspect and o the computation results related to (P_i), i=1,2,3, see Courjaret $[1]$ Bégis-Glowinski, Glowinski-Lions-Trémolières, loc. cit.

<u>Remark 1.10</u> : The (P_3) problem appears to be a particular case of general class of non differentiable optimization problems considered in Cea-Glowinski-Nedelec $[1]$.

II.   <u>Application to an optimal control problem.</u>

II.1   : <u>Problem formulation</u> :

Let $\Omega \subset R^2$ be an open, bounded, connected domain and $\Gamma = \partial \Omega$ be the boundary of Ω . Let us define u and u_{ad} by :

$$(2.1) \qquad u = L^2(\Omega)$$

$$(2.2) \quad u_{ad} = [v| v \in u, 0 \leqslant v(x) \leqslant v_0 \text{ a.e,} \int_\Omega v(x)dx = M] \text{ with,}$$

obviously $\qquad M < v_0 \text{ measure } (\Omega)$.

From $(2.1),(2.2)$ u_{ad} is a closed, convex, bounded subset of u
The positive constant g and $f \in L^2(\Omega)$ being given, we define a mapping $v \longrightarrow y(v) : u \longrightarrow V$, where V is, as in Part I,

$$(2.3) \quad \begin{cases} y(v) = \text{optimal solution, v being given, of :} \\ \underset{y \in V}{\text{Min}} \left[\int_\Omega |\Delta y|^2 \, dx - 2\int_\Omega (f+v)y \, dx + 2g \int_\Gamma \left|\frac{\partial y}{\partial n}\right| \, d\Gamma \right] \end{cases}$$

From Part I, there is <u>one and only one</u> $y(v)$ solution of (2.3)
The optimal control problem we want then to solve is :

$$(2.4) \qquad (\pi) \quad \begin{cases} y(v) \text{ solution of } (2.3) \\ \underset{v \in u_{ad}}{\text{Min}} \int_\Gamma \left|\frac{\partial y}{\partial n}(v)\right|^2 \, d\Gamma \end{cases}$$

II. 2 : Existence results.

In Lions $[2]$ it is proved (*), using compactness techniques, that problem (π) has at least one solution. About the
uniqueness, it is quite easy to get examples with an infinity of
optimal solutions.

II. 3 : A sub-optimization method. An existence theorem. Numerical results.

The computation (using efficient methods) of (π) solution appears to be a difficult problem because the mapping

$$v \longrightarrow \int_\Gamma \left| \frac{\partial y}{\partial n}(v) \right| d\Gamma : \, u \longrightarrow R$$

is, usually, a non convex, non differentiable one. In the context
of this paper, we shall limit ourselves to a sub-optimization
approach.

II.3.1: Formulation of the sub-optimization problem.

The control v being given, we know, from Part I, that
$y(v)$ is characterized by :

$$(2.5) \qquad \left\{ \begin{array}{l} \Delta^2 y = f + v \\ y|_\Gamma = 0 \\ \Delta y = - g\lambda \\ \lambda = \rho_\Lambda \left(\lambda + \rho \frac{\partial u}{\partial n}\right) \qquad \forall \, \rho \geqslant 0 \end{array} \right.$$

so we have for (π) the equivalent formulation :

$$(2.6) \qquad \left\{ \begin{array}{l} \Delta^2 y = f + v \\ y|_\Gamma = 0 \\ \Delta y|_\Gamma = - g\lambda \\ \lambda = \rho_\Lambda \left(\lambda + \rho \frac{\partial u}{\partial n}\right) \qquad \forall \, \rho \geqslant 0 \\ \displaystyle \operatorname*{Min}_{v \in u_{ad}} \int_\Gamma \left| -\frac{\partial u}{\partial n} \right|^2 d\Gamma \end{array} \right.$$

(*)   In fact, this is not proved for (π), but for a very
       close problem.

Removing, for a moment, the fourth relation (2.6), and λ being given, we consider the optimal control problem defined by :

$$(2.7) \qquad (\pi_\lambda) \quad \begin{cases} \Delta^2 y = f + v \\[4pt] y|_\Gamma = 0 \\[4pt] \Delta y|_\Gamma = - g\lambda \\[4pt] \underset{v \in \mathcal{U}_{ad}}{\text{Min}} \int_\Gamma \left| \frac{\partial y}{\partial n} \right|^2 d\Gamma \end{cases}$$

Using Lions[1], we can show quite easily that (π_λ) admits at least one solution and that any solution u is characterized using the adjoint state function p , by :

$$(2.8) \qquad \begin{cases} \Delta^2 y = f + u \quad \text{on} \ \Omega \\[4pt] y|_\Gamma = 0 \\[4pt] \Delta y|_\Gamma = - g\lambda \\[4pt] \Delta^2 p = 0 \quad \text{on} \ \Omega \\[4pt] p|_\Gamma = 0 \\[4pt] \Delta p|_\Gamma = \frac{\partial y}{\partial n} \\[6pt] \int_\Omega p(v-u) \, dx \geqslant 0 \quad \forall \ v \in \mathcal{U}_{ad} \\[4pt] u \in \mathcal{U}_{ad} \end{cases}$$

Reintroducing the missing relation $\lambda = P_\Lambda \left(\lambda + \rho \frac{\partial y}{\partial n} \right)$, we can now define the sub-optimization problem as finding (y,p,u,λ) solution of :

$$(2.9) \qquad (\pi') \quad \begin{cases} (2.8) \ \ \text{relations} \\[6pt] \lambda = P_\Lambda \left(\lambda + \rho \frac{\partial y}{\partial n} \right) \quad \forall \ \rho \geqslant 0. \end{cases}$$

II.3.2 : <u>An existence theorem</u> :

       About (π'), we have the following :

<u>Theorem 2.1</u> :

       The sub-optimal problem (π') admits at least one solution.

We just shall sketch the proof : using appropriate Galerkin approximations of $\mathcal{U}$ and $\mathcal{U}_{ad}$, $L^2(\Gamma)$ and Λ , we can reduce (π')

to a finite dimensional problem for which we can prove the existence of a solution by using Brouwer fixed-point theorem. We get the final result using compactness technique and taking the limit in the relations :

$$(2.10) \quad \begin{cases} \int_\Omega p_h(v_h - u_h) \, dx \geq 0 \quad \forall \ v_h \in \mathcal{U}_{ad,h} \\ u_h \in \mathcal{U}_{ad,h} \end{cases}$$

$$(2.11) \quad \begin{cases} \int_\Gamma \frac{\partial y_h}{\partial n} (u_h - \lambda_h) d\Gamma \leq 0 \quad \forall \ u_h \in \Lambda_h \\ \lambda_h \in \Lambda_h \end{cases}$$

where y_h, p_h, u_h, λ_h, are solutions of the approximated problem of (π') and $\mathcal{U}_{ad,h}$, Λ_h the approximations of $\mathcal{U}_{ad}$ and Λ we mentionned above, relations (2.10) approximating the two last relations (2.8) and relations (2.11) the relation :

$$(2.12) \quad \begin{cases} \int_\Gamma \frac{\partial y}{\partial n} (\mu - \lambda) \, d\Gamma \leq 0 \quad \forall \mu \in \Lambda \\ \lambda \in \Lambda \end{cases}$$

which is equivalent to $\lambda = P_\Lambda \left(\lambda + \rho \frac{\partial y}{\partial n} \right) \quad \forall \ \rho \geq 0$.

<u>Remark 2.1</u> : When $g = 0$, problems (π) and (π') are identical ; there is also identity when g is big enough, as in that case $y(v)$ is solution, $\forall \ v \in \mathcal{U}_{ad}$, of the (P_1) problem of the first Part (see Theorem 1.5).

<u>II.3.3</u> : <u>An algorithm for solving (π')</u> :

To solve (π'), we use the following algorithm :

. start from an arbitrary λ^0

. λ^n being given, solving $(\pi_\lambda n)$ (cf. (2.7)), we get

$(y^{n+1}, p^{n+1}, u^{n+1})$

. compute λ^{n+1} by :

$$\lambda^{n+1} = P_{\Lambda} \left(\lambda^{n} + \rho \frac{\partial y^{n+1}}{\partial n} \right), \quad \rho \text{ small enough and so on.}$$

<u>Convergence</u> : the convergence of the previous algorithm is not proved at this moment.

<u>II.3.4</u> : <u>Numerical example</u> :

. $\Omega =] \, 0,1 \, [\times] \, 0,1 \, [$

$\mathcal{U}_{ad} = [\, v \in L^{2}(\Omega) \mid 0 \leqslant v \leqslant 200 \quad \text{a.e. on} \Omega, \int_{\Omega} v(x)dx = 100]$

. $g = 1$

. $f = -200$

. <u>Approximation</u> : we use finite difference method and normal derivative approximation as in I.5, with $h = 1/20$. with $\rho = 5$ the result appears to be stable for about 300 iterations, corresponding to 5 minutes on CII 10 070.
For more details about computation, numerical results and other methods for solving (π), see Bégis-Glowinski, loc. cit.

References

[1] D. Bégis and R. Glowinski. To appear.
[1] A. Bossavit. Electricité de France. Internal report and Numerical Analysis seminar 1970-1971, Paris VI University.
[1] J. Cea , R. Glowinski and J.C. Nedelec. Minimisation de fonctionnelles non différentiables. Proceedings of the Dundee Numerical Analysis Symposium 1971. To appear, Springer (1971).
[1] B. Courjaret. Thesis 3d Cycle, to appear.
[1] G. Duvaut and J.L. Lions. <u>Sur les Inéquations en Mécanique et en Physique</u>. Dunod. Paris (1971).
[1] L.W. Ehrlich. Solving the Biharmonic Equation as coupled finite difference equations. Siam J. 8, (June 1971).
[1] R. Glowinski, J.L. Lions and R. Trémolières. <u>Analyse Numérique des Inéquations Variationnelles en Mécanique et en Physique</u>, To appear. Dunod, Paris (1972).
[1] Golub, Dorr and Walker. C.S. 293, Stanford University (1970)
[1] J.L. Lions. <u>Contrôle optimal de systèmes gouvernés par des équations aux dérivées partielles</u>. Paris, Dunod, Gauthier-Villars, (1968). English translation, Springer Verlag, (1971).

[2] J.L. Lions. <u>Some aspects of the optimal control of distri-ted systems</u>. Regional Conference Series in Applied Mathematics. Siam Publication,   (1972).

[1] J. Smith. The coupled Equation Approach to the Numerical solution of the Biharmonic Equation by Finite Differences. Siam Publication 7, (1970).

[1] H. Uzawa. In Arrow,Hurwicz, Uzawa, studies in linear and non linear programming, Stanford University Press.

--- § ---

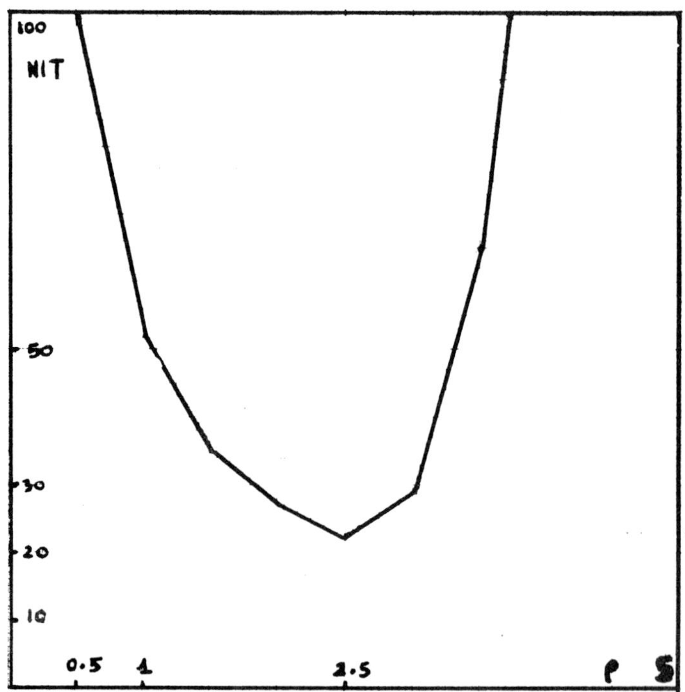

Fig. 1.3

NON LINEAR EVOLUTION EQUATIONS
IN METHEMATICAL ECONOMICS

Claude HENRY

Ecole Polytechnique
Paris, France

Introduction

In mathematical economics autonomous systems of differential equations are often considered whose solutions may
not become negative. It therefore can happen that the right
hand side of such a system is not everywhere continuous, a
case which is not covered by classical existence theorems
like Peano's theorem.

The impossibility to quote the classical existence
theorems has been considered a serious obstacle in many models of economic theory. Owing to recent developments in
mathematics (see for example $[1]$, $[2]$ and $[7]$), this obstacle can now be overcome.

This paper is devoted to existence and other related
theorems for autonomous systems of differential equations
whose multivalued right-hand side presents discontinuities
on the boundary of the domain of definition of the solutions. The existence theorems will be applied to prove the
existence of at least one solution for a system of differential equations describing the time path of prices in a
Walrasian tâtonnement adjustment process ; it will also be
applied to prove the existence of at least one solution for
a system of differential equations which formalizes a collective consumptions'planning procedure proposed by
J.H. DREZE and D. de la VALLEE POUSSIN (see $[3]$) and by
E. MALINVAUD (see $[5]$).

1. A Walrasian tâtonnement adjustment process.

Consider a pure exchange economy E with n private
goods. Let p denote the price-system in E, p being a point

in R^n_+. Let also z denote the excess demand in E, i.e. the excess of the sum of the individual demands by all consumers over the total resources ; as a consequence of usual assumptions on the consumption sets and on the preferences of the consumers, the excess demand z is a point in a subset Z (p) of R^n, with

$$ Z : R^n_+ \longmapsto \mathcal{P} R^n : p \longmapsto Z (p) $$

being an upper-semicontinuous correspondence such that, for every price-system p, Z (p) is convex (see for example [6]).

The following system of differential equations then describes the time path of prices in a Walrasian tâtonnement adjustment process defined in E (b is a fixed positive number) : for every $i \in \{1, \ldots, n\}$,

$$ \frac{dp_i}{dt} = \begin{cases} bz_i, & \text{if } p_i > 0 \\ \max \{bz_i, 0\}, & \text{if } p_i = 0 \end{cases} \tag{1.1} $$

with $z = (z_1, \ldots, z_n) \in Z (p)$.
Does this system possess at least one solution ? the answer is positive, owing to theorem 3.1 proved in section 3.

2. A collective consumptions'planning procedure.

Following E. MALINVAUD in [5], let us consider an economy with m consumers, n public goods and one private good used as numéraire. Let $x \in R^n_+$ denote the vector whose ith coordinate x_i is the total quantity of public good i available for consumption ; let $y = y (x)$ denote the maximum quantity of numéraire which is simultaneously available for consumption, and y^j the share of y allotted to consumer j. We will make the following assumptions :
1) y is continuously differentiable on R^n_+.
2) Every consumer j has a utility function U^j, which depends on x and y^j, and which is continuously differentiable on R^{n+1}_+.
3) $\forall j \in \{1, \ldots, m\}, \forall (x, y^j) \in R^{n+1}_+, \frac{d}{dy^j} U^j (x, y^j) > 0$;

4) $\forall j \in \{1, \ldots, m\}, \forall i \in \{1, \ldots, n\}, \forall (x, 0) \in R^{n+1}_+,$

$\pi_i^j (x,0) = 0$, where π_i^j denotes $\dfrac{dU^j}{dx^i} \Big/ \dfrac{dU^j}{dy^j}$.

In the planning procedure considered here, the collective consumptions x_i and the shares y^j of the numéraire are the indicators sent to the consumers, whose proposals are the marginal rates of substitution π_i^j. Taking those proposals into account, the planning bureau revises the indicators along the following equations : (2.1)

$$\star \ \forall i \in \{1,\ldots,n\}, \ \frac{dx_i}{dt} = \begin{cases} b \left(\sum_{j=1}^{m} \pi_i^j - \emptyset_i \right), & \text{if } x_i > 0 \\[2ex] \max \left\{ b \left(\sum_{j=1}^{m} \pi_i^j - \emptyset_i \right), 0 \right\}, & \text{if } x_i = 0 \end{cases}$$

where $\emptyset_i = -\dfrac{dy}{dx_i}$, and b is a fixed positive number ;

$$\star \ \forall j \in \{1,\ldots,m\}, \ \frac{dy^j}{dt} = \sum_{i=1}^{n} \frac{dx_i}{dt} \left(-\pi_i^j + \varrho^j \left(\sum_{k=1}^{m} \pi_i^k - \emptyset_i \right) \right),$$

where ϱ^j, $j=1, \ldots, m$, are m fixed positive numbers such that $\sum_{j=1}^{m} \varrho^j = 1$.

It is worth pointing out that, $\forall j \in \{1, \ldots, m\}$, $\dfrac{dy^j}{dt}$ is the sum of a compensation $-\sum_{i=1}^{n} \dfrac{dx_i}{dt} \pi_i^j$ for the revision of the collective consumptions x_i, and a share

$\varrho^j \left(\sum_{i=1}^{n} \dfrac{dx_i}{dt} \sum_{k=1}^{m} \pi_i^k - \sum_{i=1}^{n} \dfrac{dx_i}{dt} \emptyset_i \right)$ of the social surplus emerging from that revision. The existence of at least one solution for system (2.1) will be derived from theorem 3.4.

3. Existence and related theorems.

Let C be a n-dimensional closed convex set and $F : C \times R^m \longmapsto \mathcal{P} R^n$ an upper-semicontinuous correspondence such that, for every point (x,y) in $C \times R^m$, $F(x,y)$ is convex ; suppose furthermore that there exists a positive number α such that, $\forall (x,y) \in C \times R^m$,

$$\sup_{w \in F(x,y)} \| w \| < \alpha \left(1 + \| (x,y) \|^{1/2} \right). \qquad (3.1)$$

Let also, $\forall j \in \{1, \ldots, m\}$, $\theta^j : C \times R^m \longmapsto R^n$ be a continuous function on $C \times R^m$; suppose furthermore that there exist positive numbers α^j, $j \in \{1, \ldots, m\}$, such that $\forall (x,y) \in C \times R^m$,

$$\| \theta^j (x,y) \| < \alpha^j (1 + \| (x,y) \|^{1/2}). \qquad (3.2)$$

Then, using $\text{Proj}_{\pi_C(x)} F(x,y)$ for the projection of the set $F(x,y)$ onto the supporting cone $\pi_C(x)$ to C at point x, and the notation $w.\theta$ for the scalar product of the n-vector w by the nxm tensor θ, the following theorem holds :

Theorem 3.1 : (x^o, y^o) being any point in $C \times R^m$ and T any positive number, the following system of differential equations

$$\frac{dx}{dt} = w$$

$$\text{with } w \in \text{Proj}_{\pi_C(x)} F(x,y) \qquad (3.3)$$

$$\frac{dy}{dt} = w. \, \theta (x,y)$$

has at least one solution
    $(x,y) : [0,T] \longmapsto C \times R^m : t \longmapsto (x(t), y(t))$;
it is a solution in the following sense :
1) (x,y) is absolutely continuous on $[0, T]$;
2) $x(0) = x^o$ and $y(0) = y^o$;
3) $\forall t \in [0, T]$, $x(t) \in C$;
4) there exists a function $w : [0, T] \longmapsto R^n : t \longmapsto w(t)$
    with $w(t) \in \text{Proj}_{\pi_C(x(t))} F(x(t), y(t))$, such that, for

    almost every t in $[0, T]$,
    $$\frac{d}{dt} x(t) = w(t)$$
and $\frac{d}{dt} y(t) = w(t) \, . \, \theta (x(t), y(t))$.

Proof of theorem 3.1 : let $H(x,y)$ denote $\text{Proj}_{\pi_C(x)} F(x,y)$

and extend H and θ to the whole of R^{n+m} in the following manner : $\forall (x,y) \in R^{n+m}$, $H(x,y) = H(\bar{x},y)$ and $\theta(x,y) = \theta(\bar{x},y)$, where $\bar{x}$ is the projection of x onto C.

We will now consider the correspondence $K : R^{n+m} \longmapsto R^{n+m}$, which is the least upper-semi-continuous extension of the right-hand side in (3.3) such that the image of every point (x,y) in R^{n+m} be a compact convex subset $K(x,y)$ of R^{n+m}. Denoting by B the euclidean unit ball in

R^{n+m}, let :

$$(x,y) + \delta B = \{ (x',y') \mid \|(x',y') - (x,y)\| \leqslant \delta \} ;$$

$$H ((x,y) + \delta B) = \bigcup_{(x',y') \in (x,y) + \delta B} H (x',y') ;$$

$$\hat{H} ((x,y) + \delta B) = \text{convex hull of } H ((x,y) + \delta B) ;$$

$$G (x,y) = \bigcap_{\delta \in]0, + \infty [} \hat{H}((x,y) + \delta B) ;$$

then we have

$$K (x,y) = \{(w, w. \theta (x,y) \mid w \in G (x,y)\}. \qquad (3.4)$$

From (3.1) and (3.2) we easily derive the existence of a positive number β such that, $\forall (x,y) \in R^{n+m}$,

$$\sup_{(x',y') \in K (x,y)} \|(x',y')\| < \beta (1 + \|(x,y)\|). \qquad (3.5)$$

Hence VALADIER's existence theorem (see [7]) can be applied to the following system of differential equations :

$$(\frac{dx}{dt} ,\frac{dy}{dt}) \in K (x,y). \qquad (3.6)$$

Therefore there exist a function $(x,y) : [0, T] \longrightarrow R^{n+m}$: $t \longmapsto (x(t), y(t))$, and a function $w : [0,T] \longrightarrow R^{n}$: $t \longmapsto w(t)$ with $w(t) \in G (x(t), y(t))$, such that
1) (x,y) is absolutely continuous on $[0, T]$;
2) $x(0) = x^{0}$ and $y(0) = y^{0}$;
4') for almost every t in $[0, T]$,

$$\frac{d}{dt} x (t) = w (t)$$

and $\frac{d}{dt} y (t) = w (t). \theta (x (t), y (t))$.

It remains to show that, $\forall t \in [0, T]$, $x (t) \in C$, and that, for almost every t in $[0, T]$, $w (t) \in H (x(t), y(t))$; this can be done along the same lines as in [4]. Q.E.D.

Theorem 3.2 : let $S (x^{0}, y^{0})$ denote the set of all the solutions of (3.3) ; it is a compact subset of $C_{u}([0, T], R^{n+m})$, the space of continuous functions from $[0, T]$ to R^{n+m} with the uniform convergence topology.

Theorem 3.3 : let A denote a compact subset of R^{n+m} ; the correspondence $S : A \longmapsto \wp C_{u} ([0, T], R^{n+m})$: $(x^{0}, y^{0}) \longmapsto S (x^{0}, y^{0})$ is upper semi-continuous.
As every solution of (3.6) is a solution of (3.3) and conversely (see the proof of theorem 3.1), both theorems 3.2 and 3.3 immediately result from their analogous in [7].

Theorem 3.4 : if the correspondence F boils down to a continuous function f, then, for every t in
$[0, T]$, w (t) = $\text{Proj}_{\pi_C(x(t))}$ f (x(t), y(t)) and w (t). θ(x(t),
y(t)) are the derivatives on the right of x and y respecti-
vely, where (x,y) is any solution of (3.3) in the sense of
theorem 3.1.
Proof of theorem 3.4 : from theorem 3.1 we know that, for
every t in $[0, T]$ and every Δ t > 0,

$$x (t + \Delta t) = x(t) + \int_t^{t+\Delta t} w(s)\, ds \qquad (3.7)$$

and $y (t + \Delta t) = y(t) + \int_t^{t+\Delta t} w(s).\, θ (x(s), y(s))\, ds \qquad (3.8)$

Hence for every t in $[0, T]$ such that w(t) = f(x(t),
y(t)), the result is immediately derived from (3.7), (3.8)
and the continuity of f, x and y. Consider now t_0 in $[0, T]$
such that f $(x(t_0), y(t_0)) \notin \pi_C(x(t_0), y(t_0))$; as f, x and
y are continuous functions, there exists $t > 0$ such that,
for every t in $[t_0, t_0 + \Delta t]$, we also have f $(x(t), y(t))$
$\notin \pi_C (x(t), y(t))^0$; hence, when t is growing from t_0 to
$t_0 + \Delta t$, x(t) remains on the boundary of C and
w = $\text{Proj}_{\pi_C(x)}$ f (x,y) is a continuous function of t ; from

(3.7) and (3.8) it then results that $w(t_0)$ and w (t_0) .
θ $(x(t_0), y(t_0))$ are the derivatives on the right of x and
y at $t = t_0$. Q.E.D.
Theorem 3.1 can be directly applied to (1.1) ; as this
system is of the general form

$$\frac{dx}{dt} \in \text{Proj}_{\pi_C(x)} F (x),$$

it is interesting to note that the existence of at least one
solution could also have been derived from an existence
theorem recently proved by H. ATTOUCH and A. DAMLAMIAN (see
$[1]$) for systems of the form

$$\frac{dx}{dt} + A (x) + B (t,x) \ni 0$$

where A is maximal monotone and B upper semi-continuous.
Things are a little more complicated as far as system
(2.1), which formalizes the planning procedure, is concerned. Indeed in (2.1) no component of y may become negative;
in order to quote theorem 3.4 we thus have to extend the

right-hand side of (2.1) from $R^n_+ \times R^m_+$ to $R^n_+ \times R^m_+$; this can be done in the same way as for the extensions considered in the proof of theorem 3.1. From assumption 4 on the procedure, $\forall \, j \in \{1, \ldots, m\}$, $y^j = 0$ implies that

$$\sum_{i=1}^{n} \frac{dx_i}{dt} \left(-\pi_i^j + \rho^j \left(\sum_{k=1}^{m} \pi_i^k - \phi_i \right) \right) =$$

$$\rho^j \sum_{i=1}^{n} \frac{dx_i}{dt} \left(\sum_{k=1}^{m} \pi_i^k - \phi_i \right) \geqslant 0 \; ;$$

from the way we have extended the right-hand side of (2.1) to $R^n_+ \times R^m_+$, it is then clear that the solution we are supplied with by application of theorem (3.4) always remains in $R^n_+ \times R^m_+$, and is therefore a planning path suitable to the proposed procedure.

References.

[1] ATTOUCH, H. et A. DAMLAMIAN, Perturbations multivoques semi-continues supérieurement d'équations d'évolution dans les espaces de Hilbert, to appear in Annales de l'Institut Fourier.

[2] BREZIS, H., Semi-groupes non linéaires et applications, Symposium sur les problèmes d'évolution, Istituto Nazionale d'Alta Matematica, Roma, May 1970.

[3] DREZE, J. and D. de la VALLEE POUSSIN, A tâtonnement process for public goods. The Review of Economic Studies 38 (1971) 133-150.

[4] HENRY, C., An existence theorem for a class of differential equations with multivalued right-hand side, to appear in J. of Math. Analysis and Applications.

[5] MALINVAUD, E., The theory of planning for individual and collective consumptions, Symposium on the Problems of the National Economy Modelling, Novosibirsk, June 1970.

[6] QUIRK, J. and R. SAPOSNIK, "Introduction to general equilibrium theory and welfare economics", Mc Graw Hill, New York, 1968.

[7] VALADIER, M., Existence globale pour les équations différentielles multivoques, C.R.A.S. 272 (15 février 1971) 474-477.

INTERACTIVE GRAPHICAL SOLUTION OF BOUNDARY VALUE
PROBLEMS USING LINEAR PROGRAMMING*

J. B. Rosen

Computer, Information, and Control Sciences
University of Minnesota
Minneapolis, Minnesota

1. Introduction

The solution of generalized approximation problems by
the use of mathematical programming has been the subject of
a number of papers recently (for an early discussion see
[4]). In order to permit a user to formulate and solve a
problem rapidly, an interactive graphical system has been
developed for solving such problems. This system can be
used to approximate discrete data or functions in one or
two variables and to approximate the solution to linear
ordinary or partial differential equation boundary value
problems [5,9,6]. Auxiliary conditions on the approximation
may also be imposed. For two-dimensional boundary value
problems an approximate solution is obtained in terms of a
convenient spline basis using linear programming to
minimize a weighted sum of the interior and boundary errors
in the maximum norm. A specific problem is formulated
using an interactive graphics terminal, by means of function
keys and a light pen, and taking advantage of symbolic
notation where possible. The solution to the original
problem together with certain error information is then
obtained and may be graphically displayed in various forms
allowing the user to modify his formulation on line.
The formulation and method are discussed in the next
Section, and error estimates are obtained in Section 3. The
system and its use are briefly described and illustrated by
an example in Sections 4 & 5.

---

*This research was supported in part by the National
Science Foundation under Grant GJ 0362.

2. Formulation and Approximation Method

We consider a general class of linear boundary value problems with sufficiently smooth solutions. Specifically, let Ω denote a bounded domain in R^n with boundary $\partial\Omega$, and let $L[u]$ denote the linear differential operator of order $2m$, defined by

$$L[u] \equiv \sum_{|\beta| \leq 2m} a_\beta(x) D^\beta u(x) \tag{2.1}$$

where the coefficients $a_\beta(x)$ are bounded on the closure $\overline{\Omega}$. Similarly, let $B_\ell[u]$ denote a linear boundary differential operator of order $0 \leq m_\ell \leq 2m-1$, $\ell = 1,\ldots,m$ where the coefficients of B_ℓ are bounded on $\partial\Omega$ (see [3]). The boundary problem considered may be stated in the form

$$L[u] = f \quad \text{in } \Omega$$
$$B_\ell[u] = g_\ell, \quad \ell = 1,\ldots,m, \text{ on } \partial\Omega \tag{2.2}$$

where f and the g_i are defined on Ω and $\partial\Omega$, respectively. We assume that f and the g_i are sufficiently smooth (together with other required assumptions on the domain Ω and the operators B_ℓ and L) to insure the existence of a unique solution u in the Sobelev space $W_\infty^{p+1}(\Omega)$, $p \geq 2m$. For details of notation see, for example, Chap. 3 of [11]. We generate a finite-dimensional subspace of $W_\infty^p(\overline{\Omega})$ by choosing a single function $\phi(x)$, an element of W_∞^p with compact support. The subspace is then generated by linear combinations of the functions $\phi(\frac{x}{\Delta} - i)$, where Δ is a scalar parameter and i denotes an n-tuple of integers. Typically we choose $\phi(x)$ to be a tensor product of B-splines of degree p, in which case Δ is the uniform knot size of the B-spline. For a detailed discussion see [8] and [10]. The relevant results of Strang and Fix are summarized in Chap. 5 of [11]. Specifically, we consider an approximation to the solution u of the form

$$S(\Delta,\alpha;x) = \sum_{i=1}^{N} \alpha_i \phi(\frac{x}{\Delta} - i) \tag{2.3}$$

The number of terms N in this summation (i.e., the number

of elements α_i in the coefficient vector α) will be proportional to Δ^{-n}.

A suitable vector $\alpha = \hat{\alpha}$ of coefficients is computed by solving a linear programming problem. This vector minimizes a weighted sum of the error in the differential equation in Ω and the errors in the boundary conditions on $\partial\Omega$. All errors are measured in the maximum norm over a discrete grid. A method based on minimizing a weighted sum of errors measured in the L_2 norm has recently been considered in detail by Bramble and Schatz [2,3]. Let $||\cdot||_X = \sup\limits_{x \in X} |\cdot|$, and consider a discrete grid $\overline{\Omega}_h \subset \overline{\Omega}$ such that

$$\max_{x \in \overline{\Omega}} \quad \min_{y \in \overline{\Omega}_h} \quad |x-y| \leq h \tag{2.4}$$

We define $\Omega_h \equiv \overline{\Omega}_h \cap \Omega$ and $\partial\Omega_h \equiv \overline{\Omega}_h \cap \partial\Omega$, and choose h so that $h \leq \Delta/2$.

For any choice of Δ and h, the linear programming problem which determines $\hat{\alpha}$ may now be stated as follows:

$$\min_{\alpha} \quad \psi(\Delta,h;\alpha) \tag{2.5}$$

where

$$\psi(\Delta,h;\alpha) \equiv K_0(\Delta) ||L[S(\Delta,\alpha)] - f||_{\Omega_h}$$

$$+ \sum_{\ell=1}^{m} K_\ell(\Delta) ||B_\ell[S(\Delta,\alpha)]-g_\ell||_{\partial\Omega_h} \tag{2.6}$$

The weights $K_\ell(\Delta) > 0$, are given by $K_0(\Delta) = K_0\Delta^{2m}$, $K_\ell(\Delta) = K_\ell\Delta^{m_\ell}$, $\ell = 1,\ldots,m$, where usually $K_\ell = 1$, $\ell = 1,\ldots,m$, and K_0 is a positive constant determined by L and Ω.

In order to discuss the computational aspects of this more easily we let

$$\sigma_{0i}(\Delta;x) \equiv \Delta^{2m}L[\phi(\tfrac{x}{\Delta} - i)]$$

$$\sigma_{\ell i}(\Delta;x) \equiv \Delta^{m_\ell}B_\ell[\phi(\tfrac{x}{\Delta} - i)], \quad \ell = 1,\ldots,m \tag{2.7}$$

Note that the functions $\sigma_{\ell i}(\Delta,x)$, $\ell = 0,\ldots,m$ are readily obtained in symbolic form in terms of the function $\phi(x)$

and the operators L and B_ℓ. Furthermore, they are easily evaluated numerically for any $x \in \overline{\Omega}$. As $\Delta \to 0$, we have $\sigma_{\ell i}(\Delta;x) = \sigma_{\ell i}(x) + \mathcal{O}(\Delta)$, $\ell = 0,\ldots,m$, where $\sigma_{\ell i}(x)$ is independent of Δ. Using (2.7) we can now write

$$\psi(\Delta,h;\alpha) = K_0 \left|\left| \sum_{i=1}^{N} \sigma_{0i}\alpha_i - \Delta^{2m}f \right|\right|_{\Omega_h} +$$

$$+ \sum_{\ell=1}^{m} K_\ell \left|\left| \sum_{i=1}^{N} \sigma_{\ell i}\alpha_i - \Delta^{m_\ell}g_\ell \right|\right|_{\partial\Omega_h} \tag{2.8}$$

The formulation of (2.5) as a linear program in standard primal form has been discussed in detail elsewhere [1,5,7] and will not be given here. It will only be remarked that the matrix generated for the linear program consists of $\bar{N} = N + m + 1$ rows and a number of columns equal to the number of grid points in $\overline{\Omega}_h$. A typical element of a column corresponding to an interior grid point $x_j \in \Omega_h$ is given by $\sigma_{0i}(\Delta;x_j)$, and similarly for a column corresponding to a boundary point. The optimal solution selects a nonsingular (and usually well-conditioned) basis of $\bar{N}$ columns and solves the corresponding system of $\bar{N}$ linear equations for the minimizing vector α. The selected basis columns correspond to those points in $\overline{\Omega}_h$ at which the maximum interior and boundary errors are attained.

3. Error Estimates

For a given problem (2.2) and a specific choice of function $\phi(x)$, the minimizing $\hat{\alpha}$ determined by (2.5) will depend on Δ, that is $\hat{\alpha} = \hat{\alpha}(\Delta)$. For simplicity we assume that $h = h(\Delta) \leq \Delta/2$. The minimum weighted error ψ can then be considered as a function of Δ only, that is $\psi(\Delta,h(\Delta),\hat{\alpha}(\Delta)) = \hat{\psi}(\Delta)$. We now investigate the behavior of $\hat{\psi}(\Delta)$ as a function of Δ. We assume that $\phi(x) \in W_\infty^p$ satisfies the following condition. For any $u \in W_\infty^{p+1}(\overline{\Omega})$, there exists an $\alpha = \alpha(u)$ such that

$$||u - S(\Delta,\alpha(u))||_{W_\infty^k(\overline{\Omega})} \leq C\Delta^{p+1-k}||u||_{W_\infty^{p+1}(\overline{\Omega})} \tag{3.1}$$

$$k = 0,\ldots,p$$

where C is independent of Δ and u, and

$$||\cdot||_{W^s_\infty(\Omega)} \equiv \sum_{|\beta|\leq s} ||D^\beta \cdot||_{L_\infty(\Omega)} \tag{3.2}$$

The existence of functions $\phi(x)$ satisfying this condition has recently been investigated [2, 10, 11]. In particular, the tensor product of splines with uniform knot size has this property.

<u>Theorem 1.</u> <u>For</u> $u \in W^{p+1}_\infty(\overline{\Omega})$ <u>we have</u>

$$\psi(\Delta, h(\Delta); \hat{\alpha}(\Delta)) \leq C\Delta^{p+1}||u||_{W^{p+1}_\infty(\overline{\Omega})} \tag{3.3}$$

<u>where</u> C <u>is independent of</u> Δ <u>and</u> u.
To prove this we let $\psi(\Delta, 0; \alpha)$ denote the weighted error with sup norms $||\cdot||_\Omega$ and $||\cdot||_{\partial\Omega}$ replacing the corresponding norm on the discrete grids Ω_h and $\partial\Omega_h$. Since $\overline{\Omega}_h \subset \overline{\Omega}$, $\psi(\Delta, h(\Delta); \alpha) \leq \psi(\Delta, 0; \alpha)$ for any α. For $\alpha = \alpha(u)$, by assumption (3.1) and the bounded coefficients in L, we have

$$||L[S(\Delta, \alpha(u))] - f||_\Omega = ||L[S(\Delta, \alpha(u))] - L[u]||_{W^0_\infty(\Omega)}$$

$$\leq C_1||S(\Delta, \alpha(u)) - u||_{W^{2m}_\infty(\Omega)} \leq C_2\Delta^{p+1-2m}||u||_{W^{p+1}_\infty(\overline{\Omega})}$$

$$\tag{3.4}$$

where C_1 and C_2 are independent of Δ and u. Similar estimates hold for the boundary terms. Then taking into account the weights $K_\ell(\Delta)$, $\ell = 0,\ldots,m$, we obtain

$$\psi(\Delta, h(\Delta); \alpha(u)) \leq \psi(\Delta, 0; \alpha(u)) \leq C\Delta^{p+1}||u||_{W^{p+1}_\infty(\overline{\Omega})} \tag{3.5}$$

But from (2.5), $\psi(\Delta, h(\Delta); \hat{\alpha}(\Delta)) \leq \psi(\Delta, h(\Delta); \alpha(u))$, which gives (3.3).

Theorem 1 gives an a priori estimate for the weighted error on the discrete grid. We also need an estimate for $\psi(\Delta, 0; \hat{\alpha}(\Delta))$ in terms of the computed quantity $\psi(\Delta, h(\Delta); \hat{\alpha}(\Delta))$, to give a posteriori error bounds. Such an estimate is given by the following.

<u>Theorem 2.</u> <u>Let</u> $\phi(x)$ <u>be a tensor product of B-splines of</u>

<u>degree</u> p, <u>and</u> <u>assume</u> f, $g_\ell \in W^{p+1}$. <u>Then</u> <u>for</u> $h \leq \Delta/p$,

$$\psi(\Delta,0;\hat{\alpha}(\Delta)) \leq C_1\psi(\Delta,h;\hat{\alpha}(\Delta)) + C_2(f,g_\ell)h^{p+1} \qquad (3.6)$$

<u>where</u> C_1 <u>and</u> C_2 <u>are</u> <u>independent</u> <u>of</u> Δ and h, <u>but</u> C_2 <u>depends</u> <u>on</u> <u>the</u> <u>norms</u> $||f||_{W^{p+1}_\infty(\Omega)}$ <u>and</u> $||g_\ell||_{W^{p+1}_\infty(\partial\Omega)}$. The proof of Theorem 2 is given in [8.]. It should also be remarked that the coefficients C_1 and C_2 can be easily computed in many cases.

Finally, it should be noted that when the boundary value problem (2.2) has an appropriate monotone property, estimates for the error $||u - S(\Delta,\alpha)||_{\overline{\Omega}}$ can be obtained in terms of $\psi(\Delta,0;\alpha)$. See for example [2,11]. For such cases, one can often obtain estimates of the following kind.

$$||S(\Delta,\alpha) - u||_{\overline{\Omega}} \leq C\psi(\Delta,0;\alpha) \qquad (3.7)$$

4. Interactive Graphical System

An interactive graphical system was developed earlier to solve approximation problems in one dimension, including approximation to discrete data, to a given function, and to ordinary differential equation problems, both initial and multipoint boundary value [5,9]. This system also allows the user to impose certain auxiliary conditions on the approximation obtained. Any, or all of the following conditions are available: lower (upper) bounds on the coefficients, lower (upper) bounds on the approximation and its first and second derivative at specified points.

This system at the University of Wisconsin has now been extended [6] to handle second order linear boundary value problems in two dimensions, of the type discussed above. The system uses an Adage Graphics Terminal (AGT/10) and a Univac 1108 linked by a high speed data channel. Full use is made of the capabilities of the AGT in specifying two-dimensional problems. The user formulates the problem with the help of a light pen to describe the domain $\overline{\Omega}$ and function keys to define the operators L and B_ℓ in symbolic form. The domain may be specified in terms of polygons, and is not required to be simply connected (for example, a rectangle with an interior triangle deleted). The interior Ω of the domain

is specified by means of the discrete set of grid points
Ω_h. This is done by deleting all exterior points using
the light pen (see Fig. 1 showing this for a triangular
domain). The degree and knot size of the product spline
basis are also selected using the light pen, and the
matrices and cost rows for the linear program are then
automatically generated. The linear program is solved on
the 1108, giving the coefficients of the spline approximation to the solution. Contour plots of this approximate
solution and related error information may then be displayed at the user's option (see Fig. 2). The user may
also select any cross-section of the domain and display
an error curve corresponding to this cross-section. If
the results obtained are not satisfactory, he may then
modify his earlier choices and repeat the process.

It should be emphasized that this system is not best
suited to obtain highly accurate solutions to problems
with standard domains and boundary conditions which can
easily be solved by finite difference methods. One of its
main advantages is that it provides an interactive
graphical environment in which it is convenient to specify
problems with relatively complex domains and boundary
conditions for which it would be difficult to formulate
a good finite difference equivalent. Quick access to
graphical results allows the user to modify various parameters and observe the effect on the approximate solution
and corresponding error contours. It also permits rapid
computational exploration to determine values for the
constants in the error estimates for specific cases.

<h2 align="center">5. <u>Computational Example</u></h2>

The use of the interactive graphical system will be
illustrated by a selected computational example. This
example also illustrates the effect of the spline degree
p and the boundary weights K_ℓ on the error in the approximation. A hyperbolic equation with variable coefficient
is considered on the unit square $\overline{\Omega} = \{x\epsilon[0,1], y\epsilon[0,1]\}$.
Specifically we let

$$L[u] \equiv u_{yy} - (1+x)u_{xx}, \qquad f(x,y) = L[e^{2xy}]$$

and consider

189

$$L[u] = f(x,y) \text{ in } \Omega$$

with boundary conditions

$$u = 1 \text{ on } \Gamma_1, \Gamma_2; \quad u = e^{2y} \text{ on } \Gamma_3; \quad u_y = 2x \text{ on } \Gamma_1;$$

where $\Gamma_1 = \{x\epsilon[0,1], y = 0\}$, $\Gamma_2 = \{x=0, y\epsilon[0,1]\}$, and $\Gamma_3 = \{x=1, y\epsilon[0,1]\}$. The solution is of course given by $u = e^{2xy}$. We first chose bicubic splines (p=3) with a knot size $\Delta = 1/3$ (giving a total of 36 coefficients), and boundary weights such that $K_\ell(1/3) = 1$, $\ell = 0,1,2,3,4$. A value of h = 0.1 for $\overline{\Omega}_h$ was used for all cases. The calculation gave $\psi = 0.213$ and $||S(\hat{\alpha})-u||_{\overline{\Omega}} = .036$ for this case.

More accurate results were obtained using biquintic splines (p=5) with a knot size $\Delta = 1$. This is equivalent to approximating u by $P_5(x)Q_5(y)$, where P_5 and Q_5 are 5th degree polynomials, so that we again have 36 coefficients to be determined. Boundary weights were again all set to $K_\ell = 1$. The results gave $\psi = .053$ and $||S(\hat{\alpha})-u||_{\overline{\Omega}} = .0022$, a substantial improvement over the results obtained with cubic splines. A contour plot of the error $L[S(\hat{\alpha})]-f$ for this case is shown in Fig. 3 and in Fig. 4 a contour plot of $S(\hat{\alpha})-u$ is given. These Figs. are (negative) photographs of the AGT screen and show the displays as they appear (at the user's request) immediately after the linear pro- gramming problem has been solved. The contour lines increase in brightness (on the AGT screen) corresponding to increasing values of the contour.

For comparison, the solution e^{2xy} was approximated directly by the same biquintic spline basis. That is, we obtained the best uniform approximation to e^{2xy} with this basis. The corresponding error contour plot is shown in Fig. 5. The uniform error has now been reduced to 1.6×10^{-4}. Note also that twice as many oscillations appear in each direction as compared with the approximation obtained using the boundary value problem (Fig. 4). Finally, another solution was obtained with the same biquintic spline basis, but with $K_\ell = 2$, $\ell = 1,2,3$; $K_4 = 1$. That is, a boundary weight of 2, where the value of u was specified. This change decreased $||S(\hat{\alpha})-u||_{\overline{\Omega}}$ from .0022 to .00084.

An indication of the terminal and computer time

required is given by noting that this example required a total of about 20 minutes at the terminal, including the time required to input the problem statement. A total of about 2 minutes of 1108 time was required to solve the linear programming problems and carry out all other necessary computation. Complete details of the system and a variety of examples are contained in LaFata's thesis [6].

References

1. Barrodale, I., and A. Young, "Computational experience in solving linear operator equations using the Chebyshev norm" in <u>Numerical Approximation to Functions and Data</u>, J. G. Hayes (Ed.), Athlone Press, London, 1970, pp. 115-142.

2. Bramble, J. H., and A. H. Schatz, "Rayleigh-Ritz-Galerkin methods for Dirichlet's problem using subspaces without boundary conditions" Comm. Pure Appl. Math. <u>23</u> (1970), pp. 653-675.

3. Bramble, J. H. and A. H. Schatz, "Least squares methods for $2m^{th}$ order elliptic boundary-value problems", Math. of Comp. <u>25</u> (1971), pp. 1-32.

4. Collatz, L., "Approximation in partial differential equations", <u>On Numerical Approximation</u>, R. E. Langer, Ed., Univ. of Wisc. Press, Madison, 1959. pp. 413-422.

5. LaFata, P., and J. B. Rosen, "An interactive display for approximation by linear programming" Comm. ACM <u>13</u>, (1970), pp. 651-659.

6. LaFata, P. "An interactive graphical system for generalized approximation", Ph.D. thesis, Computer Science Dept., Univ. of Wisc., Sept., 1971.

7. Rosen, J. B., "Approximate solution and error bounds for quasilinear boundary value problems", SIAM J. Numer. Anal. <u>7</u>, (1970), pp. 80-103.

8. Rosen, J. B., "Minimum error bounds for multidimensional spline approximation", J. Computer and System Sciences <u>5</u> (1971), pp. 430-452.

9. Rosen, J. B., and P. LaFata, "Interactive graphical spline approximation to boundary value problems" In <u>Proc. ACM '71</u>, Association for Computing Machinery, New York, 1971, pp. 466-481.

10. Strang, G and G. Fix, <u>An Analysis of the Finite Element Method</u>, Prentice-Hall. To appear.

11. Varga, R. S., "Functional analysis and approximation theory in numerical analysis" Regional Conf. Series in Appl. Math. $\underline{3}$, SIAM, 1971.

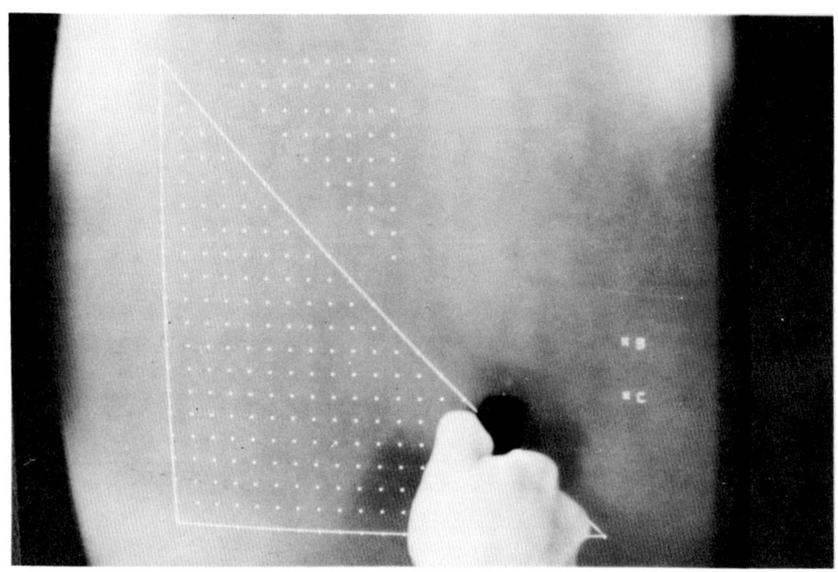

Fig. 1. Use of light pen to delete grid points exterior to domain.

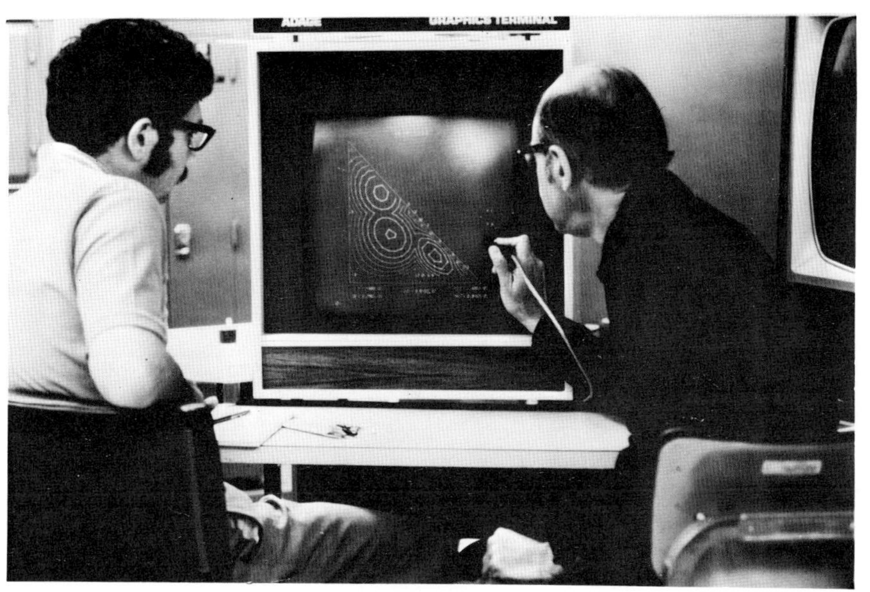

Fig. 2. Interactive modification using light
pen with graphics terminal.

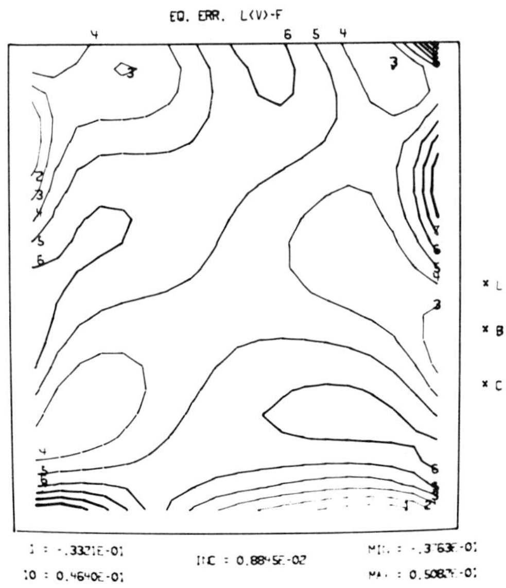

Fig. 3. Contour plot of differential equation
error.

193

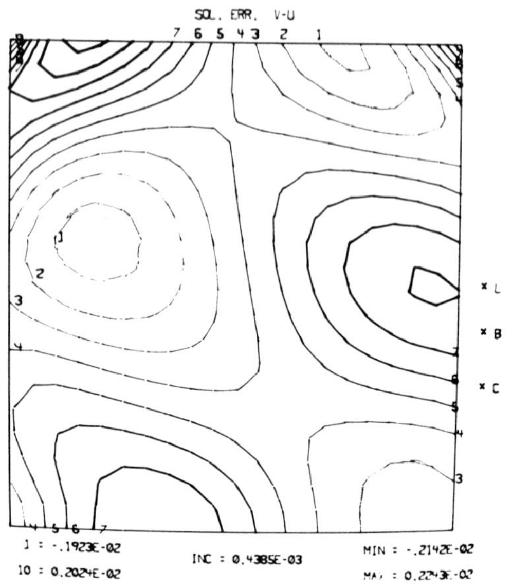

Fig. 4. Contour plot of solution error.

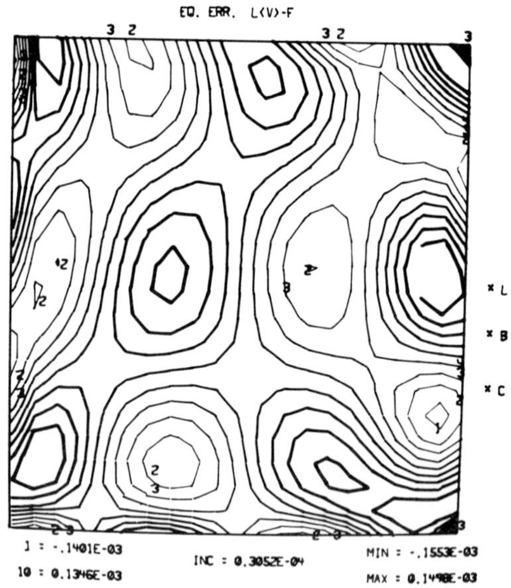

Fig. 5. Contour plot of error in direct
         approximation.

OPTIMAL CONTROL

EXISTENCE THEOREMS IN PROBLEMS OF OPTIMAL CONTROL WITHOUT PROPERTY (Q).

Leonard D. Berkovitz*

Purdue University
Lafayette, Indiana 47907

Introduction

In [4] L. Cesari obtained very general theorems for the existence of optimal controls and generalized optimal controls in problems where the state of the system is governed by ordinary differential equations. In [2] we obtained related results using a method that differed from Cesari's in many important respects. In [4] and [2] the property (Q), introduced into optimal control problems by Cesari, played a crucial role. In this paper we shall use the method of [2] to obtain a compactness and semicontinuity theorem for problems in which property (Q) is replaced by a condition involving a generalized modulus of continuity. Existence theorems for ordinary and relaxed optimal controls are immediate consequences of this theorem. For details see [2]. The results of this paper are also related to the paper of Jacobs [7].

To facilitate reference and comparison with [2] we shall use the same notation as in [2]. This will be summarized in the next paragraph.

Notation & Definitions

We shall use single letters to denote vectors, we shall use subscripts to denote components of vectors. The letter t will denote a real variable, which we call time, the letter x will denote a vector $x = (x^1, \cdots, x^n)$ in R^n,

*The research reported in this paper was carried out under National Science Foundation Grant GP-28392.

which we call the state variable, the letter y will denote a real number, which we call the output variable, and the letter w will denote a vector in R^m, which we call the control variable. By $x_1 \geq x_2$ we mean that every component of x_1 is greater than or equal to the corresponding component of x_2. The euclidean norm of a vector x will be denoted by $|x|$.

Let f be a real function $(t,x,w) \to f(t,x,w)$ defined on $R \times R^n \times R^m$ with range in R^1 and let g be a function $(t,x,w) \to g(t,x,w)$ defined on $R \times R^n \times R^m$ with range in R^n. Let R be a subset of (t,x)-space. Let Ω denote a mapping that assigns to each point (t,x) in R a subset $\Omega(t,x)$ of R^m. Let B be a set of points (t_1,x_1,t_2,x_2) in R^{2n+2} with $t_2 > t_1$.

An absolutely continuous function $\phi = (\phi^1,\cdots,\phi^n)$ defined on an interval $[t_1,t_2]$ is said to be an <u>admissible trajectory</u> if there exists a measurable function $u = (u^1,\cdots,u^m)$ defined on the same interval such that the following hold:

$\quad$ (i) $\quad (t,\phi(t)) \in R$ for all t in $[t_1,t_2]$

$\quad$ (ii) $\quad (t_1,\phi(t_1),t_2,\phi(t_2)) \in B$

(2.1) (iii) $\quad \phi'(t) = g(t,\phi(t),u(t))$ a.e. in $[t_1,t_2]$

$\quad$ (iv) $\quad u(t) \in \Omega(t,\phi(t))$ a.e. in $[t_1,t_2]$

$\quad$ (v) $\quad f(t,\phi(t),u(t))$ is in $L_1[t_1,t_2]$.

The function u is said to be an <u>admissible control</u> and the pair (ϕ,u) is said to be an <u>admissible pair</u>. By virtue of (2.1)-(v) we can define an <u>absolutely continuous output-function</u> θ corresponding to each admissible pair (ϕ,u) as follows:

$$(2.2) \qquad \theta(t) = \int_{t_1}^{t} f(s,\phi(s),u(s))ds \qquad t_1 \leq t \leq t_2.$$

Output functions and output variables arise in applications and in the reduction of control problems in Lagrange formulation to control problems in Mayer formulation. We define $\theta(t_2)$ to be the <u>output of the system</u> or simply the <u>output</u>.

Assumptions

In this section we list two sets of assumptions for our theorems. One of the statements in A requires the introduction of a function Q^+ that assigns to each point (t,x) in R a subset $Q^+(t,x)$ of R^{n+1} as follows:

$$Q^+(t,x) = \{(\eta,\xi): \eta \geq f(t,x,w), \; \xi = g(t,x,w), \; w \in \Omega(t,x)\},$$

where η is a scalar and ξ is an n-vector.

ASSUMPTION A. (1) R is a compact subset of $R \times R^n$. (2) The set B is compact. (3) For each (t,x) in R the set $\Omega(t,x)$ is closed. (4) The set $D = \{(t,x,w): (t,x) \in R, \; w \in \Omega(t,x)\}$ is closed. (5) For each (t,x) in R the set $Q^+(t,x)$ is closed and convex. (6) For all (t,x,w) in D, $f(t,x,w) \geq 0$.

For a discussion of these assumptions see [2]. In particular it follows from (2.1)-(i) and Assumption A(1) that all of the intervals $[t_1,t_2]$ corresponding to admissible pairs (ϕ,u) are contained in a fixed compact interval I.

Assumption E, which follows, is so labeled to facilitate comparison with [2]. It replaces Assumption B of [2], which utilized Cesari's property (Q).

ASSUMPTION E. (1) The function $F = (f,g)$ is continuous. (2) The sets $\Omega(t,x)$ are independent of x; i.e. for a given t in I, $\Omega(t,x) = \Omega(t,x')$ for all x and x' such that (t,x) and (t,x') are in R. (3) There exists a nondecreasing function μ defined on $[0,\infty)$ such that $\lim_{\delta \to 0} \mu(\delta) = 0$ and a non negative function L

defined on $I \times R^m$ such that

$$(3.1) \qquad |F(t,x,w)-F(t,x',w)| \leq L(t,w)\mu(|x-x'|)$$

for all (t,x,w) and (t,x',w) in $\mathcal{D}$.

Note that if F is uniformly continuous on $\mathcal{D}$, which occurs if $\mathcal{D}$ is compact, then (3.1) holds with $L \equiv 1$ and μ the modulus of continuity. If F is Lipschitz in x then (3.1) holds with $\mu(\delta) = \delta$ and L equal to the Lipschitz constant.

<u>A compactness and semi-continuity theorem</u>

Let Z denote the set of all continuous functions z defined on subintervals $[t_1,t_2]$ of I. We extend the definition of z to a function $\tilde{z}$ defined on all of I by setting $\tilde{z}(t) = z(t_1)$ if $a \leq t \leq t_1$ and $\tilde{z}(t) = z(t_2)$ if $t_2 \leq t \leq b$. If $t \in [t_1,t_2]$, then $\tilde{z}(t) = z(t)$. For z in Z defined on $[t_1,t_2]$ and $z_o \in Z$ defined on $[t_{01},t_{02}]$, we define

$$\rho(z,z_o) = |t_{01}-t_1| + |t_{02}-t_2| + \max |\tilde{z}_o(t)-\tilde{z}(t)|,$$

where the max is taken over all t in I. It is easily checked that ρ is a metric and that Z is complete under this metric. A sequence of functions $\{z_k\}$ defined on intervals $[t_{1k},t_{2k}]$ converges in this metric to a function z defined on an interval $[t_1,t_2]$ if and only if $t_{ik} \to t_i$, $i = 1,2$ and $\tilde{z}_k \to \tilde{z}$ uniformly on I.

THEOREM 1. Let Assumptions A and E hold. Let A_o be a set of admissible pairs (ϕ,u) such that the trajectories ϕ are equi-absolutely continuous, such that for the admissible controls u corresponding to the trajectories ϕ

$$(4.1) \qquad \int_{t_1}^{t_2} L(t,u(t))dt < A,$$

where A is a constant, and such that the outputs $\theta(t_2)$ are bounded. Then there exists a sequence $\{(\phi_k, u_k)\}$ in A_o, a real number γ, and an admissible pair (ϕ, u) in A with the following properties: (i) $\phi_k \to \phi$ in the ρ metric. (ii)

$$(4.2) \qquad \gamma = \lim_{k \to \infty} \theta_k(t_{2k}) \geq \theta(t_2),$$

where θ_k is the output function associated with (ϕ_k, u_k) and θ is the output function associated with (ϕ, u).

In [4] and [5] Cesari and Cesari, LaPalm and Nishiura give very general conditions guaranteeing equi-absolute continuity of the components of ϕ.

As in [2] the assumption that F is continuous (Assumption E(1)) can be replaced by Assumption C, which reads as follows. For each t in I, the function F is a continuous function of (x,w) on R^{n+m} and for each (x,w) in R^{n+m} the function F is measurable with respect to t in I.

In the following example property (Q^*) of [2] fails, and hence property (Q) of [4] fails, but the hypotheses of Theorem 1 of the present paper are fulfilled. The example is essentially that used by Brunovsky ([3], Example 2, p. 184) to show that property (Q) is not a necessary condition for lower closure.

EXAMPLE: Let $x = (x^1, x^2)$, let w be a real number, let $\Omega(t,x) = R^1$, let $f \equiv 0$ and let $g \equiv (w, x^1 w)$. Let $B = \{(t_1, x_1, t_2, x_2): t_1 = 0, \ t_2 = 1, \ x_1^1 = 0, \ 0 \leq x_1^2 \leq 1, \ (t_2, x_2) \in R\}$, and let $R = \{(t,x): 0 \leq t \leq 1, \ |x^i| < M\}$, where M is a large positive constant. For each (t,x) in R,

$$Q^+(t,x) = \{(\eta, \xi) = (\eta, \xi^1, \xi^2): \ \eta \geq 0, \ \xi^1 = w, \ \xi^2 = x^1 w\}.$$

It is readily verified that Assumptions A, E(1), and E(2) are satisfied. Assumption B(1) of [2], which is the same as E(1), is therefore also satisfied. On the other hand, for all $\delta > 0$, cl co $Q^+(N_x(t,x,\delta)) = \{(\eta,\xi): \eta \geq 0, \xi \in R^2\}$. Hence property (Q^*) fails and Assumption B(2) of [2] fails. On the other hand $F = (f,g)$ satisfies Assumption E(3) with $L = |w|$ and $\mu(|x-x'|) = |x^1-x'^1|$.

Let A be the set of admissible pairs (ϕ,u). The set A is not void. Let A_o be any subset of A such that the admissible pairs (ϕ,u) in A_o have the property that the functions ϕ are equi-absolutely continuous. The functions u then have equi-absolutely continuous integrals, and hence so do the functions $|u|$. Hence, since we are integrating over $[0,1]$ there is a constant A such that

$$\int_o^1 |u|dt \leq A$$

for all u such that $(\phi,u) \in A_o$. Thus the hypotheses of Theorem 1 of this paper are fulfilled, while the hypotheses of Theorem 1 of [2], which involve property (Q^*), are not.

Proof of Theorem 1

As in the proof of Theorem 1 of [2] we shall exploit the weak convergence in L_1 of a sequence of derivatives ϕ_k' and Mazur's theorem which states that a strongly closed convex set in a Banach space is weakly closed. In the proof we shall select subsequences of various sequences. Unless stated otherwise, we shall relabel the sequence with the labeling of the original sequence. In order to minimize technicalities we shall assume that t_1 and t_2 are fixed. The situation in which t_1 and t_2 are not fixed requires minor technical changes in the arguments for which the reader is referred to [2]. We note that if t_1 and t_2 are fixed then convergence in the ρ-metric is just uniform convergence on $[t_1, t_2]$. As in [2] we break up the argument into several steps.

STEP 1. There is a sequence $\{(\phi_k, u_k)\}$ of elements in A_o, a real γ and points x_{01} and x_{02} in R^n such that for $i = 1, 2$, $(t_i, x_{0i}) \in R$, $(t_1, x_{01}, t_2, x_{02}) \in B$ and

$$(5.1) \qquad \phi_k(t_i) \to x_{0i} \qquad \theta_k(t_2) \to \gamma.$$

Since the set of outputs $\theta(t_2)$ corresponding to pairs (ϕ, u) in A_o is bounded, it follows that there is a real number γ and a sequence $\{(\phi_k, u_k)\}$ of admissible pairs in A_o such that $\theta_k(t_2) \to \gamma$. Since B is compact there is a subsequence of this sequence such that $(t_1, \phi_k(t_1), t_2, \phi_k(t_2))$ converges to a point $(t_1, x_{01}, t_2, x_{02})$ in B. Moreover, since $(t_i, \phi_k(t_i)) \in R$, $i = 1, 2$, it follows that $(t_i, x_{0i}) \in R$.

STEP 2. There exists an absolutely continuous function ϕ defined on $[t_1, t_2]$ and a subsequence $\{\phi_k\}$ such that $\phi_k \to \phi$ uniformly and $\phi_k' \to \phi'$ weakly in $L_1[I]$. Moreover, ϕ satisfies (2.1)-(i) and (2.1)-(ii).

Since the graphs of all trajectories in A_o lie in the compact set R, the functions ϕ_k are uniformly bounded. By hypothesis, they are equi-absolutely continuous. Hence, by Ascoli's theorem there is a subsequence $\{\phi_k\}$ and a function ϕ such that $\phi_k \to \phi$ uniformly on $[t_1, t_2]$. Moreover, ϕ is absolutely continuous so that ϕ' is in L_1 and

$$\phi(t) = x_{01} + \int_{t_1}^{t} \phi'(s)\,ds.$$

From the relation

$$\phi_k(t) = \phi_k(t_1) + \int_{t_1}^{t} \phi_k'(s)\,ds$$

and the convergence of ϕ_k to ϕ it follows that for all t in $[t_1,t_2]$,

$$\int_{t_1}^{t} \phi_k'(s)\,ds \to \int_{t_1}^{t} \phi'(s)\,ds.$$

Since the functions ϕ_k are equi-absolutely continuous, their derivatives ϕ_k' have equi-absolutely continuous integrals. Hence (See [1], page 136) $\phi_k' \to \phi'$ weakly in L_1.

Since ϕ is the uniform limit of functions for which (2.1)-(i) and (ii) hold, these conditions hold for ϕ.

STEP 3. There exists a function λ that is integrable on $[t_1,t_2]$ such that $(\lambda(t),\phi'(t)) \in Q^+(t,\phi(t))$ for a.e. t in $[t_1,t_2]$ and such that

(5.2)
$$\int_{t_1}^{t_2} \lambda(s)\,ds \le \gamma.$$

Since $\phi_k' \to \phi'$ weakly in L_1, we obtain the following statement from a corollary to Mazur's theorem (Corollary, Theorem 2.9.3, p. 36[6]). For each integer j there exists an integer n_j, a set of integers $i = 1,\cdots,k$, where $k = k(j)$ depends on j, and a set of numbers $\alpha_{ij},\cdots,\alpha_{kj}$ satisfying

(5.3)
$$\alpha_{ij} \ge 0, \quad i = 1,\cdots,k \qquad \sum_{i=1}^{k} \alpha_{ij} = 1$$

such that $n_{j+1} > n_j + k(j)$ and

$$(5.4) \qquad \int_{t_1}^{t_2} \left| \phi' - \sum_{i=1}^{k} \alpha_{ij} \phi'_{n_j+i} \right| dt < 1/j$$

Let

$$(5.5) \quad \psi_j(t) = \sum_{i=1}^{k} \alpha_{ij} \phi'_{n_j+i}(t) = \sum_{i=1}^{k} \alpha_{ij} g(t, \phi_{n_j+i}(t), u_{n_j+i}(t)).$$

In terms of ψ_j, (5.4) says that $\psi_j \to \phi'$ in L_1. Hence there is a subsequence $\{\psi_j\}$ such that

$$(5.6) \qquad \psi_j(t) \to \phi'(t) \quad \text{a.e.}$$

We suppose that (5.5) is now this subsequence. Corresponding to the sequence (5.6) we define a sequence λ_j as follows:

$$(5.7) \qquad \lambda_j(t) = \sum_{i=1}^{k} \alpha_{ij} f(t, \phi_{n_j+i}(t), u_{n_j+i}(t)),$$

where for each j the numbers α_{ij}, the indices n_{j+i} and the functions ϕ_{n_j+i} and u_{n_j+i} are as in (5.5).

Up to now the proof was the same as in Theorem 1 of [2]. The rest of the argument in this Step is different from that in [2]. The reader is cautioned to keep in mind the order in which various subsequences are chosen.

Define sequences of functions σ_j and ω_j corresponding to ψ_j and λ_j as follows:

$$(5.8) \qquad \begin{aligned} \sigma_j(t) &= \sum_{i=1}^{k} \alpha_{ij} g(t, \phi(t), u_{n_j+i}(t)) \\ \omega_j(t) &= \sum_{i=1}^{k} \alpha_{ij} f(t, \phi(t), u_{n_j+i}(t)). \end{aligned}$$

The functions σ_j and ω_j are measurable. Let $M_k = \max \{ |\phi_k(t) - \phi(t)| : t \in [t_1, t_2] \}$. Since ϕ_k converges uniformly to ϕ, $M_k \to 0$ as $k \to \infty$. Let $\hat{g}_q = g(t, \phi(t), u_q(t))$ and let $g_q = g(t, \phi_q(t), u_q(t))$.

Since the σ_j are measurable we get, using (5.8), Assumption E(3), and (4.1),

$$\int_{t_1}^{t_2} |\sigma_j - \psi_j|\, dt \leq \sum_{i=1}^{k} \alpha_{ij} \int_{t_1}^{t_2} |\hat{g}_{n_j+i} - g_{n_j+i}|\, dt$$

$$\leq \sum_{i=1}^{k} \alpha_{ij} \mu(M_{n_j+i}) \int_{t_1}^{t_2} L(t, u_{n_j+i}(t))\, dt$$

$$\leq A \sum_{i=1}^{k} \alpha_{ij} \mu(M_{n_j+i}).$$

Thus $\sigma_j - \psi_j$ is in L_1. Since $M_k \to 0$ and $\mu(\delta) \to 0$ as $\delta \to 0$ we get that $\sigma_j - \psi_j \to 0$ in L_1. A similar argument shows that $\omega_j - \lambda_j$ is in L_1 and $\omega_j - \lambda_j \to 0$ in L_1. Hence there exists subsequences such that

$$(5.9) \qquad \sigma_j(t) - \psi_j(t) \to 0 \quad \omega_j(t) - \lambda_j(t) \to 0 \quad \text{a.e.}$$

We henceforth take the functions in (5.5), (5.7), and (5.8) to be the functions in these subsequences.

Define

$$(5.10) \qquad \lambda(t) = \lim \inf \lambda_j(t).$$

Since $f \geq 0$ it follows that $\lambda \geq 0$. Hence by Fatou's theorem we get, on setting $f_{n_j+i} = f(t, \phi_{n_j+i}(t), u_{n_j+i}(t))$,

$$\int_{t_1}^{t_2} \lambda\, dt \leq \lim \inf \sum_{i=1}^{k} \alpha_{ij} \int_{t_1}^{t_2} f_{n_j+i}\, dt.$$

$$= \lim \inf \sum_{i=1}^{k} \alpha_{ij} \theta_{n_j+i}(t_2).$$

The inequality (5.2) now follows from (5.1) and (5.3). Since $\lambda \geq 0$ we get that λ is in L_1 and is finite a.e.

We now show that for a.e. t; $(\lambda(t),\phi'(t)) \in Q^+(t,\phi(t))$.

For each integer k define a set E_k as follows:
$E_k = \{t: u_k(t) \notin \Omega(t,\phi_k(t))\}$. By $(2.1)-(iv)$, meas $E_k = 0$.
Let E denote the union of the sets E_k and let T_2 denote the set of points in $[t_1,t_2]$ that do not belong to E. Let $T_1 = \{t: \lambda(t) < \infty, \psi_j(t) \to \phi'(t)\}$, and let $T' = T_1 \cap T_2$. Then meas $T' = t_2-t_1$. Let T denote the set of points in T' at which (5.9) holds. Then meas $T = t_2-t_1$.

Let t be a fixed but arbitrary point in T. Since $\psi_j(t) \to \phi'(t)$, it follows from (5.9) that $\sigma_j(t) \to \phi'(t)$. From the definition of λ it follows that there is a subsequence $\lambda_j(t)$, which will in general depend on t, such that $\lambda_j(t) \to \lambda(t)$. By virtue of (5.9), $\omega_j(t) \to \lambda(t)$. For the corresponding subsequence $\sigma_j(t)$ we still have $\sigma_j(t) \to \phi'(t)$. Since $t \in T \subset T'$, it follows that for all j and i

$$u_{n_j+i}(t) \in \Omega(t,\phi_{n_j+i}(t)) = \Omega(t,\phi(t)),$$

the last equality being a consequence of Assumption $E(2)$. Hence

$$(f(t,\phi(t),u_{n_j+i}(t)),g(t,\phi(t),u_{n_j+i}(t)) \in Q^+(t,\phi(t)).$$

Since $Q^+(t,\phi(t))$ is convex, the points $(\psi_j(t),\sigma_j(t))$ belong to $Q^+(t,\phi(t))$. Since $Q^+(t,\phi(t))$ is closed and $(\omega_j(t),\sigma_j(t)) \to (\lambda(t),\phi'(t))$, we get that $(\lambda(t),\phi'(t)) \in Q^+(t,\phi(t))$. Since t was an arbitrary point of T, we have that $(\lambda(t),\phi'(t)) \in Q^+(t,\phi(t))$ a.e.

STEP 4. There exists a measurable function u defined on $[t_1,t_2]$ such that for almost all t:

(i) $\phi'(t) = g(t,\phi(t),u(t))$; (ii) $u(t) \in \Omega(t,\phi(t))$;
(iii) $\lambda(t) \geq f(t,\phi(t),u(t))$.

The existence of a function v satisfying the conclusion of Step 4 is a restatement of $(\lambda(t),\phi'(t)) \in Q^+(t,\phi(t))$. To show that there exists a measurable function u with the same properties as v one uses a standard argument involving the McShane-Warfield extension of Filippov's Lemma [8]. For details see [2].

STEP 5.   Completion of Proof

Statements (i) and (ii) of Step 4 assert that (ϕ,u) satisfies (2.1)-(iii) and (iv). Since ϕ and u are measurable and f is continuous or satisfies Assumption C, it follows that $f(t,\phi(t),u(t))$ is measurable. Since λ is integrable and f is bounded from below, it follows from (iii) of Step 4 that $f(t,\phi(t),u(t))$ is integrable. Hence (ϕ,u) satisfies (2.1)-(v). From this and from Step 2 it follows that (ϕ,u) is admissible and that $\phi_k \to \phi$ uniformly in the ρ metric. It therefore only remains to prove (4.2).

The equality in (4.2) was established in Step 1. Let θ be the output corresponding to (ϕ,u). Then from (iii) of Step 4 and (5.2) it follows that $\lambda \geq \theta(t_2)$ and (4.2) is thereby established.

REFERENCES

1. Banach, S., _Théorie des Opérations Linéaires_, Warszawa, 1932.
2. Berkovitz, L. D., Existence theorems in problems of optimal control, to appear _Studia Mathematica_ 1972.
3. Brunovsky, P., On the necessity of a certain convexity condition for lower closure of control problems, SIAM J. Control 6(1968), 174-185.
4. Cesari, L., Existence theorems for optimal controls of the Mayer type, _SIAM J. Control_ 6(1968), 517-552.
5. Cesari, L., J. R. LaPalm, and T. Nishiura, Remarks on some existence theorems for optimal control, _J. Optimization Theory Appl._ 3(1969), 296-305.

6. Hille, E., and R. S. Phillips, _Functional Analysis_ and
   Semi-Groups, Revised Ed., American Mathematical
   Society, Providence, 1957.
7. Jacobs, M. Q., Attainable sets in systems with unbounded controls, _J. Differential Equations_ 4(1968),
   408-423.
8. McShane, E. J., and R. B. Warfield, Jr., On Filippov's
   implicit functions lemma, _Proc. Amer. Math. Soc._
   18(1967), 41-47.

THE BI-CANONICAL SYSTEMS

Christian MARCHAL

Office National d'Etudes et de Recherches Aérospatiales
O.N.E.R.A.
92 - Châtillon - France.

Summary

The bi-canonical systems are the differential systems
which satisfy the conditions of canonicity in the generalized meaning of Contensou and in the generalized meaning of
Pontryagin ; they appear very often in the optimization
problems.
The conditions of bi-canonicity are given in the first
part and the properties of bi-canonical systems are developed in the second and in the third parts ; these properties
are useful as well for the theoretical studies as for the
numerical computations.

Les Systèmes bi-canoniques

Résumé

Les systèmes bi-canoniques sont ceux qui satisfont aux
conditions de canonicité au sens de Contensou généralisé et
au sens de Pontryagin généralisé ; ces systèmes apparais-
sent très souvent dans les problèmes d'optimisation.
Les conditions de bi-canonicité sont présentées dans
la première partie et les propriétés des systèmes bi-canoniques sont développées dans la seconde et dans la troisiè-
me partie ; ces propriétés sont utiles aussi bien pour les
études théoriques que pour les applications numériques pra-
tiques.

1. Introduction

This paper is an extension of the reference $\left[11\right]$.

The theories of optimization have been developed in three main directions :

I) The local analytical studies summarized by the question "a solution being given is it possible to improve it by infinitesimal modifications ?" [1-7], question which leads to the "maximum principle" of Pontryagin [2].
II) The general analytical studies such as the studies of Contensou [8-10].
III) The studies of methods allowing the use of computers to solve the problems of optimization.

We shall consider especially the two first points which have led to the notions of "canonicity in the generalized meaning of Pontryagin" and "canonicity in the generalized meaning of Contensou"[11] ; when a differential system is canonical in both meanings it is bi-canonical, it has of course all the properties related to these canonicities[11] and, furthermore, it has some particular properties we are going to consider.

First Part

The Conditions of Bi-Canonicity

Let us consider as usual a differential system S defined by :

1) A parameter of description t (usually the time) ;

2) A "state vector" $\vec{X}$ related to all other interesting parameters (including the performance index which must be a function of the terminal values $\vec{X_o}$, t_o , $\vec{X_f}$ and t_f ; if necessary, when for instance the performance index is related to an integral taken along the trajectory, one must add in the state vector a component related to this integral).

3) The maneuverability of this system S can be given by a maneuverability domain $\mathcal{D}(\vec{X}, t)$:

$$(1) \qquad \vec{V} = \text{velocity vector} = \frac{d\vec{X}}{dt} \in \mathcal{D}(\vec{X}, t)$$

4) If the maneuverability is given by a control, for instance by a control vector $\vec{u}$:

$$(2) \qquad \vec{V} = f(\vec{X}, \vec{u}, t) \; ; \; \text{with} \; \vec{u} \in \mathcal{D}(\vec{X}, t)$$

then the control domain $\mathscr{D}(\vec{X}, t)$ is obviously related to the maneuverability domain $D(\vec{X}, t)$ by :

$$(3) \qquad D(\vec{X}, t) = f\left[\vec{X}, \mathscr{D}(\vec{X}, t), t\right]$$

The question of the optimal control is only a question of minor importance immediately solved when, for a given problem, the optimal trajectory $\vec{X}(t)$ is determined.

We shall consider only cases of admissibility of the continuous type [11] , by the definition of the admissibility of this type (the largest definition allowing the use of the relation (5)) the admissible trajectories $\vec{X}(t)$ are the absolutely continuous functions of the time t which verify almost always :

$$(4) \qquad \vec{V}(t) = \frac{d\vec{X}(t)}{dt} \in D\left[\vec{X}(t), t\right].$$

The absolute continuity involves that $\vec{X}(t)$ is almost always derivable and it involves also, for any t_0 and t_1, the very important relation :

$$(5) \qquad \vec{X}(t_1) = \vec{X}(t_0) + \int_{t_0}^{t_1} \vec{V}(t)\, dt$$

This relation is even generally the very definition of $\vec{X}(t_1)$.

In order to define the conditions of the bi-canonicity let us remember the following points :

A) A property is __almost always__ true in a given set of the $\vec{X}, t$ space if it is true everywhere in that set except at most at points corresponding to a set of measure zero of the t axis.

B) The differential system of interest is analysed in a "zone of interest" of the $\vec{X}, t$ space, this zone is defined by a closed interval of time $\left[T_0, T_f\right]$ finite or infinite.

C) An admissible trajectory $\vec{X}(t)$ is defined on a closed and finite interval of time $\left[t_0, t_f\right]$, t_0 being the initial time and t_f the final time (with of course $T_0 \leqslant t_0 \leqslant t_f \leqslant T_f$).

If $t_o < t_f$ the trajectory of interest is a non-zero trajectory.

D) As usual we shall call "Hamiltonian" the scalar product $\vec{P}.\vec{V}$:

$$(6) \qquad H = \vec{P}.\vec{V}$$

$\vec{P}$ being the "adjoint vector" or "vector of Pontryagin".

We shall call "Generalized Hamiltonian" the function of $\vec{P}$, $\vec{X}$ and t defined by :

$$(7) \qquad H^*(\vec{P}, \vec{X}, t) = \sup_{\vec{V} \in \mathcal{D}(\vec{X}, t)} \vec{P}.\vec{V}$$

H^* is then equal to the scalar product $\vec{P}.\vec{V}$ drawn on the figure one.

If $\vec{V}$ is given in terms of a control vector $\vec{u}$ as in (2), H^* is directly given by :

$$(8) \qquad H^*(\vec{P}, \vec{X}, t) = \sup_{\vec{u} \in \mathcal{D}(\vec{X}, t)} \vec{P}.\vec{V}(\vec{X}, \vec{u}, t)$$

Let us note that H^* remains the same after the convexisation and the closure of the maneuverability domain $\mathcal{D}(\vec{X}, t)$; the convexisation of $\mathcal{D}(\vec{X}, t)$ corresponds of course to the "relaxation" of the control [12].

Using the reference [11] the conditions of bi-canonicity are then easy to define, we obtain two cases : either the differential system of interest has no non-zero admissible trajectory (this case is of course uninteresting) or any point of the zone T_o, T_f of interest belongs to at least one non-zero admissible trajectory (in other words the "useful space" is the whole zone of interest).

This last case happens if and only if the four following conditions are satisfied :

1) $T_o < T_f$

2) The maneuverability domain $\mathcal{D}(\vec{X}, t)$ is almost always closed, convex, finite and non empty.

214

3) $H^*(\vec{P}, \vec{X}, t)$ is a measurable function of $\vec{P}, \vec{X}$ and t .

4) For any bounded set B of the zone of interest there exist an integrable function $g(t)$ such that almost always :

$$(9) \quad \left.\begin{array}{l}(\vec{X_1}, t) \in B \\ \\ (\vec{X_2}, t) \in B\end{array}\right\} \Rightarrow \left|H^*(\vec{P_2}, \vec{X_2}, t) - H^*(\vec{P_1}, \vec{X_1}, t)\right| \leqslant g(t) \cdot \left[\left\|\vec{P_2} - \vec{P_1}\right\| + \left\|\vec{P_1}\right\| \cdot \left\|\vec{X_2} - \vec{X_1}\right\|\right]$$

Remark I.
We have made no difference between the partial canonicity and the total canonicity in the generalized meaning of Contensou because a system partially canonical in the generalized meaning of Pontryagin (and hence "partially bi-canonical") is also always totally canonical, into both directions, in the generalized meaning of Contensou.

Remark II.
If the velocity vector $\vec{V}$ is given in terms of a control vector $\vec{u}$ as in (2), the condition (9) is equivalent to :

For any bounded set B of the zone of interest there exist an integrable function $h(t)$ such that almost always :

$$(10) \quad \left.\begin{array}{l}(\vec{X_1}, t) \in B \\ \\ (\vec{X_2}, t) \in B \\ \\ \vec{u_1} \in \mathcal{D}(\vec{X_1}, t)\end{array}\right\} \Rightarrow \left\{\begin{array}{l}\text{Firstly} : \left\|\vec{V}(\vec{X_1}, \vec{u_1}, t)\right\| \leqslant h(t) \\ \\ \text{Secondly} : \text{there exist } \vec{u_2} \in \mathcal{D}(\vec{X_2}, t) \text{ such} \\ \\ \text{that} \left\|\vec{V}(\vec{X_2}, \vec{u_2}, t) - \vec{V}(\vec{X_1}, \vec{u_1}, t)\right\| \leqslant h(t) \cdot \left\|\vec{X_2} - \vec{X_1}\right\|\end{array}\right.$$

If the control domain $\mathcal{D}(\vec{X}, t)$ depends only on t it is generally sufficient to choose $\vec{u_2} = \vec{u_1}$ in order to verify the condition (10).

Remark III.
The condition (10) involves that $D(\vec{X}, t)$ is almost always finite, it is also generally easy to verify that $D(\vec{X}, t)$ is almost always non empty but it is more difficult to verify that $D(\vec{X}, t)$ is almost always closed and especially that it is almost always convex. However the convexisation and the closure of the maneuverability domain $D(\vec{X}, t)$, i.e. the "relaxation" of the control [12] , are very often justified (i.e. the admissible trajectories obtained after convexisation and closure of $D(\vec{X}, t)$ are limit of appropriate sequen-

ces of admissible trajectories obtained before convexisation and closure) indeed, if the condition (9) or the condition (10) is verified, it is sufficient that the relation $\vec{V} \in \mathcal{D}(\vec{X}, t)$ be measurable in the restricted meaning of Borel [13] , as it is of course the case in all ordinary problems.

Let us note that this last condition involves the measurability of $H^*(\vec{P}, \vec{X}, t)$ in terms of $\vec{P}, \vec{X}$ and t .

<u>Remark IV.</u>
The bi-canonicity doesn't involve any kind of continuity of the maneuverability with respect to the time, it can even be completely discontinuous with respect to t ; but the condition (9) involves that, after convexisation and closure, the maneuverability domain $\mathcal{D}(\vec{X}, t)$ is a continuous function of $\vec{X}$.

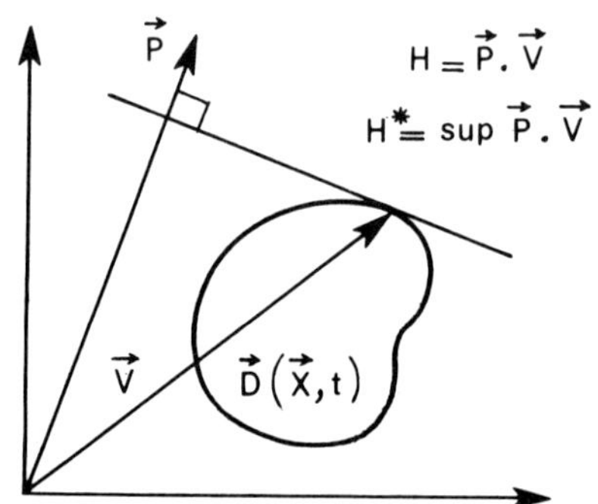

Figure 1 - The Generalized Hamiltonian H^* and the maneuverability domain $\mathcal{D}(\vec{X}, t)$.

Second Part

Properties of Bi-Canonical Systems

The bi-canonical systems have of course all the properties associated to either type of canonicity and they have also some particular properties.

Let us remember the main ordinary properties of canonical systems [11] .

2.1. Main properties associated with the total canonicity in the generalized meaning of Contensou.

A) The set of admissible trajectories of the differential system S of interest is "complete" and even "locally compact".

B) The system S can be considered as the limit for $\varepsilon \to 0$ of "jump systems" S_ε , but these jump systems are quite complicated (this property is very important for the numerical applications).

C) The useful space being there all the zone T_0 , T_f of interest, as long as the accessible domain from a given point $(\vec{X}_1 , t_1)$ remains finite its sections by the $t = $ constant spaces remain closed, connex and upper semi-continuous with respect to $\vec{X}_1$ and t_1 .

etc...

2.2. Main properties associated with the canonicity in the generalized meaning of Pontryagin.

A) If an admissible trajectory $\vec{X}(t)$ defined for $t_0 \leqslant t \leqslant t_f$ is locally optimal it is also locally extremal (i.e. it is locally optimal between any instants t_1 and t_2 such that $: t_0 \leqslant t_1 \leqslant t_2 \leqslant t_f$), and if it is optimal it is also extremal.

B) The maneuverability domain $\mathcal{D}(\vec{X}, t)$ being almost always finite and non empty, the generalized Hamiltonian $H^*(\vec{P}, \vec{X}, t)$ can always be used as $\mathcal{L}(\vec{P}, \vec{X}, t)$ function [11] and hence the generalization of the Pontryagin maximum principle leads to :

If $\vec{X}(t)$ is a locally extremal trajectory defined for $t_0 \leqslant t \leqslant t_f$ there exists at least one adjoint absolutely continuous function $\vec{P}(t)$ defined on the same interval of time and such that :

I) $\qquad t \in \left[t_0 , t_f \right] \implies \vec{P}(t) \neq 0$

II) If, at the points $(\vec{P}(t), \vec{X}(t), t)$, $H^*(\vec{P}, \vec{X}, t)$ is almost always continuously differentiable with respect to $\vec{P}$ and $\vec{X}$, $\vec{P}(t)$ and $\vec{X}(t)$ must verify almost always :

$$(11) \quad \frac{d\vec{X}(t)}{dt} = \frac{\partial H^*(\vec{P}, \vec{X}, t)}{\partial \vec{P}} \; ; \; \frac{d\vec{P}(t)}{dt} = - \frac{\partial H^*(\vec{P}, \vec{X}, t)}{\partial \vec{X}}$$

$\vec{X}(t)$ is then either an "ordinary trajectory of Pontryagin" (if there exist only one independent function $\vec{P}(t)$) or a "singular trajectory of type II".

III) If, at the points $(\vec{P}(t), \vec{X}(t), t)$, $H^*(\vec{P}, \vec{X}, t)$ is not almost always continuously differentiable with respect to $\vec{P}$ and $\vec{X}$ the equations (11) become ($\vec{P}$, $\vec{X}$ being considered as a single vector with $2n$ components) :

$$(12) \quad \begin{cases} \left[\dfrac{d\vec{X}(t)}{dt} \; ; \; - \dfrac{d\vec{P}(t)}{dt} \right] \in \mathcal{D}_H \left[\vec{P}(t), \vec{X}(t), t \right] \; ; \; \text{almost always.} \\[2mm] \mathcal{D}_H\,(\vec{P}, \vec{X}, t) \text{ being the smallest closed and convex set} \\ \quad \text{containing the gradient vectors } \dfrac{\partial H^*(\vec{P}, \vec{X}, t)}{\partial (\vec{P}, \vec{X})} \\ \quad \text{obtained at points} (\vec{P}+\delta\vec{P}, \vec{X}+\delta\vec{X}, t) \text{where : A) } H^* \text{ is dif-} \\ \quad \text{ferentiable with respect to } (\vec{P}, \vec{X}) \; ; \; \text{B) } \delta\vec{P} \text{ and } \delta\vec{X} \text{ are} \\ \quad \text{infinitely small.} \end{cases}$$

Remark I.
It is easy to verify that the conditions (11) or (12) involve the "condition of the maximum" i.e. that, taking account of the definition of H^* given in (7) or (8), the velocity vector $\vec{V}(t)$ is almost always such that :

$$(13) \quad \vec{P}(t) \cdot \vec{V}(t) = H^* \left[\vec{P}(t), \vec{X}(t), t \right]$$

Remark II.
If, for any function $\vec{P}(t)$ satisfying (12) the domain $\mathcal{D}_H \left[\vec{P}(t), \vec{X}(t), t \right]$ is almost always such that the choice of $\frac{d\vec{X}(t)}{dt}$ imposes $\frac{d\vec{P}(t)}{dt}$, then the given trajectory is either a "singular trajectory of type I" (if there exist only one independent function $\vec{P}(t)$) of a "singular trajectory of type II".

The singular trajectories of type I appear very often into the optimization problems when the maneuverability domains $\mathcal{D}(\vec{X}, t)$ are not strictly convex (or when they are the result of the convexisation of non-convex domains) ; it is to these trajectories that can be applied the optimalty test of singular extremals (or "generalized Legendre Clebsch conditions") [3,4,6,11] .

<u>Remark III.</u>
If the domain $\mathcal{D}_H\left[\vec{P}(t), \vec{X}(t), t)\right]$ is not almost always such that the choice of $\dfrac{d\vec{X}(t)}{dt}$ imposes $\dfrac{d\vec{P}(t)}{dt}$ then the given trajectory $\vec{X}(t)$ is a "singular trajectory of type III" (for instance reference [11] chapter 5.3.3.3.) ; the use of appropriate $\mathcal{L}(\vec{P}, \vec{X}, t)$ functions improves the analysis and may even lead to the conclusion of non-optimality of the trajectory $\vec{X}(t)$ of interest.

<u>Remark IV.</u>
Let us put $\vec{P}_0 = \vec{P}(t_0)$ and $\vec{P}_f = \vec{P}(t_f)$. As usual the different couples $(\vec{P}_0, \vec{P}_f)$ associated to a given trajectory $\vec{X}(t)$ can be used to describe the "neighbouring surely accessible domain" ; for instance if $\vec{X}(t)$ is an "ordinary trajectory of Pontryagin" or a "singular arc of type I" there is only one adjoint function $\vec{P}(t)$ and the neighbouring surely accessible domain is the half space of couples $(\vec{\delta X}_0, \vec{\delta X}_f)$ define by :

$$(14) \qquad \vec{P}_0 \cdot \vec{\delta X}_0 \geqslant \vec{P}_f \cdot \vec{\delta X}_f$$

More generally let us consider an ordinary problem of optimization under terminal constraints and let $\vec{X}(t)$ be a solution of this problem.

Let us remember that the performance index is a function of the terminal values $\vec{X}_0, t_0, \vec{X}_f, t_f$ and let us call $\mathcal{K}$ the cone (in the $\vec{\delta X}_0, \vec{\delta X}_f$ space) of directions : 1) authorized, 2) improving the performance index ; we obtain two cases :

A) <u>If $\mathcal{K}$ is convex and is not a sub-space of the space of the couples $(\vec{\delta X}_0, \vec{\delta X}_f)$,</u> a necessary condition of optimality of the trajectory $\vec{X}(t)$ is that there exist a corresponding adjoint couple $(\vec{P}_0, \vec{P}_f)$ such that :

$$(15) \qquad (\vec{\delta X}_0, \vec{\delta X}_f) \in \mathcal{K} \implies \vec{P}_0 \cdot \vec{\delta X}_0 \leqslant \vec{P}_f \cdot \vec{\delta X}_f$$

Such couples $(\vec{P_0}, \vec{P_f})$ must exist for any function $L(\vec{P},\vec{X},t)$ if $\vec{X}(t)$ is a "singular arc of type III" (see remark III above).

B) If K is not convex or is a sub-space of the space of the couples $(\delta\vec{X_0}, \delta\vec{X_f})$ a similar condition must be verified with respect to any part of K satisfying the above conditions A.

The question of the continuity of the cone K is important (see in [11] page 71 the strict definition of the cone K).

Remark V.
If the generalized Hamiltonian $H^*(\vec{P},\vec{X},t)$ is continuous at $(\vec{P_0},\vec{X_0},t_0)$ and $(\vec{P_f},\vec{X_f},t_f)$, the parameter t can be considered as an ordinary parameter : the corresponding adjoint values are as usual :

$$ P_{t_0} = -H^*(\vec{P_0},\vec{X_0},t_0) \qquad \text{and} \qquad P_{t_f} = -H^*(\vec{P_f},\vec{X_f},t_f) $$

and the inequation (14) becomes :

$$ (16) \qquad P_0 \cdot \delta X_0 - H_0^* \, \delta t_0 \;\geqslant\; P_f \cdot \delta X_f - H_f^* \, \delta t_f $$

the "neighbouring surely accessible domain" can be then analysed around the values t_0 and t_f .

2.3. Particular properties of bi-canonical systems

The particular properties of bi-canonical systems are very important for the numerical applications : they allow to built systematical methods of research of the optimal trajectories ; we shall consider the four main properties.

2.3.1. Sufficient conditions of existence of an optimal trajectory in a given problem of optimization.

These conditions may be summarized by : "not to get into trouble with the discontinuities of the performance index, with the closure of the authorized domain and with infinity".

A problem of optimization is usually presented in the following way :

A differential system S and a performance index $f(\vec{X}_0, t_0, \vec{X}_f, t_f)$ being given, what is the admissible trajectory $\vec{X}(t)$ which gives the maximum value to the performance index various terminal constraints being imposed ? (for instance $\vec{X}_0, t_0$ and t_f are given, $\vec{X}_f$ being free).

Let us call $\mathcal{D}$ the domain of authorized couples $\left[(\vec{X}_0, t_0)(\vec{X}_f, t_f)\right]$ and $\mathcal{R}$ the domain of connectable couples (i.e. such that there exist an admissible trajectory starting at $(\vec{X}_0, t_0)$ and going to $(\vec{X}_f, t_f)$).

If $\mathcal{D} \cap \mathcal{R} = \emptyset$ the problem is impossible and without interest, hence a first obvious condition is that the problem be possible : there must exist at least one admissible trajectory such that : $\left[(X_0, t_0)(X_f, t_f)\right] \in \mathcal{D}$

If $\mathcal{D} \cap \mathcal{R} \neq \emptyset$ let us put :

$$(17) \qquad \left\{ \quad f_M = \underbrace{\sup}_{\left[(X_0, t_0), (X_f, t_f)\right] \in \mathcal{D} \cap \mathcal{R}} f(\vec{X}_0, t_0, \vec{X}_f, t_f) \right.$$

f_M is the maximum value we are looking for.

If f_M corresponds to no admissible trajectory there is no optimal solution for the problem of interest, this may happen in three cases :

A) When the performance index has discontinuities (for instance :
$f(\vec{X}_0, t_0, \vec{X}_f, t_f) = t_f -$ integer part of t_f), hence the second condition is that the performance index be a continuous function, or at least an upper semi-continuous function of $\vec{X}_0, t_0, \vec{X}_f$ and t_f .

B) When the authorized domain $\mathcal{D}$ is not closed (for instance : $f = t_f$ and $t_f < 1$), hence the third condition is that the authorized domain $\mathcal{D}$ be closed.

C) When the sequence of admissible trajectories $\vec{X}_n(t)$ such that $f_n \longrightarrow f_M$ when $n \longrightarrow \infty$ doesn't belong to a bounded set of the $\vec{X}, t$ space ; for instance if in a problem where

the final time is free the optimality leads to infinite
duration. A more complicated case is the case of trajecto-
ries "through infinity" : $n \to \infty \Rightarrow sup \| \vec{X}_n (t) \| \longrightarrow \infty$, $\vec{X}$
"goes to infinity and come back in a finite time". (see for
instance [11] chapter 5.6.).

Hence the fourth condition is that the infinitely remo-
te parts of $\mathcal{D}$ make no trouble (they are unreachable or
they correspond to uninteresting values of the performance
index) and furthermore that the phenomenum of trajectories
"through infinity" is either impossible or at least hazard-
less.

If these four conditions are satisfied the problem has
at least one optimal solution, but there remains one obstacle : this solution can correspond to a local maximum of
$f(\vec{X}_o, t_o , \vec{X}_f , t_f)$ in $\mathcal{D}$ (the problem is then rather a study
of $\mathcal{D}$ and f than a problem of optimization), in that case
the "optimal solution" is generally not an "optimal trajectory" in the ordinary meaning of these words [11] and the
maximum principle cannot be applied.

If this last obstacle is overcome there is an optimal
solution and this solution is an optimal and even an extremal trajectory for which it is justified to apply the generalized maximum principle (chapter 2.2.B. or reference [11]
chapter 5.3.3.).

2.3.2. Existence of a sequence of very simple "jump
systems" the limit of which is the differential system of
interest.

The sequence of "jump systems" presented in [11] page
20, is complicated and we shall consider a simpler one in
which, for a given "time jump", the "jump domains" are
always smaller.

Let us define the jump system S_ε by :

1) Time jump :

$$(18) \qquad\qquad t_{i+1} = t_i + \varepsilon$$

2) State jump :

(19)
$$\vec{X}_{i+1} - \vec{X}_i \in \Delta\,(\vec{X}_i\,,\,t_i)$$

$\Delta\,(\vec{X}_i\,,\,t_i)$ being the convex, bounded and closed jump domain defined by :

(20) $\vec{u} \in \Delta(\vec{X}_i,t_i) \iff$
$$\begin{cases} \exists W(t)\text{defined for } t_i < t < t_{i+1}, \text{ and such that:} \\ \text{I} \quad \vec{W}(t) \in \mathcal{D}\,(\vec{X}_i\,,\,t)\,, \qquad \text{almost always} \\ \text{II} \quad \vec{u} = \int_{t_i}^{t_{i+1}} \vec{W}(t)\,.\,dt\,. \end{cases}$$

Let us demonstrate that the reachable domains obtained by these jump systems S_ε from an initial point $(\vec{X}_0\,,\,t_0)$ tend towards the reachable domain corresponding to the system S when $\varepsilon \to 0$.

The systems S_ε defined by (18), (19) and (20) having a maneuverability smaller than or equal to that of the jump systems of [11] their corresponding reachable domains are also smaller or equal and hence the only point to demonstrate is that if $\vec{X}(t)$ is an arbitrary admissible trajectory of the system S , starting at $(\vec{X}_0\,,t_0)$ and leading to $(\vec{X}_f\,,\,t_f)$, there exists a sequence of jump trajectories going to $\vec{X}(t)$ when $\varepsilon \to 0$.

This point is easy ; let us choose for instance the systems S_{ε_j} starting at the same instant t_0 and such that :

(21)
$$\varepsilon_j = t_{i+1} - t_i = \frac{t_f - t_0}{j}$$

and let us choose at any instant in $\mathcal{D}(\vec{X}_i,t)$ the vector $\vec{W}(t)$ of the relation (20) as close as possible to $\vec{V}(t)$.

Let us take account of (9) in a sufficient neighbourhood of the trajectory $\vec{X}(t)$, we obtain :

(22)
$$\begin{cases} \|\vec{V}(t)\| < g\,(t) \\ t_i < t < t_{i+1} \implies \|\vec{V}(t) - \vec{W}(t)\| < g\,(t)\,.\,\|\vec{X}(t) - \vec{X}_i\| \end{cases}$$

and hence for the jump trajectory of the system S_{ε_j} :

$$(23) \begin{cases} t_i = t_0 + \dfrac{i}{j}\,(t_f - t_0) \\[2mm] \|\vec{X}_{i+1} - \vec{X}(t_{i+1})\| \leqslant \|\vec{X}_i - \vec{X}(t_i)\| \cdot \left(1 + \displaystyle\int_{t_i}^{t_{i+1}} g(t)\,dt\right) + \dfrac{1}{2}\left[\displaystyle\int_{t_i}^{t_{i+1}} g(t)\,dt\right]^2 \end{cases}$$

and then, at the final time :

$$(24) \quad \begin{array}{l} t_j = t_f \\[2mm] \|\vec{X}_j - \vec{X}_f\| \leqslant \dfrac{1}{2}\displaystyle\sum_{i=0}^{j-1}\left[\displaystyle\int_{t_i}^{t_{i+1}} g(t)\,dt\right]^2 \cdot \exp\displaystyle\int_{t_0}^{t_f} g(t)\,.\,dt \end{array}$$

which involves as we looked for :

$$(25) \qquad j \longrightarrow \infty \implies \vec{X}_j \longrightarrow \vec{X}_f$$

2.3.3. As an easy consequence of the previous property we obtain that the reachable domain from a given point is a continuous function of this point at least as long as this reachable domain remains finite.

Remark

It is possible to demonstrate that this last property remains true, after convexisation and closure of the maneuverability domains, if only the conditions of canonicity in the generalized meaning of Pontryagin [11] are satisfied and as long as the reachable domain remains finite and remains in the open set of the $\vec{X}, t$ space where these conditions are verified (this involves in particular that the maneuverability domains be almost always finite).

2.3.4. The fourth particular property of bi-canonical systems is related to the relations (12) i.e. to the generalized equations of Pontryagin.

These relations constitute a differential system with respect to the new "state vector" $(\vec{X}, -\vec{P})$, the maneuverability domain of this system is of course $\mathcal{D}_H\,(\vec{P}, \vec{X}, t)$ and it is easy to verify that the conditions of total canonicity (into both directions) in the generalized meaning of Contensou are satisfied. Hence, for instance, the admissible trajectories of the vector $(\vec{X}, -\vec{P})$ (which, if $\vec{P} \neq 0$, correspond to the extremals of the system of interest) can be obtained by the method of the "jump systems" (ref. [11]

224

page 20).

Third Part

Use of the Equation $\dfrac{dH^*}{dt} = \dfrac{\partial H^*}{dt}$

The main properties of canonical systems in the generalized meaning of Pontryagin, for instance those concerning the test of singular extremals or the chattering arcs, can be applied with no change to the bicanonical systems.

It is also the case for the two very useful equations :

(26)
$$\frac{dH^*}{dt} = \frac{\partial H^*}{\partial t}$$

(27)
$$H^*(t_1) = H^*(t_0) + \int_{t_0}^{t_1} \frac{\partial H^*}{\partial t}\, dt$$

These equations may be utilized anywhere (in the generalized form given in (30)) if and only if the following condition is satisfied :

For any bounded set B of the $\vec{X}, t$ space (or of the zone of interest defined by the closed interval of time $[T_0, T_f]$) there exist an absolutely continuous function $f(t)$ such that :

(28)
$$\left.\begin{array}{l}(\vec{X}, t_1) \in B \\[2ex] (\vec{X}, t_2) \in B\end{array}\right\} \Rightarrow \left| H^*(\vec{P}, \vec{X}, t_2) - H^*(\vec{P}, \vec{X}, t_1) \right| \leq \left\| \vec{P} \right\| . \left| f(t_2) - f(t_1) \right|$$

If the velocity vector $\vec{V}$ is given in terms of a control vector $\vec{u}$ as in (2), the condition (28) is equivalent to :

For any bounded set B of the zone of interest there exist an absolutely continuous function $k(t)$ such that :

225

$$(29) \quad \left. \begin{array}{l} (\vec{X}, t_1) \in B \\[4pt] (\vec{X}, t_2) \in B \\[4pt] \vec{u}_1 \in \mathscr{D}(\vec{X}, t_1) \end{array} \right\} \Rightarrow \left\{ \begin{array}{l} \exists\, \vec{u}_2 \ \text{ belonging to } \mathscr{D}(\vec{X}, t_2) \text{ and such} \\ \qquad \text{that :} \\[8pt] \left\| \vec{V}(\vec{X}, \vec{u}_2, t_2) - \vec{V}(\vec{X}, \vec{u}_1, t_1) \right\| \leqslant \left| k(t_2) - k(t_1) \right| \end{array} \right.$$

If the control domain $\mathscr{D}(\vec{X}, t)$ depends only on $\vec{X}$ it is generally sufficient to choose $\vec{u}_2 = \vec{u}_1$ in order to verify the condition (29) ; of course this condition (or the condition (28)) associated with (9) involves the continuity of $H^*(\vec{P}, \vec{X}, t)$ everywhere.

If the conditions (28) or (29) are satisfied the equations (26) and (27) may be used in the following generalized form very similar to the relations (12) :

$\vec{X}(t)$ being an arbitrary extremal and $\vec{P}(t)$ being a corresponding adjoint function we must verify :

$$(30) \quad \left\{ \begin{array}{l} \left[\dfrac{d\vec{X}(t)}{dt} \,;\, -\dfrac{d\vec{P}(t)}{dt} \,;\, \dfrac{dH^*(\vec{P}(t), \vec{X}(t), t)}{dt} \right] \in D_{H t}\left[\vec{P}(t), \vec{X}(t), t \right]; \ \text{almost always.} \\[12pt] D_{H t}\left[\vec{P}(t), \vec{X}(t), t \right] \text{ being the smallest closed and convex set} \\[8pt] \text{(in the space of the velocities of } \ \vec{X}, \vec{P} \quad \text{ and } H^* \text{)} \\[8pt] \text{containing the gradient vectors } \dfrac{\partial H^*(\vec{P}, \vec{X}, t)}{\partial(\vec{P}, \vec{X}, t)} \text{ obtained at} \\[8pt] \text{points } (\vec{P} + \vec{\delta P}, \vec{X} + \vec{\delta X}, t) \text{ where :} \\ \qquad\qquad \text{A) } H^* \text{ is differentiable with respect} \\ \qquad\qquad\qquad \text{to } \vec{P}, \vec{X} \text{ and } t \ ; \\ \qquad\qquad \text{B) } \vec{\delta P} \text{ and } \vec{\delta X} \text{ are infinitely small.} \end{array} \right.$$

In particular if the maneuverability doesn't depends on the time, H^* is a function of $\vec{P}$ and $\vec{X}$ only, the condition (28) is satisfied and the conditions (30) lead to : $H^*\left[\vec{P}(t), \vec{X}(t), t \right]$ is constant

<u>Remark</u>
Let us consider the "jump systems" given in (18), (19) and (20) and the limit of which is the system of interest.

If the condition (28) is satisfied these jump systems

can be simplified, indeed it is possible to choose as jump domain $\Delta\,(\vec{X_i}, t_i)$ the domain such that :

$$(31) \qquad \vec{V} \in D\,(\vec{X_i}, t_i) \Longleftrightarrow \varepsilon\,\vec{V} \in \Delta\,(\vec{X_i}, t_i)$$

This last point is certainly particularly useful for the numerical applications.

Conclusion

The bi-canonical systems appear very often in the problem of optimization, they don't involve the continuity of the maneuverability (or of the control) with respect to the time and hence they are very general.

These systems have many properties related to either type of canonicity (canonicity in the generalized meaning of Contensou or in the generalized meaning of Pontryagin) and they have also many particular properties very useful for the numerical applications.

It would be very interesting and probably easy to extend this study to the case of admissibility of the discontinuous type.

References

1. G.A. Bliss, "Lectures on the calculus of variations" The University of Chicago Press, Chicago, (1946).
2. L.S. Pontryagin, V.G. Boltyanski, R.V. Gamkrelidze and E.F. Mischenko, "The mathematical theory of optimal processes", Interscience Publishers, John Wiley and Sons, Inc. New York (1962).
3. P. Contensou, "Conditions d'Optimalité pour les Domaines de Manoeuvrabilité à Frontière Semi-Affine", Colloquium on Methods of Optimization, Novossibirsk, URSS, (1968).
4. H.M. Robbins, "Optimality of intermediate thrust arcs of rocket trajectories", AIAA. J-3-1094-1098 (1965).
5. D.F. Lawden, "Optimal trajectories for space navigation", Butterworths Mathematical texts, Butterworths, London (1963).

6. H.J. Kelley, R.E. Kopp, H.G. Moyer, "Singular extremals". "Topics in Optimization" Chapter 3 (George Leitmann Editor) Academic Press New York - London (1967).

7. J.V. Breakwell, "The optimization of trajectories" S.I.A.M. Journal, 7, 215-247 (1959).

8. P. Contensou, "Note sur la cinématique générale du mobile dirigé", Association technique maritime et aéronautique, 1946 vol 45 n° 836.

9. P. Contensou, "Application des méthodes de la mécanique du mobile dirigé à la théorie du vol plané", Association technique maritime et aéronautique, 1950, vol 49, n° 958.

10. P. Contensou, "Etude théorique des trajectoires optimales dans un champ de gravitation. Application au cas d'un centre d'attraction unique", Astronautica Acta VIII fasc. 2-3 (1962).

11. C. Marchal, "Theoretical research in deterministic optimization", Publication ONERA n° 139 (1971).

12. J. Warga, "Necessary conditions for minimum in relaxed variational problems", J. Math. Anal. Applied 4 (1962) p 111-128.

13. P.R. Halmos, "Measure Theory", D Van Nostrand Company, Inc., Princeton New Jersey. Toronto, London, New York (August 1966).

THE NUMERICAL SOLUTION OF DISCRETE DYNAMIC
COMBAT TYPE GAMES

Richard E. Kopp

Grumman Aerospace Corporation
Bethpage, New York 11714

Abstract

This paper discusses one approach to the solution of two-player discrete dynamic games. Each player has prescribed dynamics and a capture set. The game is formulated as a zero sum game with a simultaneous move structure, thus allowing for mixed strategies. The computer algorithm utilizes a "backing up process," reducing the solution of the multistage dynamic game to the solution of many two-player matrix games. Results obtained for specific examples are discussed. The optimal strategy (a probability distribution in the case of mixed strategies) and the optimal value (expected value for mixed strategies) are calculated for the complete game state for each discrete time step. A second computer program allows the game to be played with the aid of a computer driven graphic display.

Introduction

As the performance of combat aircraft improves and weapon systems become more exotic, the logic associated with aerial combat becomes less evident. This is not meant to imply that after proper training and experience one cannot eventually arrive at optimum or near optimum maneuvering strategies, but rather to emphasize that the cost of training and experience and the uncertainty associated with the strategies are ever increasing. Of even more significance is the subjectivity involved in comparing one aircraft/weapon system with another aircraft/weapon system. Today, this is usually accomplished by means of elaborate computer simulation of the two aircraft/ weapon systems with strategies for each player previously stipulated. These strategies are for the most part heuristic, not necessarily optimal, and many times biased.

This paper discusses an approach to the numerical solution of aerial combat type problems in which we treat aerial combat as a two-player dynamic game. The numerical results to date are admittedly sparse and the mathematical models somewhat less than what is required for real world application. However, our results thus far are encouraging in that our approach appears to be viable.

The continuous version of the dynamic two-player game falls into the theory of differential games. A host of literature has appeared on this subject beginning with some of the pioneering work of Isaacs and Scarf (Refs. 1-4), and in fact one of the classical works in differential games today is Isaacs' book (Ref. 5). More recent treatises on the subject are the works of Berkovitz (Refs. 6-8), Leitmann (Refs. 9-10), Blaquiere (Ref. 10), Ho (Ref. 11), Pontryagin (Refs. 12-13), and Breakwell (Ref. 14). However, in all these works, a general approach to a numerical solution of the combat type game is conspicuously absent. At best, solutions have been obtained by Baron, Isaacs, and Breakwell (Refs. 15, 5, and 14), and others to specific games using somewhat heuristic methods of analysis suitable for the specific example under consideration.

The underlying cause of the absence of a general method of solution to combat-type differential games is the lack of continuity in a global sense of the optimal game value function with respect to the game state. Where continuity exists, the solution is determined by a min-max version of the Bellman-Hamilton-Jacobi equation. These regions of continuity must in turn be pieced together to satisfy a global min-max condition on the game, and can result in very peculiar phenomena occurring at joining regions.

To circumvent the difficulty associated with continuity, we have in this paper formulated the aerial combat problem in the framework of a discrete two-player dynamic game, discrete in both time and state. This is done with full appreciation of the fact that the problems of convergence of the discrete game to the continuous game pose an open mathematical question (Ref. 16). In fact, we anticipate that some light may be shed on that question through numerical experimentation and comparison with known solution properties of the continuous game. Our approach to the solution of the discrete game reduces to solving many two-player matrix games for which the theory is well developed.

Discrete Dynamic Games

For the sake of clarity, let us consider a two-player discrete dynamic game, played on an xy grid with each player having three degrees of freedom and a single control variable.

The dynamics for the game in relative coordinates is given by

$$x_{i+1} = f(x_i, y_i, \psi_i, u_i, v_i)$$

$$y_{i+1} = g(x_i, y_i, \psi_i, u_i, v_i)$$

$$\psi_{i+1} = h(x_i, y_i, \psi_i, u_i, v_i)$$

where x and y are the relative position and ψ the relative orientation of player 2 to player 1. The control choices for players 1 and 2 are u and v, respectively.

Each player has a capture set, ξ, associated with his vehicle. The capture set is most conveniently defined with respect to the vehicle's coordinates and is specified as a set of points, ξ_1 and ξ_2, for each vehicle.

In the discrete dynamic game, there are at least two possible move structures. The first, and the one used by Isaacs in his analysis of discrete games, is an alternate move structure in which one player moves at a time. This always results in a pure strategy when perfect information concerning the state of each player is available prior to the player's move. In the second move structure, each player moves simultaneously. This can result in mixed strategies even when perfect state information is available prior to the player's move. It is the latter move structure that is considered in this paper because it is believed that this more closely resembles the real world combat situation.

Finally, a discussion of the game value function is in order. Intuitively, the objectives of each player are quite obvious in an aerial combat situation — capture the opponent's vehicle. With this notion in mind, the different outcomes are: Vehicle 1 captures vehicle 2 without being captured himself; a draw in which neither vehicle captures the other; a sacrifice in which both vehicles capture each other; and vehicle 2 captures vehicle 1 without being captured himself. (Capture of vehicle 2 by vehicle 1 means that the state of vehicle 2 enters the set ξ_1.) Very few people would argue against the premise that player 1 would prefer the sequence of outcomes in the order stipulated, with player 2's preference reversed, with the draw and sacrifice outcomes interchanged. However, if this premise is accepted, a nonzero sum game results, for which the solution is somewhat subjective and can often lead to paradoxical situations. However, if we are willing to violate the aforementioned premise, the combat problem can be formulated as a zero-sum game for which an intuitively unique solution definition exists. This I believe can be justified by considering one player to be more "aggressive" than the other. Further we give relative weightings for the different outcomes, which leads to a game of degree rather than a game of kind, which again I believe more closely resembles a real world combat situation.

Computer Algorithms

In summary, the aerial combat problem has been formulated as a discrete, two-player simultaneous move, zero-sum, dynamic game, discrete in both state and time. The computer algorithm developed to solve this game relies on a backing up procedure similar to dynamic programming. The termination of the game is defined by the occurrence of specific outcomes that can be determined by examining the game state after a move has occurred. For example, a logical definition of termination would be when one vehicle captures the other vehicle. As will be seen in one of the examples, this need not be the only criterion for game termination. The particular algorithm developed here also requires the game to terminate after a finite number of moves — the termination in this case resulting in a draw outcome, that is, neither vehicle capturing the other one. As will also be demonstrated with the aid of an example, this does not imply that the solution of an infinite move game cannot be obtained by the algorithm.

For the purpose of exposition let us assume that only the four outcomes mentioned previously can occur at game termination and that a weighting or measure P_i, $i = 1, \ldots, 4$, reflecting the strength of preference, is given for each outcome. Thus at game termination there is associated a value $P = P_i$ for each state which would constitute initial inputs to the computer program. A flow diagram for the computer program is shown in Fig. 1.

Let us consider first a one move game, that is, for all game initial states where neither vehicle is in the other vehicle's capture set and the game terminates after a single move. For all these states a game matrix A is evaluated by calculating the forward dynamics (forward indicating time increasing) for all control choices of each vehicle and examining the outcomes. The resulting game matrix A is first tested for a pure strategy by seeing if it possesses a saddle point. The absence of a saddle point indicates a mixed strategy solution which is solved for by using a Brown-Robinson algorithm (Ref. 17). A convergence proof for this algorithm is included as an appendix for completeness of this paper.

It is clear that the game matrix A need not be calculated for those states where only a draw outcome occurs for all control combinations. This can be achieved in the computer program by slightly modifying the above mentioned procedure. At game termination, for all states where either vehicle is in the other vehicle's capture set, calculate, using inverse dynamics (inverse indicating time decreasing) and all control combinations, that set of states from which an outcome other than a draw could occur. Then calculate the game matrix A for the subset of those states where neither vehicle is in the other vehicle's capture set.

The generalization to an N move game is straightforward. The values of the one move game are used to calculate the game matrices for the two move game, and so forth. It is seen that for each state at each move the value of the game and the strategy must be stored. If in the backing up procedure the game value and strategy for the N − 1 move game is the same as the game value and strategy for the N move game then we have the solution to the infinite (countable) move game.

Examples

As a first example we consider a case where both vehicles move on a rectangular grid, as shown in Fig. 2. Both vehicles have the same dynamics and capture set. Each vehicle has an xy coordinate position and four possible orientations which are designated by -1, 0, +1, +2. The dynamics are given by the rule that one vehicle advances one grid point with a move in the direction in which it is presently oriented, with three possible changes in orientations under the control of the player: a 90° counter clockwise rotation, 0° rotation, or 90° clockwise rotation designated by control choices -1, 0, +1. The capture set for each vehicle is the grid point the vehicle is occupying and the next two grid points in the direction of the vehicle's orientation. In Fig. 2 the apex of the triangle designates the vehicle's position and orientation and the solid line the vehicle's capture set. Thus the state of the game is given by three numbers that indicate the relative xy position and orientation of vehicle 2 with respect to vehicle 1.

The relative weightings given to the four possible outcomes are:

1.   Vehicle 1 captures vehicle 2, $P = +1$

2.   Vehicle 2 captures vehicle 1, $P = -1$

3.   Sacrifice — mutual capture, $P = 0.5$

4.   Draw — no capture, $P = 0$

After 3 moves the game value and strategy repeat themselves, thus giving the game value and strategy for an infinite move game. In Figs. 3-5, the value of the game is given for different initial conditions. The letter K at a grid point indicates that vehicle 1 will capture vehicle 2 in 3 moves or less, giving a value P to the game of +1. Similarly, C indicates a value of $P = -1$, and S a value $P = 0.5$. The letter M indicates a mixed outcome, the expected value of the game being 0.43 in Fig. 3 and 0.25 in Figs. 4 and 5. The absence of a letter indicates a draw condition with a game value of 0. The figure for an orientation +1 is not given since it is a mirror image of Fig. 4.

As a second example, another termination condition was added to make the game somewhat more interesting. If vehicle 2 moves outside a specified engagement region around vehicle 1 the game terminates with a value of 0.5. A scenario could be constructed around this outcome to account for loss of visual or radar contact, thus terminating the aerial combat duel. If vehicle 1 were on an escort type mission he could then successfully complete his mission. Other scenarios of course could be rationalized. In any event this addition makes for a more interesting combat problem. The draw outcome in this case was given a value of 0.25. The engagement region for this problem is given by $-6 < x < 6$ and $-5 < y < 6$.

After 11 moves the game value and strategy reached the solution for an infinite move game as is shown in Figs. 6-8. In this case the numerical values at the initial conditions indicate the expected value of the game multiplied by 100. The lack of exact symmetry where expected is due to the finite number of "fictitious" plays (30) used in the Brown-Robinson algorithm.

In the spirit of the scenario mentioned, the results of example 2 could be interpreted as follows: If vehicle 2 has the option of choosing his initial position at the beginning of combat (first visual or radar contact), he should select his position and orientation to maximize his advantage. Since he is able to advance one grid unit at a move, the periphery of

the engagement region for orientations consistent with initial
encounters should be examined for minimum expected values of
the game. For example, initial states ±4, 5, 2; ±2, 5, 2
and 0, 5, 2 as shown in Fig. 6 are relatively good initial
conditions for vehicle 2.

Simulation

A second computer program was written so that the discrete two-player game could be simulated on an interactive
graphic computing system. Options were provided so that an
analyst could control the move of either one of the vehicles
with the computer optimally selecting the control of the other
vehicle. Where mixed strategies are involved, the optimum
play is chosen with the aid of a random number generator using
the appropriate probability distribution given by the optimal
strategy. Although extensive experiments were not conducted,
in all cases the analyst was unable to move optimally, even
after a learning period, thus indicating that example 2 is not
a transparent, trivial example.

An option in the computer program also allows the computer to play the role of both vehicles optimally. In Figs. 9
and 10 are two photographs of the computer output display for
optimal play by the computer of both vehicles. The initial
state of the game is 1, 4, 0 in both cases and is designated
in the figures by a zero next to the triangular symbol for vehicle 2. A relative coordinate system is used with vehicle 1
at the origin oriented in the upward direction. Below the
grid is printed the optimal strategy, expected value of the
game, and the control choices for each move of the game. For
example, the first line of printout in Fig. 9 is the initial
state 1, 4, 0 of vehicle 2. The first set of three numbers
of the second line is the optimal strategy for vehicle 1 given
as a probability distribution on the three control choices;
the second set of three numbers is the strategy for vehicle 2;
the next number is the expected value of the game for the current game state; and the last two numbers are the actual control choices played by the computer. Each line thereafter is
the same information for the then current state of the game.
After 10 moves a draw situation is seen to exist and all
moves thereafter would be pure strategies. In Fig. 10 is another optimal play of the game that ends in mutual capture
after 5 moves.

Conclusions

The amount of numerical evidence to date is admittedly
small and the models somewhat academic. However, the computa-

tional approach discussed appears viable, and the results are
encouraging. The discretizing in both space and time avoids
the annoying question of continuity. Mixed strategies that
can occur in the simultaneous move structure offer no obstacle. The "backing up" approach allows the multistage game to
be solved by finding the solution to a number of matrix games
for which there is a well developed theory.

Considerably more experimentation with parametric studies
is required on some of these simple models prior to examining
a model that is closer to a real world situation. In this way
we can examine numerically the convergence of the discrete
model to the continuous case as the grid size is made smaller,
which is an open mathematical question. More efficient computer coding is possible and estimates of the computing time
for more sophisticated models can be made.

Appendix

Let the game matrix A be an $n \times m$ matrix and the
mixed strategies for players 1 and 2 be designated by the
vectors u and v, respectively, u being an n vector
and v an m vector. The components of the vectors u and
v are the distributions of pure strategies for players 1 and
2. Thus each of the components of the vectors u and v is
greater than or equal to zero and less than or equal to one,
and the sum of the components of each vector equals one.
Designate the corresponding sets of u and v vectors by U
and V, respectively. The value of the game is P and is
given by

$$P = \underset{v \in V}{\text{Max}}[\underset{u \in V}{\text{Min}}\ u^T A V] = \underset{u \in U}{\text{Min}}[\underset{v \in V}{\text{Max}}\ u^T A V] \tag{1}$$

Consider the following algorithm for determining u and
v. Arbitrarily select a vector v belonging to V and designate it by v_0. Determine the sequence of vectors u_i and
v_i by the following operations:

$$\underset{u\epsilon U}{\underline{\text{Min}}} \; u^T A v_0 = u_1^T A v_0 = P_1 \leq P$$

$$\underset{v\epsilon V}{\underline{\text{Max}}} \; u_1^T A v = u_1^T A v_2 = P_2 \geq P$$

$$\underset{u\epsilon U}{\underline{\text{Min}}} \; \tfrac{1}{2} u^T A(v_0 + v_2) = \tfrac{1}{2} u_3^T A(v_0 + v_2) = P_3 \leq P$$

$$\vdots$$

$$\underset{u\epsilon U}{\underline{\text{Min}}} \; \frac{1}{n} u^T A \sum_{i=0}^{n-1} v_{2i} = \frac{1}{n} u_{2n-1}^T A \sum_{i=0}^{n-1} v_{2i} = P_{2n-1} \leq P \qquad (2)$$

$$\underset{v\epsilon V}{\underline{\text{Max}}} \; \frac{1}{n} \sum_{i=0}^{n-1} u_{2i+1}^T A v = \frac{1}{n} \sum_{i=0}^{n-1} u_{2i+1}^T A v_{2n} = P_{2n} \geq P \qquad (3)$$

Therefore

$$P_{2n} \geq P \geq P_{2n-1} \quad . \qquad (4)$$

Let

$$\bar{v}_{2n} = \frac{1}{n+1} \sum_{i=0}^{n} v_{2i} \qquad (5)$$

$$\bar{u}_{2n} = \frac{1}{n+1} \sum_{i=0}^{n} u_{2i+1} \quad . \qquad (6)$$

From (2), (3), (5), and (6), we obtain

$$P_{2n} \geq \bar{u}_{2n-1}^T A \bar{v}_{2n-1} \geq P_{2n-1} \quad . \qquad (7)$$

We now define an error bound sequence, E_{2n}, for which there exists an n sufficiently large so that the sequence becomes monotone decreasing with a limit value equal to zero.

$$E_{2n} = P_{2n} - P_{2n-1} = \frac{1}{n}\left\{\sum_{i=0}^{n-1} u_{2i+1}^T Av_{2n} - u_{2n-1}^T A \sum_{i=0}^{n-1} v_{2i}\right\} \tag{8}$$

which gives with the aid of inequalities (2) and (3) the recursive relationship

$$E_{2n+2} = \frac{n}{n+1} E_{2n} + \frac{1}{n+1}\left\{u_{2n+1}^T A(v_{2n+1} - v_{2n}) - C(n)\right\} \tag{9}$$

where C(n) is a bounded function of n. In the limit as n approaches infinity

$$\lim_{n \to \infty} E_{2n+2} = \lim_{n \to \infty} \frac{1}{n+1} E_2 = 0 \ . \tag{10}$$

Therefore from (4), (7), and (10)

$$\lim_{n \to \infty} \bar{u}_{2n-1}^T Av_{2n-1} = P \ . \tag{11}$$

Acknowledgments

The author acknowledges with appreciation the many helpful discussions he has had with M. Falco and J. Mendelsohn as well as with other colleagues within the Research Department of the Grumman Aerospace Corporation. He also extends thanks to M. Falco and J. Mendelsohn for their careful scrutiny of this manuscript.

References

1. R. Isaacs, <u>Rand Corp.</u> Report P-257 (1957).

2. R. Isaacs, <u>Naval Res. Logistics Quart.</u> 2, 1 and 2 (1955).

3. R. Isaacs, <u>Rand Corp. Reports</u> RM-1391, RM-1399, RM-1411, RM-1486 (1954).

4. H. E. Scarf, <u>Annals Math. Study 59</u>, Princeton Univ. Press, pp. 393-405 (1957).

5. R. Isaacs, _Differential Games_, John Wiley (1965).

6. L. D. Berkovitz and W. H. Fleming, _Contribution to the Theory of Differential Games_, III, Princeton Univ. Press, pp. 413-435 (1957).

7. L. D. Berkovitz, _Advances in Game Theory_, Annals Math. Study, 52, Princeton Univ. Press, pp. 175-194 (1964).

8. L. D. Berkovitz, _Journal SIAM on Control_ 5, 1, pp. 1-24 (1967).

9. G. Leitmann, _Journal of Optimization and Applications_ 2, 4, pp. 220-225 (1968).

10. A. Blaquiere, F. Gerard, and G. Leitmann, _Quantitative and Qualitative Games_, Academic Press (1963).

11. Y. C. Ho, _SIAM Journal Control_ 4, 3, pp. 421-428 (1966).

12. L. S. Pontryagin, _SIAM Journal Control_ 3, 1, pp. 49-52 (1966).

13. L. S. Pontryagin, _Math. Theory of Control_, Academic Press, pp. 330-334 (1967).

14. J. V. Breakwell and A. W. Merz, First International Conference on the Theory and Applications of Differential Games, pp. III-1—III-5 (1969).

15. S. Baron et al., _NASA Report_ CR-1626 (1970).

16. W. H. Fleming, _Annals Math. Study 52_, Princeton Univ. Press, pp. 195-210 (1964).

17. J. Robinson, _Annals Math. Study 54_, Princeton Univ. Press, pp. 296-301 (1951).

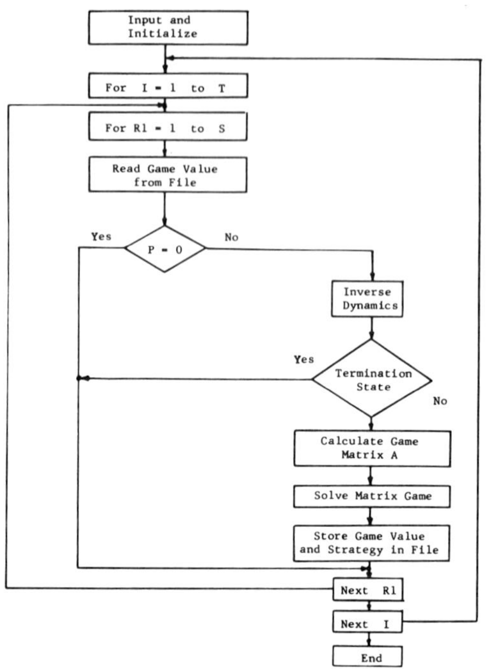

Fig. 1   Flow Chart for Computer Solution of Discrete Dynamic Games — I designates number of game moves; R1 enumerates states for each game move; P is value of game

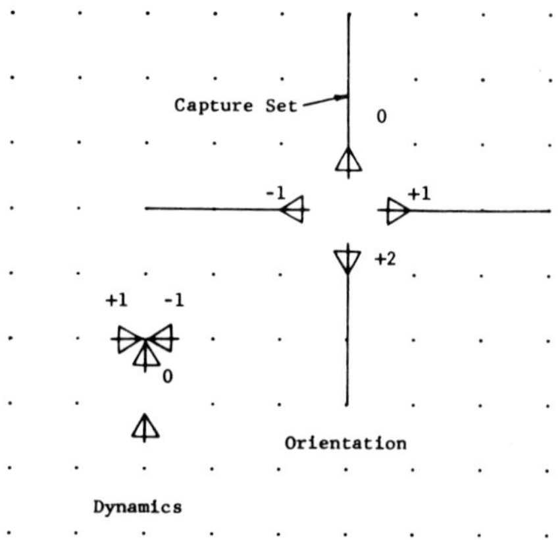

Fig. 2   Example 1, Dynamics, Orientation, and Capture Set of Each Vehicle

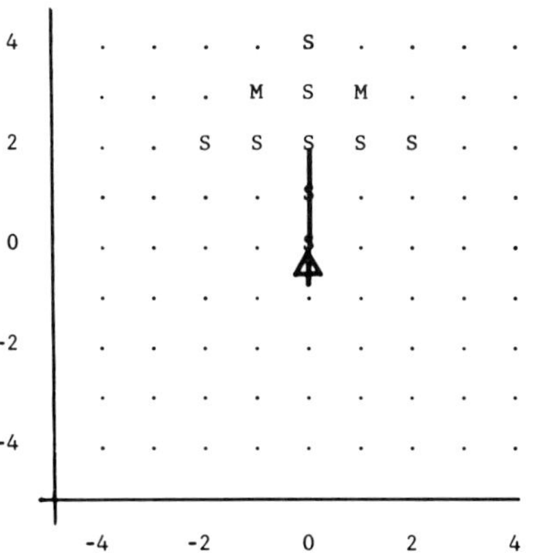

Fig. 3   Example 1: Optimum Outcome versus Game State;
Vehicle 2 Oriented ↓

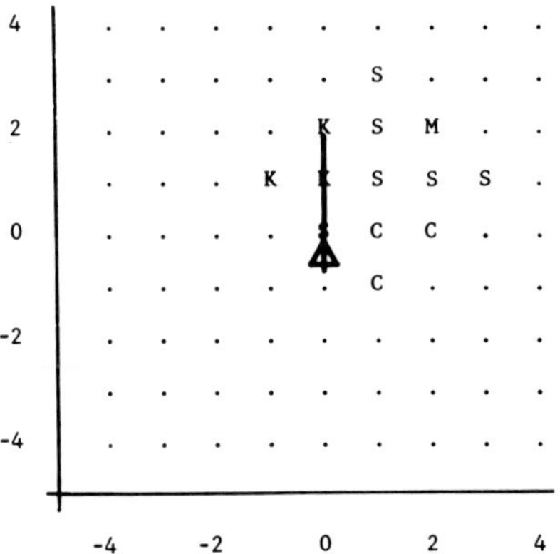

Fig. 4   Example 1: Optimum Outcome versus Game State;
Vehicle 2 Oriented ←

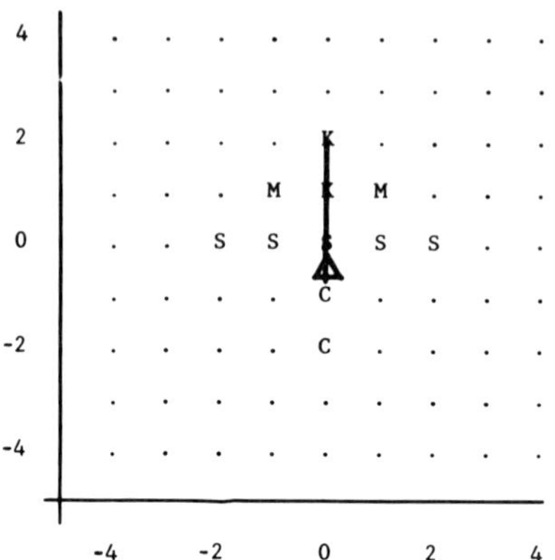

Fig. 5   Example 1: Optimum Outcome versus Game State; Vehicle 2 Oriented ↑

	-6	-5	-4	-3	-2	-1	0	1	2	3	4	5	6
6	E	E	E	E	E	E	E	E	E	E	E	E	E
5	E	E	E	E	E	E.	E	E	E	E	E	E	E
4	E	E	48	7	21	9	19	9	21	6	49	E	E
3	E.	E	5	17	7	15	6	15	8	18	6	E	E
2	E	E	16	.	.	.	K	.	.	.	17	E	E
1	E	E	7	.	.	24	K	22	.	.	8	E	E
0	E	F.	22	.	S	S	S	S	S	.	22	E	E
-1	E	E	8	15	18	.	C	.	18	14	7	E	E
-2	E	E	17	.	15	7	C	8	14	.	16	E	E
-3	E	E	5	.	.	14	6	15	.	.	5	E	E
-4	E	E	48	.	.	.	15	.	.	.	49	E	E
-5	E	E	E	E	E	E	E	E	E	E	E	E	E
-6	E	E	E	E	E	E	E	K	E	R	E	F.	R

Fig. 6   Example 2: Optimum Outcome versus Game State; Vehicle 2 Oriented ↓

	-6	-5	-4	-3	-2	-1	0	1	2	3	4	5	6
6	E	E	E	E	E	E	E	E	E	E	E	E	E
5	E	E	E	E	6	17	8	19	7	16	5	49	E
4	E	E	E	E	18	12	16	10	18	6	9	6	E
3	E	E	E	E	.	.	.	S	.	.	.	17	E
2	E	E	E	E	.	.		S	10	.	.	8	E
1	E	E	E	E	.	K		S	S	S	.	22	E
0	E	E	E	E	15	6		C	C	7	14	7	E
-1	E	E	E	E	.	17	5	C	13	15	.	16	E
-2	E	E	E	E	.	.	15	7	14	.	.	5	E
-3	E	E	E	E	.	.	.	15	.	.	.	49	E
-4	E	E	E	E	E	E	E	E	E	E	E	E	E
-5	E	E	E	E	E	E	E	E	E	E	E	E	E
-6	E	E	E	E	E	E	E	E	E	E	E	E	E

Fig. 7   Example 2: Optimum Outcome versus Game State; Vehicle 2 Oriented ←

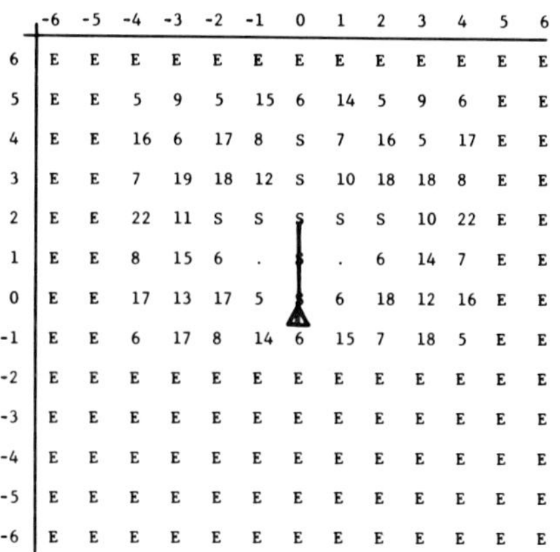

	-6	-5	-4	-3	-2	-1	0	1	2	3	4	5	6
6	E	E	E	E	E	E	E	E	E	E	E	E	E
5	E	E	5	9	5	15	6	14	5	9	6	E	E
4	E	E	16	6	17	8	S	7	16	5	17	E	E
3	E	E	7	19	18	12	S	10	18	18	8	E	E
2	E	E	22	11	S	S	S	S	S	10	22	E	E
1	E	E	8	15	6	.		.	6	14	7	E	E
0	E	E	17	13	17	5		6	18	12	16	E	E
-1	E	E	6	17	8	14	6	15	7	18	5	E	E
-2	E	E	E	E	E	E	E	E	E	E	E	E	E
-3	E	E	E	E	E	E	E	E	E	E	E	E	E
-4	E	E	E	E	E	E	E	E	E	E	E	E	E
-5	E	E	E	E	E	E	E	E	E	E	E	E	E
-6	E	E	E	E	E	E	E	E	E	E	E	E	E

Fig. 8   Example 2: Optimum Outcome versus Game State; Vehicle 2 Oriented ↑

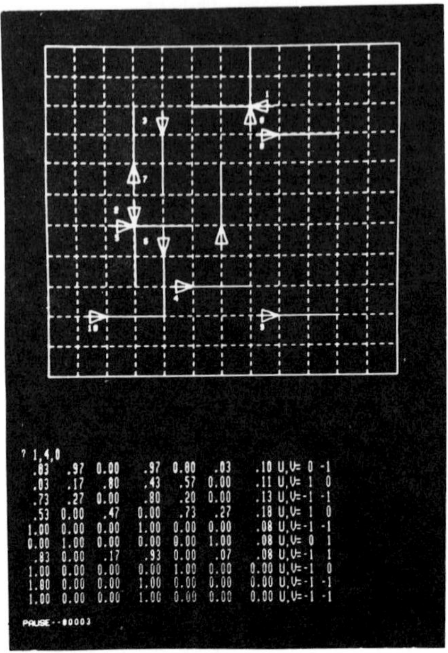

Fig. 9   Example 2: Optimal Strategy Game Simulation
Terminating in a Draw Outcome

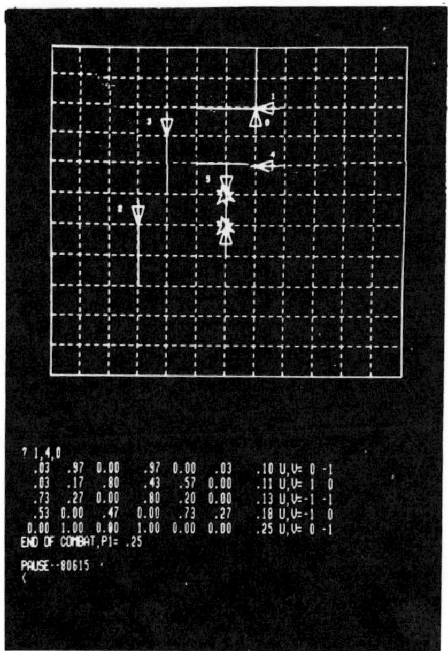

Fig. 10   Example 2: Optimal Strategy Game Simulation
Terminating in a Sacrifice Outcome

A MODIFICATION OF THE METHOD OF MULTIPLIERS FOR MATHEMATICAL PROGRAMMING PROBLEMS[1]

A. Miele[2], P.E. Moseley[3], and E.E. Cragg[3]

Rice University
Houston, Texas

Abstract

This paper considers the problem of minimizing a function $f(x)$ subject to a constraint $\varphi(x) = 0$, where f is a scalar, x an n-vector, and φ a q-vector, with $q < n$. A modification of Hestenes' method of multipliers is presented, with one idea in mind: to accelerate convergence. The modification consists of (i) imbedding Hestenes' updating rule into a one-parameter family of updating rules and determining the scalar parameter β so that the total error in the system is minimized and (ii) increasing the updating frequency so that the number of iterations in a cycle is $\Delta N = 1$ for the ordinary-gradient algorithm and $\Delta N = n$ for the conjugate-gradient algorithm. Numerical experiments show that the modified method of multipliers exhibits faster convergence than Hestenes' method of multipliers in connection with both the ordinary-gradient algorithm and the conjugate-gradient algorithm.

---

[1] This research was supported by the National Science Foundation, Grant No. GP-27271. The authors are indebted to Mr. A.V. Levy for computational assistance.

[2] Professor of Astronautics and Director of the Aero-Astronautics Group, Department of Mechanical and Aerospace Engineering and Materials Science.

[3] Graduate Student in Aero-Astronautics, Department of Mechanical and Aerospace Engineering and Materials Science.

1. Introduction

In a previous paper (Ref. 1), the problem of minimizing a function f(x) subject to a constraint $\varphi(x) = 0$ was considered, where f is a scalar, x an n-vector, and φ a q-vector, with q < n. In Ref. 1, some numerical experiments on Hestenes' method of multipliers (Ref. 2) were presented in conjunction with both the ordinary-gradient algorithm and the conjugate-gradient algorithm.

Hestenes' method is superior to the standard penalty function method in both stability and rapidity of convergence. However, for the examples considered in Ref. 1, Hestenes' method exhibits much slower convergence than the methods developed in Refs. 3-5. This being the case, we present in this paper a technique which accelerates convergence of the method of multipliers. We shall call this technique the modified method of multipliers. For the sake of identification, Hestenes' method of multipliers is called Method (I) and the modified method of multipliers is called Method (II).

2. Statement of the Problem

We consider the problem of minimizing the function

$$f = f(x) , \qquad (1)$$

subject to the constraint

$$\varphi(x) = 0 . \qquad (2)$$

In the above equations, f is a scalar, x an n-vector, and φ a q-vector[4], where q < n. It is assumed that the first and second partial derivatives of the functions f and φ with respect to x exist and are continuous; it is also assumed that the constrained minimum exists.

2.1. Exact First-Order Conditions. From theory of maxima and minima, it is known that the previous problem can be recast

---

[4] All vectors are column vectors.

as that of minimizing the <u>augmented function</u>

$$F(x, \lambda) = f(x) + \lambda^T \varphi(x), \tag{3}$$

subject to the constraint (2). Here, λ is a q-vector Lagrange multiplier and the superscript T denotes the transpose of a matrix. If

$$F_x(x, \lambda) = f_x(x) + \varphi_x(x)\lambda \tag{4}$$

denotes the gradient of the augmented function[5], the optimum solution x, λ must satisfy the simultaneous equations

$$\varphi(x) = 0 \, , \quad F_x(x, \lambda) = 0. \tag{5}$$

2.2. <u>Approximate Solutions.</u> In general, the system (5) is nonlinear; consequently, approximate methods must be employed. These are of two kinds: first-order methods (such as those discussed in subsequent sections of this report) and second-order methods. Here, we introduce the scalar quantities

$$P(x) = \varphi^T(x)\varphi(x) \quad , \quad Q(x, \lambda) = F_x^T(x, \lambda)F_x(x, \lambda) \, , \tag{6}$$

which measure the errors in the constraint and the optimum condition, respectively. We observe that $P = 0$ and $Q = 0$ for the optimum solution, while $P > 0$ and/or $Q > 0$ for any approximation to the solution. When approximate methods are used, they must ultimately lead to values of x, λ such that

$$P(x) \leq \epsilon_1 \quad , \quad Q(x, \lambda) \leq \epsilon_2 \, . \tag{7}$$

Alternatively, (7) can be replaced by

$$R(x, \lambda) \leq \epsilon_3 \, , \tag{8}$$

where

$$R(x, \lambda) = P(x) + Q(x, \lambda) \tag{9}$$

denotes the cumulative error in the constraint and the optimum

---

[5] In Eq.(4), the gradients f_x, F_x denote n-vectors and the matrix φ_x is n x q.

condition. Here, ϵ_1, ϵ_2, ϵ_3 are small, preselected numbers. Note that satisfaction of Ineq. (8) implies satisfaction of Ineqs. (7), if one chooses $\epsilon_1 = \epsilon_2 = \epsilon_3$.

3. Summary of the Method of Multipliers

The basic ideas of Hestenes' method of multipliers or Method (I) can be summarized as follows:

(a) The original, constrained minimization problem is replaced by a sequence of unconstrained minimization problems.

(b) In each element of the sequence or cycle, the vector x is viewed as unconstrained, and the augmented penalty function

$$W(x, \lambda, k) = F(x, \lambda) + kP(x), \tag{10}$$

where

$$F(x, \lambda) = f(x) + \lambda^T \varphi(x), \quad P(x) = \varphi^T(x)\varphi(x), \tag{11}$$

is minimized with respect to x for given values of λ and k. Here, the q-vector λ is the Lagrange multiplier and the scalar $k > 0$ is the penalty constant.

(c) A cycle is terminated when the following inequality is satisfied:

$$W_x^T(x, \lambda, k)W_x(x, \lambda, k) \le \epsilon_4, \tag{12}$$

where ϵ_4 denotes a small, preselected number and where the gradient of the augmented penalty function is given by

$$W_x(x, \lambda, k) = F_x(x, \lambda) + kP_x(x), \tag{13}$$

where

$$F_x(x, \lambda) = f_x(x) + \varphi_x(x)\lambda, \quad P_x(x) = 2\varphi_x(x)\varphi(x). \tag{14}$$

(d) At this point, the multiplier and the penalty constant are updated according to the simple rule

$$\lambda_* = \lambda + 2k\varphi(x), \quad k_* = k. \tag{15}$$

(e) Once λ and k are updated according to (15), the next cycle is started; that is, one returns to (b) and proceeds iteratively.

(f) The algorithm is terminated when the following inequality is satisfied:

$$R(x, \lambda) \le \epsilon_3 , \qquad (16)$$

where ϵ_3 is a small, preselected number.

(g) To start the algorithm, some assumption concerning the multiplier is necessary. The simplest assumption is

$$\lambda = 0 \qquad (17)$$

and is equivalent to stating that the augmented penalty function (10) and the standard penalty function

$$U(x, k) = f(x) + kP(x) \qquad (18)$$

are identical for the first cycle of the algorithm.

Remark 3.1. Whenever λ is replaced by λ_*, the updated error in the optimum condition (6-2) satisfies the inequality

$$Q(x, \lambda_*) \le \epsilon_4 , \qquad (19)$$

where ϵ_4 is the parameter characterizing the stopping condition (12) of a cycle. This inequality can be derived from (12) and (15-1) if one observes that

$$W_x(x, \lambda, k) = F_x(x, \lambda_*) \qquad (20)$$

and that

$$Q(x, \lambda_*) = F_x^T(x, \lambda_*)F_x(x, \lambda_*) . \qquad (21)$$

In turn, Ineq. (19) implies that

$$Q(x, \lambda_*) \le \epsilon_3 , \qquad (22)$$

if

$$\epsilon_4 = \epsilon_3 , \qquad (23)$$

that is, if the parameter characterizing the stopping condition (12) of a cycle is identical with the parameter characterizing the stopping condition (16) of the algorithm.

4. Modification of the Method of Multipliers

The method of multipliers described in Section 3 has one drawback, common to all penalty-function methods: each auxiliary optimization problem or cycle may require a large number of iterations ΔN. Therefore, the total number of iterations for convergence $N_* = \Sigma(\Delta N)$ may be large.

To reduce this difficulty, we assign a priori the number of iterations in a cycle ΔN, specifically,

$$\Delta N = 1 \tag{24}$$

for the ordinary-gradient algorithm and

$$\Delta N = n \tag{25}$$

for the conjugate-gradient algorithm. Therefore, the stopping condition (12) for a cycle is bypassed and is replaced by (24) or (25) in the hope that, by means of a shorter cycle, the method of multipliers can be improved.

Whenever (24) or (25) is employed in connection with (15), the updated total error $R(x, \lambda_*)$ may be larger or smaller than the total error $R(x, \lambda)$ prior to updating. Since this uncertainty is undesirable, we renounce Hestenes' updating rule (15) and replace it with the more general updating rule

$$\lambda_* = \lambda + 2\beta\varphi(x), \quad k_* = k , \tag{26}$$

where β is a parameter. This scalar parameter must be determined so as to produce some optimum effect.

For given values of x and λ, a change in β causes a change in the updated multiplier λ_*. Consequently, the updated error in the optimum condition

$$Q(x, \lambda_*) = F_x^T(x, \lambda_*) F_x(x, \lambda_*) \tag{27}$$

changes, and the updated total error

$$R(x, \lambda_*) = P(x) + Q(x, \lambda_*) \tag{28}$$

changes. The optimum value of β is that which gives $Q(x, \lambda_*)$, and hence $R(x, \lambda_*)$, the smallest value for given x and λ.

After combining (26-1) and (27), we obtain the relation

$$Q(x, \lambda, \beta) = [F_x(x, \lambda) + \beta P_x(x)]^T [F_x(x, \lambda) + \beta P_x(x)], \tag{29}$$

which is quadratic in β and admits the derivatives

$$Q_\beta(x, \lambda, \beta) = 2P_x^T(x)[F_x(x, \lambda) + \beta P_x(x)] , \tag{30}$$

$$Q_{\beta\beta}(x, \lambda, \beta) = 2P_x^T(x)P_x(x) ,$$

the first of which vanishes for

$$\beta = -F_x^T(x, \lambda)P_x(x)/P_x^T(x)P_x(x) . \tag{31}$$

This value of β minimizes $Q(x, \lambda, \beta)$, since $Q_{\beta\beta}(x, \lambda, \beta) > 0$ if $P_x(x) \neq 0$.

The method of multipliers with cycle stopping condition (12) replaced by (24) or (25) and with updating rule (15) replaced by (26) and (31) is called the <u>modified method of multipliers</u> or <u>Method (II)</u>.

Remark 4.1. In the light of Eq. (13), the optimum value of the parameter β, given by Eq. (31), can be rewritten as

$$\beta = k - W_x^T(x, \lambda, k)P_x(x)/P_x^T(x)P_x(x) . \tag{32}$$

This relation shows that

$$\beta = k \tag{33}$$

if

$$W_x(x, \lambda, k) = 0 , \tag{34}$$

or more in general if

$$W_x^T(x, \lambda, k)P_x(x) = 0 \ , \quad P_x(x) \neq 0 \ . \tag{35}$$

5. First-Order Algorithms

In this section, we describe some first-order algorithms to be employed with each cycle of both the method of multipliers and the modified method of multipliers. They are: the ordinary-gradient algorithm and the conjugate-gradient algorithm. Both of these algorithms make use of the augmented penalty function

$$W(x, \lambda, k) = F(x, \lambda) + kP(x) , \tag{36}$$

where

$$F(x, \lambda) = f(x) + \lambda^T \varphi(x) \ , \quad P(x) = \varphi^T(x)\varphi(x) , \tag{37}$$

and are employed with this understanding: in each cycle, the Lagrange multiplier λ and the penalty constant k are held unchanged, and the vector x is viewed as unconstrained.

5.1. <u>Ordinary-Gradient Algorithm</u>. Let x denote the nominal point, $\tilde{x}$ the varied point, Δx the displacement leading from the nominal point to the varied point, and α the stepsize. With this understanding, the ordinary-gradient algorithm is represented by

$$F_x(x, \lambda) = f_x(x) + \varphi_x(x)\lambda \ , \quad P_x(x) = 2\varphi_x(x)\varphi(x) \ ,$$

$$W_x(x, \lambda, k) = F_x(x, \lambda) + kP_x(x) , \tag{38}$$

$$p = W_x(x, \lambda, k) \ , \quad \Delta x = -\alpha p , \quad \tilde{x} = x + \Delta x \ .$$

For given nominal point x, Lagrange multiplier λ, and penalty constant k, Eqs. (38) constitute a complete iteration leading to the varied point $\tilde{x}$ providing one specifies the stepsize α.

5.2. <u>Conjugate-Gradient Algorithm</u>. Let x denote the nominal point, $\hat{x}$ the previous point, $\tilde{x}$ the varied point, Δx the displacement leading from the nominal point to the varied point, p the present search direction, $\hat{p}$ the previous search direction, γ the directional coefficient, and α the stepsize. Both p and $\hat{p}$ are n-vectors, while γ and α are scalars. With this understanding, the conjugate-gradient algorithm is represented by

$$F_x(x, \lambda) = f_x(x) + \varphi_x(x)\lambda \; , \; P_x(x) = 2\varphi_x(x)\varphi(x) \; ,$$

$$W_x(x, \lambda, k) = F_x(x, \lambda) + kP_x(x) \; ,$$

$$\gamma = W_x^T(x, \lambda, k)W_x(x, \lambda, k)/W_x^T(\hat{x}, \lambda, k)W_x(\hat{x}, \lambda, k),$$

$$\tag{39}$$

$$p = W_x(x, \lambda, k) + \gamma\hat{p} \; , \; \Delta x = -\alpha p \; , \; \tilde{x} = x + \Delta x \; .$$

For given nominal point x, Lagrange multiplier λ, directional coefficient γ, and penalty constant k, Eqs. (39) constitute a complete iteration leading to the varied point $\tilde{x}$, providing one specifies the stepsize α. For the first iteration of a cycle and every nth iteration thereafter, Eq. (39-4) is bypassed and is replaced by $\gamma = 0$.

    5.3. <u>Stepsize Determination.</u> For both the ordinary-gradient algorithm and the conjugate-gradient algorithm, the position vector at the end of the step can be written as

$$\tilde{x} = x - \alpha p \; , \tag{40}$$

where p denotes the search direction. This is a one-parameter family of varied points $\tilde{x}$, for which the augmented penalty function (36) takes the form

$$W(\tilde{x}, \lambda, k) = W(x - \alpha p, \lambda, k) = W(\alpha) \; . \tag{41}$$

A precise search to be employed in the conjugate-gradient algorithm and an approximate search to be employed in the ordinary-gradient algorithm are described below.

    <u>Precise Search.</u> We now assume that a minimum of $W(\alpha)$ exists. Then, we employ some one-dimensional search scheme (for instance, quadratic interpolation, cubic interpolation, or quasilinearization) to determine the value of α for which

$$W_\alpha(\alpha) = 0 \ . \tag{42}$$

Ideally, this procedure should be used iteratively until the modulus of the slope satisfies any of the following inequalities:

$$|W_\alpha(\alpha)| \leq \epsilon_5 \quad \text{or} \quad |W_\alpha(\alpha)| \leq \epsilon_6 \ |W_\alpha(0)| \ , \tag{43}$$

where ϵ_5 and ϵ_6 are small, preselected numbers. Of course, the value of α satisfying Ineq. (43) must be such that

$$W(\alpha) < W(0) \ . \tag{44}$$

Approximate Search. Since the rigorous determination of α might require excessive computing time, one might renounce solving Eq. (42) with a particular degree of precision and determine the stepsize in a noniterative fashion, for instance, by employing the first optimum value of α supplied by the search procedure. Of course, this value of α is acceptable only if Ineq. (44) is satisfied. Otherwise, α must be replaced by a smaller value (for example, with a bisection process) until Ineq. (44) is met.

6. Experimental Conditions and Numerical Example

In order to illustrate the theory, a numerical example was developed using a Burroughs B-5500 computer and double-precision arithmetic[6]. The algorithms were programmed in Extended ALGOL. Concerning the stepsize α, the one-dimensional search on the function $W(\alpha)$ was done in accordance with Section 5.3.

For the ordinary-gradient algorithm, an approximate search was employed. This consisted of one-step, corrected quasi-linearization, followed by a bisection process until the inequality

$$W(\alpha) < W(0) \tag{45}$$

was satisfied.

---

[6] For additional numerical examples, see Ref. 6.

For the conjugate-gradient algorithm, a precise search was employed. This consisted of multi-step, corrected quasilinearization such that, in any given step, the inequality

$$W(\alpha) < W(\alpha_o) \tag{46}$$

was satisfied, where α_o is the nominal stepsize and α is the varied stepsize. The search was terminated when the following stopping condition was satisfied:

$$W_\alpha^2(\alpha) \leq W_\alpha^2(0) \times 10^{-6} \ . \tag{47}$$

Termination of a cycle of Hestenes' method of multipliers was defined by

$$W_x^T(x, \lambda, k) W_x(x, \lambda, k) \leq 10^{-12} \ . \tag{48}$$

Conversely, termination of a cycle of the modified method of multipliers was defined by

$$\Delta N = 1 \quad \text{or} \quad \Delta N = n \ , \tag{49}$$

where ΔN denotes the number of iterations composing the cycle. Equation (49-1) applies to the ordinary-gradient algorithm, and Eq. (49-2) applies to the conjugate-gradient algorithm.

Convergence of the complete algorithm was defined as follows:

$$R(x, \lambda) \leq 10^{-12} \ . \tag{50}$$

Conversely, nonconvergence was defined by means of the inequalities

$$\text{(a)} \ \begin{cases} N \geq 1000 \text{ for the ordinary-gradient algorithm,} \\ N \geq \ \ 200 \text{ for the conjugate-gradient algorithm,} \end{cases} \tag{51}$$

or

$$\text{(b)} \qquad\qquad N_s \geq 20 \ , \tag{52}$$

or

$$\text{(c)} \qquad\qquad M \geq 0.4 \times 10^{69} \ . \tag{53}$$

Here, N is the iteration number, N_s is the number of bisections of the stepsize α required to satisfy Ineq. (45) or Ineq. (46), and M is the modulus of any of the quantities employed in the algorithm.

Example 6.1. Consider the problem of minimizing the function

$$f = (x_1 - 1)^2 + (x_1 - x_2)^2 + (x_2 - x_3)^2 + (x_3 - x_4)^4 + (x_4 - x_5)^4,$$

$$(54)$$

subject to the constraints

$$x_1 + x_2^2 + x_3^3 - 2 - 3\sqrt{2} = 0 \ , \ x_2 - x_3^2 + x_4 + 2 - 2\sqrt{2} = 0 \ ,$$

$$x_1 x_5 - 2 = 0 \ . \tag{55}$$

This function admits the relative minimum $f = 0.7877 \times 10^{-1}$ at the point defined by

$$x_1 = 1.1911 \ , \quad x_2 = 1.3626 \ , \quad x_3 = 1.4728 \ ,$$

$$(56)$$

$$x_4 = 1.6350 \ , \quad x_5 = 1.6790 \ ,$$

and

$$\lambda_1 = -0.3882 \times 10^{-1} \ , \quad \lambda_2 = -0.1672 \times 10^{-1} \ ,$$

$$\lambda_3 = -0.2879 \times 10^{-3} . \tag{57}$$

The nominal point chosen to start the algorithm is the point of coordinates

$$x_1 = x_2 = x_3 = x_4 = x_5 = 2 \ , \tag{58}$$

not consistent with (55).

7. Numerical Results and Conclusions

For the previous example and experimental conditions, Hestenes' method of multipliers and the modified method of multipliers were tested in connection with both the ordinary-gradient algorithm and the conjugate-gradient algorithm. Several values of the penalty constant k, ranging from 10^{-2} and 10^{2}, were considered. The numerical results are given in Table 1, where the number of iterations at convergence N_* is shown. In the table, Method (I) denotes Hestenes' method of multipliers and Method (II) denotes the modified method of multipliers.

From the table, one concludes that the modified method of multipliers exhibits faster convergence than Hestenes' method of multipliers in both the ordinary-gradient version and the conjugate-gradient version. However, both Method (I) and Method (II) exhibit slower convergence than the methods developed in Refs. 3-5, where constraint satisfaction to first order is enforced at each iteration.

Table 1. Number of iterations for convergence N_*
for Example 6.1.

k	Ordinary-gradient algorithm		Conjugate-gradient algorithm	
	Method (I)	Method (II)	Method (I)	Method (II)
10^{-2}	>1000	205	>200	>200
10^{-1}	>1000	196	138	113
10^{0}	>1000	874	84	43
10^{1}	>1000	>1000	81	50
10^{2}	>1000	>1000	95	55

References

1. MIELE, A., MOSELEY, P.E., and CRAGG, E.E., Numerical Experiments on Hestenes' Method of Multipliers for Mathematical Programming Problems, Rice University, Aero-Astronautics Report No. 85, 1971.
2. HESTENES, M.R., Multiplier and Gradient Methods, Journal of Optimization Theory and Applications, Vol. 4, No. 5, 1969.
3. MIELE, A., LEVY, A.V., and CRAGG, E.E., Modifications and Extensions of the Conjugate Gradient-Restoration Algorithm for Mathematical Programming Problems, Journal of Optimization Theory and Applications, Vol. 7, No. 6, 1971.
4. MIELE, A., CRAGG, E.E., IYER, R.R., and LEVY, A.V., Use of the Augmented Penalty Function in Mathematical Programming Problems, Part 1, Journal of Optimization Theory and Applications, Vol. 8, No. 2, 1971.
5. MIELE, A., CRAGG, E.E., and LEVY, A.V., Use of the Augmented Penalty Function in Mathematical Programming Problems, Part 2, Journal of Optimization Theory and Applications, Vol. 8, No. 2, 1971.
6. MIELE, A., MOSELEY, P.E., and CRAGG, E.E., A Modification of the Method of Multipliers for Mathematical Programming Problems, Rice University, Aero-Astronautics Report No. 86, 1971.

TIME/FUEL OPTIMAL AND ADAPTIVE CONTROL
FOR GRAVITY GRADIENT SPACECRAFT

FRANZ C. ZACH*

Computing & Software, Inc.,
NASA Goddard Space Flight Center
Greenbelt, Maryland

Abstract

This paper treats the time/fuel optimal control of gravity gradient spacecraft. The equations of motion of the
spacecraft are simplified such that the Maximum Principle
becomes applicable; the validity of this approximation is
proven by comparison of the simulation results of the rigorous and the simplified system. Based upon the simplified
equation, a time/fuel optimal law for spacecraft attitude
control is developed. This law is based upon definition
of a fuel savings angle, a new way of defining the trade-
off between pure time optimality and pure fuel optimality.
Chattering in the vicinity of zero angles and zero rates
is eliminated by a simple adaptive modification.

Symbols

The index i stands for yaw ($i = x$), roll ($i = y$) and pitch
($i = z$) and is omitted when no distinction in the treatment of the different axes is necessary.

---

*Part of the work for this paper was accomplished while the
author held a National Research Council Postdoctoral Re-
sident Research Associateship supported by the National
Aeronautics and Space Administration, Goddard Space Flight
Center. The author is very much obliged to Mr. W. Isley
and Mr. D. Endres for their valuable suggestions, as well
as to Computing and Software, Inc., for the support provided during the preparation of this paper.

a_j = gravity gradient coefficients, $j = x,y,z$
I_{ii} = moments of inertia, $i = x,y,z$
r = distance from earth center to spacecraft center
T_{ci} = control torque, $i = x,y,z$
T_{di} = disturbance torque, $i = x,y,z$
Γ_+ = line for switching from $u = 0$ to $u = +k$
Γ^+ = line for switching from $u = 0$ to $u = -k$
Λ_+^- = line for switching from $u = k$ to $u = 0$
Λ_-^+ = line for switching from $u = -k$ to $u = 0$
μ = earth gravitational constant
ω_0 = orbital angular rate for circular equatorial orbit

All other variables are defined wherever needed in the
text.

Introduction

Several spacecraft applications and experiments, e.g.,
laser communications and some types of earth observation,
make it necessary to acquire and maintain the S/C attitude
with a precision of .001°. Furthermore, severe constraints
on the amount of fuel available in spacecraft make it necessary to develop economical policies for attitude control. A promising approach is to combine active control
(e.g., by means of electric propulsion devices) and pas-
sive (gravity gradient) control. The configuration and
the equations of motion of a gravity gradient spacecraft
(in the case considered here, a synchronous earth satellite) will be presented in the following.

Configuration and Equations of Motion

The satellite under consideration is shown in Figure 1.

The control problem consists in bringing the attitude
angles of the spacecraft to zero. The attitude angles are
defined as the Euler angles between the x_b, y_b, z_b - system and the Attitude Coordinate System (x_o, y_o, z_o). The
axes of the latter are defined by the following directions:

x_o pointing from the earth center of mass to the spacecraft center of mass

y_o normal to x_o, in the orbit plane and in the same sense
as the vehicle motion

z_o normal to the orbit plane and such that (x_o,y_o,z_o) constitutes a right handed orthogonal coordinate system

Figure 2 shows both coordinate systems mentioned and the Euler angles used.

According to standard definitions, and considering Figure 2, we have:

pitch α about z_{o1}
roll ϕ about $-y^{\dagger}$
yaw ψ about x_b

The following equations of motion are valid for this satellite[1]:

$$[I] \begin{pmatrix} \dot{\Omega}_{xb} \\ \dot{\Omega}_{yb} \\ \dot{\Omega}_{zb} \end{pmatrix} = \begin{pmatrix} \delta_{xb} \\ \delta_{yb} \\ \delta_{zb} \end{pmatrix}, \tag{1}$$

where

$$\delta_{xb} = M_{xb} + M_{xd} \sin \rho_2 + M_{zd} \cos \rho_2 - (\Omega_{yb} H_{zb} - \Omega_{zb} H_{yb})$$
$$- P \cos \rho_2 + T_{cx} + T_{dx},$$

$$\delta_{yb} = M_{yb} + M_{xd} \cos \rho_2 \cos \rho_1 + M_{SD} \sin \rho_1 - M_{zd} \sin \rho_2 \cos \rho_1 \tag{2}$$
$$- (\Omega_{zb} H_{xb} - \Omega_{xb} H_{zb}) + P \sin \rho_2 \cos \rho_1 + T_{cy} + T_{dy}$$

and

$$\delta_{zb} = M_{zb} + M_{xd} \cos \rho_2 \sin \rho_1 - M_{SD} \cos \rho_1 - M_{zd} \sin \rho_2 \sin \rho_1$$
$$- (\Omega_{xb} H_{yb} - \Omega_{yb} H_{xb}) + P \sin \rho_2 \sin \rho_1 + T_{cz} + T_{dz}.$$

The torques M_{ib} and M_{id} are caused by the gravity gradient effect and are given in Reference 2. Also, P and M_{SD} and the equation of motion for the damper rod and the relationship between the time derivatives of the Euler angles and the Ω_{ib} (i = x,y,z) are given in Reference 2.

A digital program was written[1] which allows the simulation of the satellite motion according to the equations of motion presented.

Small Angle Approximation of the Equations of Motion

It is shown in References 2 and 3 that for small angles and negligible inner damping, the equations of motion can be approximated by:

$$\text{(yaw)} \qquad \ddot{\psi} + \omega_0^2 \frac{I_{zz} - I_{yy}}{I_{xx}} \psi - \frac{I_{zz} - I_{yy} - I_{xx}}{I_{xx}} \omega_0 \dot{\phi} = \frac{T_{cx} + T_{dx}}{I_{xx}}, \qquad (3)$$

$$\text{(roll)} \qquad \ddot{\phi} + 4\omega_0^2 \frac{I_{zz} - I_{xx}}{I_{yy}} \phi + \frac{I_{zz} - I_{xx} - I_{yy}}{I_{yy}} \omega_0 \dot{\psi} = \frac{T_{cy} + T_{dy}}{I_{yy}} \qquad (4)$$

and

$$\text{(pitch)} \qquad \ddot{\alpha} + 3\omega_0^2 \frac{I_{yy} - I_{xx}}{I_{zz}} \alpha = \frac{T_{cz} + T_{dz}}{I_{zz}}, \qquad (5)$$

It is now assumed that

$$I_{zz} \approx I_{yy} + I_{xx}, \qquad (6)$$

as is true, e.g., for the ATS-D and ATS-E spacecraft.* Thus, Equations 3-5 become

$$\ddot{\psi} + a_\psi \psi = u_\psi + d_\psi, \qquad (7)$$

$$\ddot{\phi} + a_\phi \phi = u_\phi + d_\phi \qquad (8)$$

and

$$\ddot{\alpha} + a_\alpha \alpha = u_\alpha + d_\alpha, \qquad (9)$$

where the a_i, u_i and d_i can be gained immediately by comparison with Equations 3-5.

When $d_\psi = d_\phi = d_\alpha = 0$ and u_ψ, u_ϕ, u_α are constant each of the Equations 7-9 gives circles for each of the phase plane plots $(\psi, \dot{\psi}/\sqrt{a_\psi})$, $(\phi, \dot{\phi}/\sqrt{a_\phi})$ or $(\alpha, \dot{\alpha}/\sqrt{a_\alpha})$ respectively. Simulations of the GGS under the same conditions result in circles for the phase plane plots (trajectories) for angles up to 10°. This, together with the results of Reference 2, proves the applicability of the approximations used.

---

*The case of coupled satellite axes is treated in Reference 5. However, no unique switching lines in the angle angular rate phase plane can be derived.

Derivation of the Time/Fuel Optimal Control Law

The Maximum Principle[4,5] will be applied to derive the time/
fuel optimal control for $\ddot{x} + ax = u$ in order to drive x and
$\dot{x}$ to zero, where x stands for ψ, ϕ, α of Eqs. 7 and 9.

The performance index is given by

$$P = \int_{t_o}^{t_e} \left(\lambda_t + \lambda_f \, |u| \right) dt \tag{10}$$

where t_o is the start and t_e the end of control action.
λ_t is the time weighting factor, λ_f the fuel weighting factor. E.g., $\lambda_f = 0$ leads to a pure time optimal solution.
The Hamiltonian[5] for this case is with $x = x_1$ and $\dot{x} = x_2$.

$$H = p_1 x_2 + p_2 \left(u - ax_1 \right) - \lambda_t - \lambda_f \, |u| \tag{11}$$

H is maximized at every instant of time by

$$u = k \operatorname{sgn} p_2 \quad \text{for} \quad |p_2| \geq \lambda_f \tag{12}$$

and

$$u = 0 \quad \text{for} \quad |p_2| < \lambda_f \tag{13}$$

With Equation 11 and $\quad dp_i/dt = -\partial H/\partial x_i \quad$ (see Re- (14)
ference 5), one obtains

$$p_2 = P_m \sin \left(a^{1/2} t + \varphi \right) \tag{15}$$

where P_m and ϕ are integration constants.

The final arc: Figure 3 shows p_2 and u versus time t.
From this figure it can be seen that u has to be on and off
periodically. Furthermore, it can be said that u has to be
on at the final time t_e; this is true since the state would
never reach the origin with control off because in this
case the trajectories are circles with the origin as center.
The on time before reaching the origin is between 0 and t_c,
t_c being the maximum on time. Therefore the part of the
trajectory which leads the state to the origin has to be
part of a semicircle with its center at $+k/a$ or $-k/a$ as
shown in Figure 2. The longest arc leading to the origin
is given by $\beta = \omega t_c$, where $\omega = a^{1/2}$.

Coasting before final arc: From Figure 3 it is obvious that
the state has to have a coasting period before reaching this
final arc unless the initial condition happened to be on
this final arc. The length of the coasting period is

limited by $\gamma = \omega t_{off}$. If the initial conditions are far enough from the origin that more thrust intervals than the final one leading to the origin are required to drive the state to the origin the coasting interval is given by $\gamma = \omega t_{off}$. This means that by going back in time by t_{off} the time of switching from control to coasting has been found. In the phase plane this can be depicted by drawing circular arcs with the center in the origin and starting at each point of the arcs of Figure 4. The length of these arcs for coasting is given by γ. The endpoints of these arcs again form circle arcs which are produced by rotating the circle arcs of Figure 4 by an angle counter clockwise about the origin. This is shown in Figure 5. It can be seen from this figure that each initial state lying within the hatched area can be brought to the origin in a time-fuel optimal maneuver by coasting to Γ_+ or Γ_-, respectively and then by control application.

<u>Thrusting before final coasting period</u>: If the initial state does not lie in the hatched area of Figure 5 one sees from Figures 3 and 5 that another thrusting period is required with control of the reverse sign when compared with the final control. This thrusting period again can last up to a time interval t_c. Again shorter intervals than t_c for thrustings are applicable for cases where the initial condition is located such that it can be brought to Λ_- by thrusting intervals less than t_c. In the general case of thrust application of exactly t_c duration all possible trajectories for this period again are circular arcs with an angle β and their centers at $-k/a$ or $+k/a$, respectively. This means that the geometric locus of their starting points can be constructed by rotation of Λ_- or Λ_+, respectively, in Figure 5 counterclockwise by an angle β. The result is shown in Figure 6.

<u>Complete switching lines</u>: The procedures outlined can be pursued in the same manner as above. The geometric relationship produced by rotating the circle arcs by β or γ, respectively, show that all arcs which form the switching lines have their center on straight lines as shown in Figure 7. Also, the intersection points of the arcs lie on a straight line.

The Equations for the switching lines are given in Reference 5.

For the case where $\lambda_f = 0$ the pure time optimal control law is derived. This means for the switching lines that $\gamma = 0$ which leads to the well known semicircles as switching lines for the system $\dot{x} + ax = u$ (see Fig. 8 and Refs.2,4,5).

On the other hand, for $\lambda_t = 0$ the pure fuel optimal control law is derived. In this case thrusting is only provided when the state is exactly on the $\dot{x}a^{-\frac{1}{2}}$- axis, which means an impulse at this time. This corresponds with the results derived in Reference , where it is shown that most economic control for a system given by $\ddot{x} + ax = u$ is achieved when the thruster is only in operation when the state falls on the $\dot{x}a^{-\frac{1}{2}}$-axis.

Adaptive Features and Application

The optimal control law derived has been applied to a seven degree of freedom computer simulation for gravity gradient spacecraft. The results show that very good consistency is obtained between this simulation and the simplified plant $\ddot{x} + ax = u$. However, this is only valid as long as the distance sq of the state from the origin is

$$sq = \sqrt{x_1^2 + \dot{x}_2^2/a} \geq 0.2 \text{ degrees} \qquad (16)$$

In many sapcecraft attitude control cases, much higher accuracy is required. For such requirements, the nonlinearities and cross-coupling between the different spacecraft axes become crucial.

The application of the optimal control (as derived for the linearized plant model) leads to chattering in the vicinity of the phase plane origin. Two principles can be applied which eliminate such chattering[5]. One is based upon (disturbance) identification, the other, is based upon control performance. We only apply the second method here.

Adaptive Control Based Upon Control Performance

Chattering about the origin is caused by deviations of the real trajectory from the trajectory of the plant $\ddot{x} + ax = u$. This results in missing of the origin (see Figure 9).

Because missing of the origin is more severe when the

applied control torque is higher, we developed a control
law which adapts the control level k based upon the behavior of the trajectory as it approaches the origin or
(if overshoot occurs) as it departs from it. We define
this behavior as control performance; this type of performance obviously is related to control accuracy.

Out tests have shown that the control level adjustment is
best achieved by letting (for states approaching the origin)

$$k_{db} = C_1 \frac{sq}{db} k \tag{17}$$

where

$$C_1 = \text{constant}$$
$$sq = \sqrt{x_1^2 + \dot{x}_2^2/a} \tag{18}$$

and $\quad$ db is a deadband for sq, such that k_{db} <k for
sq <db and

$$k_{db} = k \text{ for sq} \geq db.$$

This means that this control level adaptation is only applied within a certain deadband. C_1 and db are optimized
by experimentation. For states going away from the origin
we have found that

$$k_{db} = C_2(sq_{i+1} - sq_i) \cdot k \text{ for } C_2(sq_{i+1} - sq_i) \leq 1$$
and
$$k_{db} = k \text{ for } C_2(sq_{i+1} - sq_i) \geq 1 \tag{19}$$

provides very satisfying and stable control performance.
In the last equation, sq_i and sq_{i+1} are two distances of
the phase from the origin in two subsequent sampling times.
(Obviously we assume here that sq_i is measured in discrete
time fashion.) C_2 is a constant which depends on the
sampling interval; C_2 has been optimized by experimentation.

We want to emphasize that the combination of the more
mathematically derived time-fuel optimal control law and
the more empirically found control level adaptation are essential for stable working conditions; this results in much
higher pointing accuracy of the S/C (i.e., maintaining
the state close to the origin) and high fuel savings when
compared to conventional control. In particular, the
pointing accuracy can be improved from .0018° (in roll)

and $.00039^{\circ}$ (in pitch) to better than 10^{-6} degrees, while fuel expenditure is decreased by a factor of 0.1.

Concluding Comments

Time/fuel optimal control procedures have been derived for gravity gradient satellites where yaw and roll axes are decoupled. Because it seems to be impossible to derive such control laws for the complete, coupled satellite equations of motion, it is recommended to decouple the yaw and roll axes by proper choice of the satellite moments of inertia. Adaptive features have to be added to the control law to make it applicable for small angles. We want to point out that here, as in many other cases, the definition of optimality is more or less arbitrary. Exact definitions, however, provide a basis for optimal control developments. Control laws gained by purely theoretical methods are usually not sufficient for practical applications, e.g., due to uncertainties in the mathematical model, or in the knowledge of disturbances, or because the mathematical model only could be treated in an approximate form. This requires additional efforts, such as adaptive control, and, most important, extensive experimentation with either the real system or a sufficiently realistic computer simulation. Furthermore, it should be pointed out that optimal control laws, due to their trade-off characteristics, are only of practical importance if they are easy to realize. This seems to postulate a new optimality criterion which takes the simplicity of application and realization into consideration.

References

1. Barrett, C.C.,"The Development of a Mathematical Model and a Study of one Method of Orbit Adjust and Station Keeping Gravity-Oriented Satellites," TN D-3652, NASA, Washington, D.C., November 1966.
2. Zach, F.,"Time Optimal Control of Gravity Gradient Satellites," Journal of Spacecraft and Rockets, Vol. 7, No. 12, pp. 1434-1440, December 1970.
3. Frick, R.H.,"Perturbation of a Gravity Gradient Stabilization System," Report AD 623279, The Rand Corp., Santa Monica, California, Sept. 1965.
4. Pontryagin, L.S. et al.," The Mathematical Theory of

Optimal Processes," Pergamon Press, Oxford, 1964.

5. Zach, F.,"Practical Optimal Control," Springer-Verlag, Wien-New York, in publication.

6. Flugge-Lotz, I., Craig, A.J., <u>Trans. ASME, J. of Basic Engineering</u>, March 1965, pp. 39-57.

7. Isley, W.,"Optimal Control Application for Electrothermal Multijet Systems on Synchronous Earth Spacecraft," <u>J. of Spacecraft and Rockets</u>, Vol. 5, No. 12, Dec. 1968.

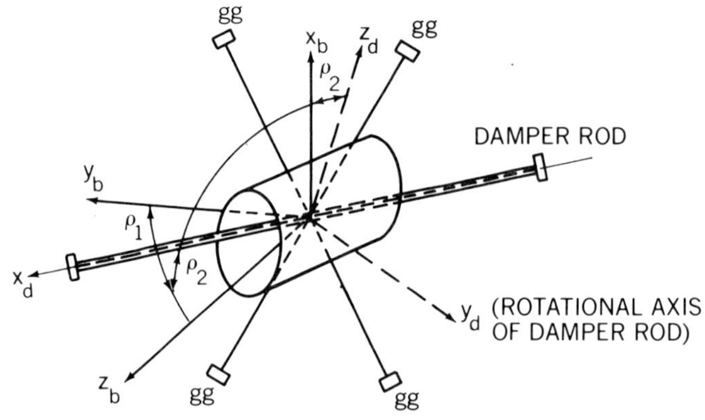

Fig. 1 Gravity Gradient Satellite. x_b and y_b lie in the plane of the four booms and are fixed to the spacecraft. x_d, y_d and z_d are fixed to the damper rod.

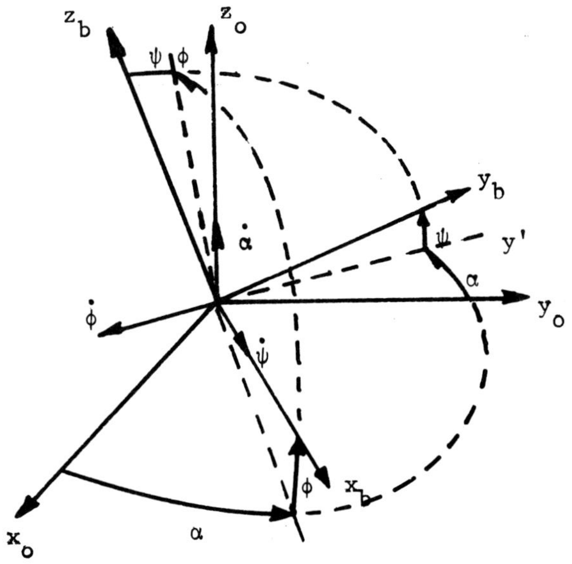

Fig. 2 Body fixed coordinate system and attitude coordinate system.

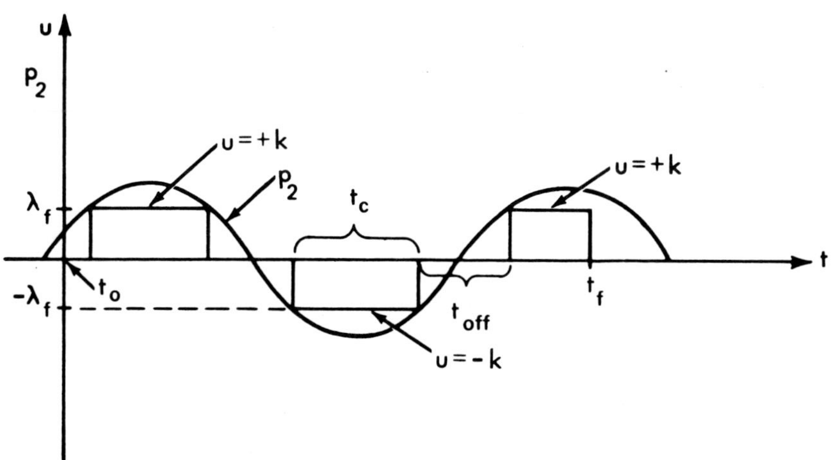

Fig. 3 Control u and auxiliary variable p_2.

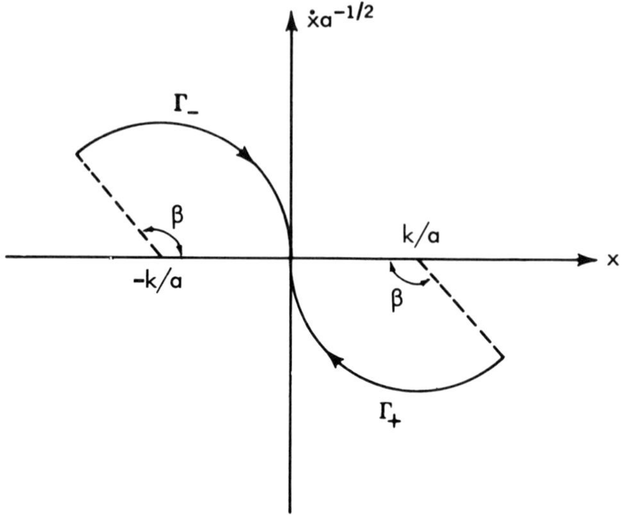

Fig. 4 Final arc.

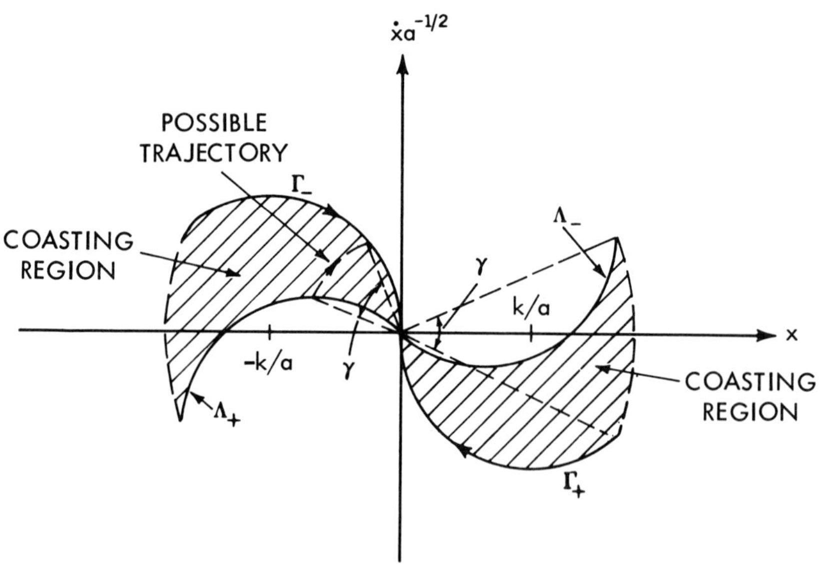

Fig. 5 Coasting regions and switching lines.

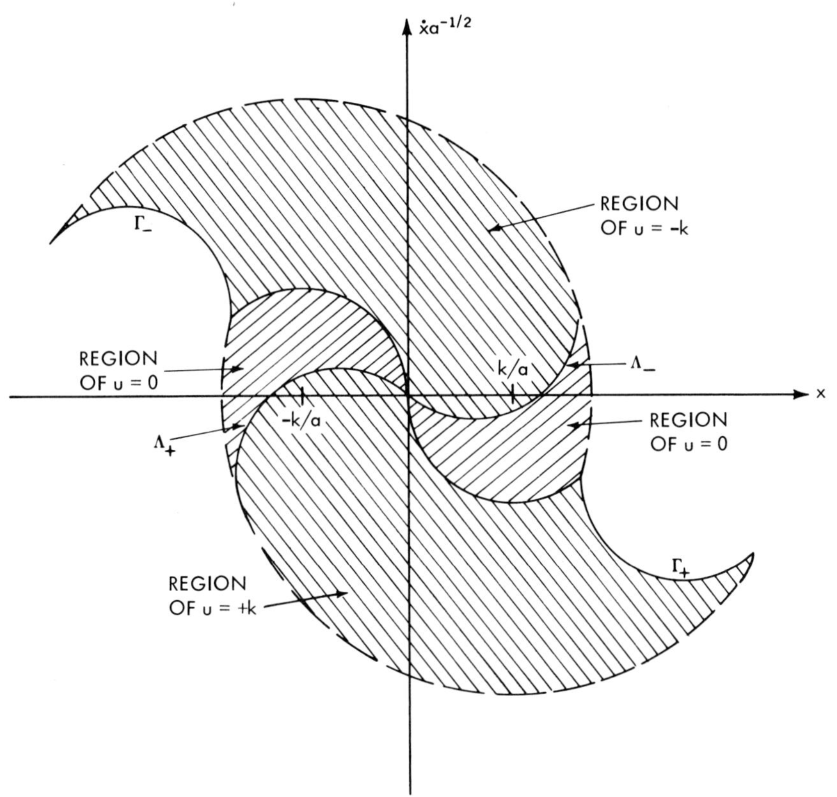

Fig. 6 Extension of Fig. 5, showing regions where u = +k, 0, -k is applied.

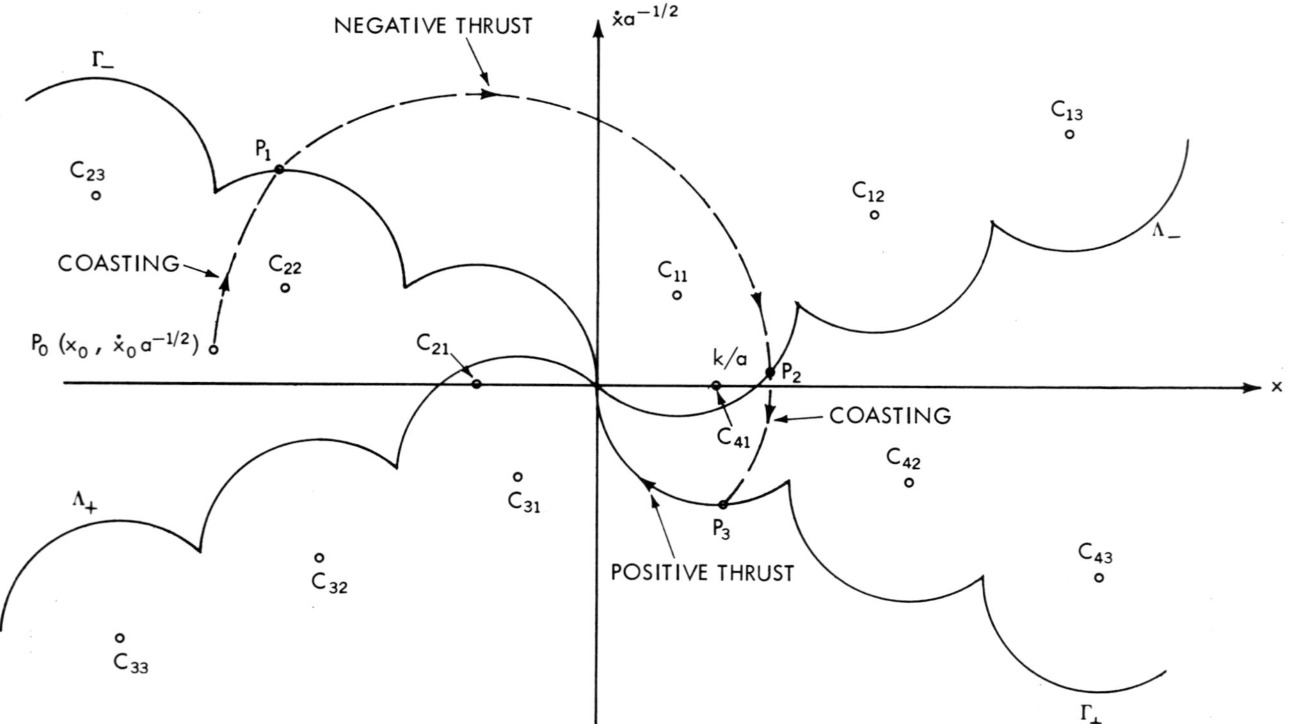

Fig. 7 Switching lines with sample optimal trajectory. C_{ik} ...centers of switching circle arcs.

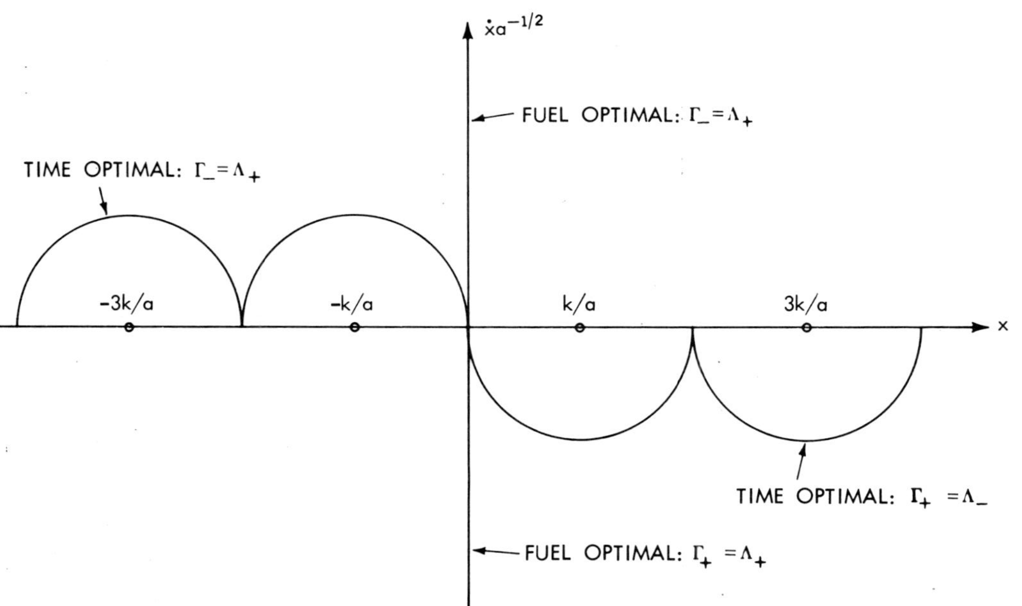

Fig. 8 Reduction of the time/fuel optimal switching lines to the pure time optimal case ($\gamma=0^\circ$) and to the pure fuel optimal case ($\gamma=180^\circ$). The switching lines for the latter fall into the $\dot{x}a^{-1/2}$ axis.

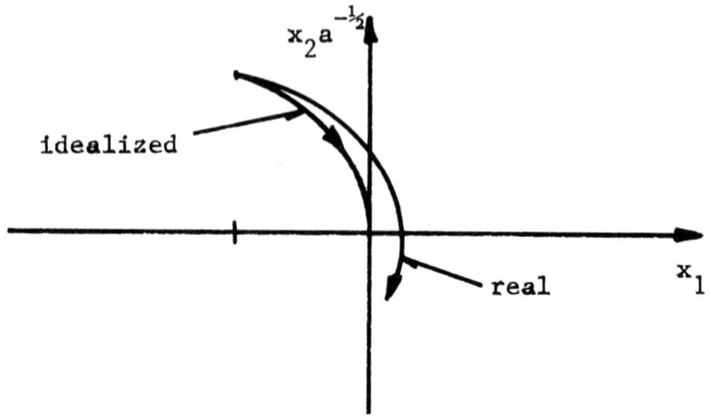

Fig. 9 Behavior of the real and idealized trajectories.

DIFFERENTIAL GAMES

SUFFICIENCY FOR OPTIMAL STRATEGIES
IN NASH EQUILIBRIUM GAMES[1]

G. Leitmann[2] and H. Stalford[3]

Introduction

Candidates for optimal strategies in Nash equilibrium
games are usually found by the constructive utilization of
necessary conditions; e.g., see Refs. 1-4 for two-person
zero-sum games and Refs. 5-7 for N-person games. However,
enabling one to assert the optimality of candidate strategies is often more difficult. One method for assuring optimality is the verification of sufficiency conditions.

A number of sufficiency conditions may be found in the
literature. These are field theorems requiring the existence of sufficiently smooth functions of the state and
time; e.g., see Isaacs' "Verification Theorem" in Ref. 1
and other versions and generalizations in Refs. 2, 4, 5, 7,
8 and 9. Here we give a simpler and sometimes more readily
verifiable sufficiency theorem; it is an adaptation of an
earlier theorem for sufficiency of optimal control, Ref. 10.

Problem Statement

Consider a system with state equation

$$\dot{x} = f(x,\ t,\ v^1,\ v^2,\ldots,v^N) \qquad (1)$$

where state variable $x \in E^n$, control variables $v^i \in E^{m_i}$, $i =$
1,2,...,N, and state velocity function f is Borel measurable
on $E^n \times E^1 \times E^{m_1} \times E^{m_2} \times \ldots \times E^{m_N}$.

---

[1]Research supported by NSF under Grant GP 24205.

[2]University of California, Berkeley, California.

[3]Naval Research Laboratory, Washington, D. C.

The playing space X is a given Lebesgue measurable subset of E^n. The initial state x^o and the target set[+] θ are given sets contained in X. Here we consider games with fixed playing time; play commences at time $t_o \geq 0$ and terminates at time t_f, $t_o \leq t_f < \infty$.

Let $\mathcal{P}^i$, $i = 1,2,..,N$, denote the spaces of all Borel measurable functions from $X \times [0,t_f]$ to E^{m_i}; space $\mathcal{P}^i$ constitutes the space of <u>strategies</u> $p^i : X \times [0,t_f] \to E^{m_i}$ for player i. A strategy p^i is admissible iff $p^i \in \mathcal{P}^i$ and

$$p^i(x,t) \in U^i(x,t) \qquad (x,t) \in X \times [0,t_f] \qquad (2)$$

for given functions

$$U^i : X \times [0,t_f] \to \text{set of all non-empty subsets of } E^{m_i} \qquad (3)$$

Thus, state constraints are introduced by specifying X and control constraints by specifying the U^i.

An absolutely continuous function $\phi_{x^o} : [t_o,t_f] \to X$ is called a trajectory iff there exist admissible strategies p^i, $i = 1,2,..,N$, such that

$$\phi_{x^o}(t) = x^o + \int_{t_o}^{t} f(\phi_{x^o}(\tau),\tau,p^1(\phi_{x^o}(\tau),\tau),...,p^N(\phi_{x^o}(\tau),\tau))d\tau$$
$$t \in [t_o,t_f] \qquad (4)$$

A strategy N-tuple $p = (p^1,p^2,..p^N)$ is <u>playable</u> at (x^o,t_o) iff it is admissible and there exists a trajectory ϕ_{x^o} such that $\phi_{x^o}(t_f) \in \theta$. Let $\mathcal{J}(x^o,t_o)$ denote the set of all strategy N-tuples that are playable at (x^o,t_o). Let $\Phi(x^o,t_o;p)$ denote the set of all trajectories generated by $p \in \mathcal{J}(x^o,t_o)$. We assume $\mathcal{J}(x^o,t_o) \neq \phi$.

The i-th player's cost, J^i, is a function of strategy N-tuple $p \in \mathcal{J}(x^o,t_o)$ and trajectory $\phi_{x^o} \in \Phi(x^o,t_o;p)$ given by

$$J^i(p^1,p^2,...,p^N;\phi_{x^o}) =$$
$$\int_{t_o}^{t_f} f_o^i(\phi_{x^o}(t),t,p^1(\phi_{x^o}(t),t),...,p^N(\phi_{x^o}(t),t))dt \qquad (5)$$

[+] θ may be all of X.

where the f_0^i are given real valued bounded Borel measurable functions with domain $E^n \times E^1 \times E^{m_1} \times \ldots \times E^{m_N}$.

Let $p^* = (p^{1*}, p^{2*}, \ldots, p^{N*}) \in \mathcal{J}(x^0, t_0)$ and $\phi^*_{x^0} \in \Phi(x^0, t_0; p^*)$. Strategy N-tuple p^* is optimal at (x^0, t_0) iff it satisfies Nash equilibrium condition

$$J^i(p^*; \phi^*_{x^0}) \leqq J^i(p^{1*}, p^{2*}, \ldots, p^i, \ldots, p^{N*}; \phi^i_{x^0})$$

$$i = 1, 2, \ldots, N) \tag{6}$$

for all $\phi^*_{x^0} \in \Phi(x^0, t_0; p^*)$, $(p^{1*}, p^{2*}, \ldots, p^i, \ldots, p^{N*}) \in \mathcal{J}(x^0, t_0)$ and $\phi^i_{x^0} \in \Phi(x^0, t_0; p^{1*}, p^{2*}, \ldots, p^i, \ldots, p^{N*})$.

A strategy N-tuple p^* is optimal on $X \times [0, t_f]$ iff it is optimal at all $(x^0, t_0) \in X \times [0, t_f]$.

Sufficiency

Let $\lambda^i : [t_0, t_f] \to E^n$, $i = 1, 2, \ldots, N$, denote absolutely continuous functions, and

$$H^i(\lambda^i(t), x, t, v^1, \ldots, v^N) = f_0^i(x, t, v^1, \ldots, v^N)$$

$$+ \lambda^{iT}(t) f(x, t, v^1, \ldots, v^N) \tag{7}$$

Also let $u^{i*}(t) = p^{i*}(\phi^*_{x^0}(t), t)$, $i = 1, 2, \ldots, N$, where $\phi^*_{x^0} \in \Phi(x^0, t_0; p^*)$.

<u>Theorem.</u> Strategy N-tuple p^* is optimal at (x^0, t_0) iff (a) $p^* \in \mathcal{J}(x^0, t_0)$, and (b) for all $\phi^*_{x^0} \in \Phi(x^0, t_0; p^*)$ there exist λ^i such that for $i = 1, 2, \ldots, N$

(i) $\qquad H^i(\lambda^i(t), \phi^*_{x^0}(t), t, u^{1*}(t), \ldots, u^{i*}(t), \ldots, u^{N*}(t))$

$\qquad - H^i(\lambda^i(t), x, t, p^{1*}(x, t), \ldots, v^i, \ldots, p^{N*}(x, t))$

$\qquad + \dot{\lambda}^{iT}(t)(\phi^*_{x^0}(t) - x) \leqq 0$

a.e. on $[t_0, t_f]$, for all $x \in X$ and for all $v^i \in U^i(x, t)$, and

(ii) $\qquad \lambda^{iT}(t_f)(\phi^*_{x^0}(t_f) - y) = 0$ for all $y \in \theta$.

<u>Proof</u>. Consider a strategy N-tuple $(p^{1*},p^{2*},\ldots,p^{i},\ldots,p^{N*})$ that is playable at (x^{o},t_{o}) and a trajectory $\phi^{i}_{x^{o}} \in \Phi(x^{o},t_{o}; p^{1*},p^{2*},\ldots,p^{i},\ldots,p^{N*})$.

By (i) of the Theorem

$$f^{i}_{o}(\phi^{*}_{x^{o}}(t),t,u^{1*}(t),\ldots,u^{i*}(t),\ldots,u^{N*}(t))$$

$$- f^{i}_{o}(\phi^{i}_{x^{o}}(t),t,u^{1}(t),\ldots,u^{i}(t),\ldots,u^{N}(t))$$

$$+ \frac{d}{dt} [\lambda^{iT}(t)(\phi^{*}_{x^{o}}(t) - \phi^{i}_{x^{o}}(t))] \leq 0$$

where $u^{i}(t) = p^{i}(\phi^{i}_{x^{o}}(t),t)$ and $u^{j}(t) = p^{j*}(\phi^{i}_{x^{o}}(t),t)$ for $j \neq i$.

On integration, and use of $\phi^{i}_{x^{o}}(t_{o}) = \phi^{*}_{x^{o}}(t_{o}) = x^{o}$ and condition (ii), equilibrium condition (6) follows, concluding the proof.

For two-person zero-sum games, $i = 1,2$, $J^{1} = -J^{2}$. For such games one has the following corollary.

<u>Corollary</u>. Strategy pair $p^{*} = (p^{1*},p^{2*})$ is optimal (saddle-point) at (x^{o},t_{o}) iff (a) $p^{*} \in \mathfrak{J}(x^{o},t_{o})$, and (b) for all $\phi^{*}_{x^{o}} \in \Phi(x^{o},t_{o};p^{*})$ there exists a λ such that

(i) $\qquad H(\lambda(t),t,\phi^{*}_{x^{o}}(t),u^{1*}(t),u^{2*}(t))$

$$- H(\lambda(t),t,x,v^{1},p^{2*}(x,t)) + \dot{\lambda}^{T}(t)(\phi^{*}_{x^{o}}(t) - x) \leq 0$$

$$H(\lambda(t),t,\phi^{*}_{x^{o}}(t),u^{1*}(t),u^{2*}(t))$$

$$- H(\lambda(t),t,x,p^{1*}(x,t),v^{2}) + \dot{\lambda}^{T}(t)(\phi^{*}_{x^{o}}(t) - x) \geq 0$$

a.e. on $[t_{o},t_{f}]$, for all $x \in X$ and for all $v^{1} \in U^{1}(x,t)$, $v^{2} \in U^{2}(x,t)$, and

(ii) $\qquad \lambda^{T}(t_{f})(\phi^{*}_{x^{o}}(t_{f}) - y) = 0$ for all $y \in \theta$.

<u>Remark 1</u>. The solutions of the adjoint equations in the necessary conditions, e.g. Eq. (12) of Ref. 6 or Eq. (93) of Ref. 7, may be used in place of functions λ^{i}. Then, if $\phi^{*}_{x^{o}}(t_{f})$ is free (i.e., $\theta = X$), the necessary transversality condition assures satisfaction of condition (ii).

<u>Remark 2</u>. If the solutions of the adjoint equations are used in place of the λ^{i}, it may be necessary to allow the λ^{i} to be piecewise absolutely continuous on $[t_{o},t_{f}]$. In

that event, condition (ii) of the Theorem is modified as follows:

$$(ii)' \quad \lambda^{iT}(t_f)(\phi_{x^o}^*(t_f) - y) + \sum_{k=1}^{r_i} [\lambda^{iT}(t_k - 0)$$

$$- \lambda^{iT}(t_k + 0)][\phi_{x^o}^*(t_k) - x] = 0$$

for all $y \in \theta$ and all $x \in X$, where the t_k, $k = 1, 2, \ldots, r_i$, are the points of discontinuity of λ^i.

<u>Remark 3</u>. While the Theorem and Corollary are given for specified playing time, <u>in principle</u> no loss of generality is incurred since a game with unspecified terminating time t_f can be transformed into one with specified terminal value of the independent variable.

Example

Here we give a simple example to illustrate the use of the sufficiency conditions. We consider a two-person zero-sum game so that the Corollary may be utilized.

The state equations are

$$\dot{x}_i = x_{i+1} \qquad i = 1, 2, \ldots, n-1$$

$$\dot{x}_n = v^1 + v^2$$

with control constraints

$$|v^j| \leq 1 \qquad j = 1, 2$$

and costs

$$J^1(p^1, p^2 ; \phi_{x^o}) = - J^2(p^1, p^2 ; \phi_{x^o})$$

$$= \int_{t_o}^{t_f} \sum_{i=1}^{n} \phi_{x^o_i}(t) dt$$

where

$$\phi_{x^o} = (\phi_{x^o_1}, \phi_{x^o_2}, \ldots, \phi_{x^o_n})$$

is a trajectory generated by admissible strategy pair (p^1, p^2) with

283

$$\phi_{x^o}(t_o) = x^o$$

and $\phi_{x^o}(t_f)$ unspecified but t_f is fixed.

The H-function is given by

$$H(\lambda(t),x,v^1,v^2) = \sum_{i=1}^{n} x_i + \sum_{i=1}^{n-1} \lambda_i(t)x_{i+1} + \lambda_n(t)(v^1 + v^2)$$

From necessary conditions we consider $\lambda : [t_o,t_f] \to E^n$ satisfying

$$\dot{\lambda}_1 = -1$$

$$\dot{\lambda}_{i+1} = -1 - \lambda_i \qquad i = 1,2,\ldots,n-1$$

with $\lambda_i(t_f) = 0$, $i = 1,2,\ldots,n$. Consequently,

$$\lambda_i(t) = \sum_{k=1}^{i} \frac{(t_f-t)^k}{k!} \qquad i = 1,2,\ldots,n$$

Since $\lambda_n(t) > 0$ for $t \in [t_o,t_f)$,

$$\left.\begin{array}{l} \underset{v^1}{\mathrm{Min}}\ H(\lambda(t),\phi^{*}_{x^o},v^1,u^{2*}(t)) \\[2em] \underset{v^2}{\mathrm{Max}}\ H(\lambda(t),\phi^{*}_{x^o},u^{1*}(t),v^2) \end{array}\right\} \Rightarrow \left\{\begin{array}{l} u^{1*}(t) \equiv -1 \\[2em] u^{2*}(t) \equiv 1 \end{array}\right.$$

Since this result is independent of initial condition (x^o, t_o), the candidates for optimal strategies are

$$p^{1*}(x,t) \equiv -1 \quad , \quad p^{2*}(x,t) \equiv 1$$

and the corresponding trajectory is given by

$$\phi^{*}_{x^o_i}(t) = \sum_{k=0}^{n-i} x^o_{i+k} \frac{(t-t_o)^k}{k!} \qquad i = 1,2,\ldots,n \ (\text{with } 0! = 1)$$

Using the above λ-function in the sufficiency conditions allows one to verify these conditions, thereby establishing the optimality of (p^{1*},p^{2*}) at (x^o,t_o). In fact, since these conditions are verified for all (x^o,t_o), the strategy pair is optimal on $E^2 \times [0,t_f]$.

Condition (i) of the Corollary reduces to

$$- \lambda_n(t)(v^1 + 1) \leqq 0 \qquad \forall \, v^1 : |v^1| \leqq 1$$

$$- \lambda_n(t)(v^2 - 1) \geqq 0 \qquad \forall \, v^2 : |v^2| \leqq 1$$

and condition (ii) follows from $\lambda(t_f) = 0$.

References

1. R. Isaacs, *Differential Games*, Wiley, 1965.
2. L. Berkovitz, A Variational Approach to Differential Games, in *Advances in Game Theory*, Princeton University Press, 1964.
3. G. Leitmann and G. Mon, Some Geometric Aspects of Differential Games, J. Astronaut. Sci. 14, No. 2, 1967.
4. A. Blaquière, F. Gérard and G. Leitmann, *Quantitative and Qualitative Games*, Academic Press, 1969.
5. J. H. Case, Toward a Theory of Many Player Differential Games, SIAM J. Control 7, No. 2, 1969.
6. A. W. Starr and Y. C. Ho, Nonzero-Sum Differential Games, J. Optimization Theory and Appl. 3, No. 3, 1969.
7. G. Leitmann, Differential Games, in *Differential Games: Theory and Applications* (1970 JACC), ASME, 1970.
8. H. Stalford, Sufficiency Conditions in Optimal Control and Differential Games, ORC 70-13, Operations Research Center, University of California, Berkeley, 1970.
9. H. Stalford and G. Leitmann, Sufficient Conditions for Optimality in Two-Person Zero-Sum Differential Games with State and Strategy Constraints, J. Math. Anal. Appl. 33, No. 3, 1971.
10. G. Leitmann and H. Stalford, A Sufficiency Theorem for Optimal Control, J. Optimization Theory and Appl. 8, No. 3, 1971.

DIFFERENTIAL GAMES WITH TIME LAG

Austin Blaquière

Laboratoire d'Automatique Théorique
Université de Paris 7

Introduction and Problem Statement

This paper is a continuation of the work developed
in (1). Here we shall consider a system whose dynamical
behavior is governed by

$$\dot{y} = f(y,u,v) \tag{1}$$

where $y \triangleq (y_1, \ldots y_n) \in D \subseteq E^n$, $y_n \equiv t$, $u \in U \subseteq E^r$,

$v \in V \subseteq E^q$, $f : D \times U \times V \longrightarrow E^n$.

Players J_P and J_E make their decisions through
choosing the values of control variables u and v, respectively, at each instant of time. These choices are governed
by strategies which J_P and J_E select from two prescribed
sets of strategies S_P and S_E , respectively.
We shall consider strategies for players J_P and J_E
to be functions of y, $p : y \longmapsto p(y)$ and $e : y \longmapsto e(y)$,
$y \in Y \subseteq D$, respectively, belonging to prescribed classes of

This research was supported in part by the
Délégation Générale à la Recherche Scientifique
et Technique under Grant 69-01-691-02. It has
been performed in collaboration with Pierre
CAUSSIN , Laboratoire d'Automatique Théorique.

functions, and such that $p(y) \in K_u \subseteq U$, $e(y) \in K_v \subseteq V$, for all $y \in Y$.

__Assumption 1.__ Whatever $p',p'' \in S_P$ and $e',e'' \in S_E$, and whatever $y^j \in Y$, functions $p''' : y \longmapsto p'''(y)$ and $e''' : y \longmapsto e'''(y)$, $y \in Y$, where

$$p''' (y) = p'(y) \text{ and } e'''(y) = e'(y) \text{ for } y_n \leqslant y_n^j$$

$$p''' (y) = p''(y) \text{ and } e'''(y) = e''(y) \text{ for } y_n > y_n^j$$

are strategies; that is $p''' \in S_P$ and $e''' \in S_E$.

Target θ is a given set of points in D. We shall assume that the state of the system, for given time lag d, $d > 0$, or as we shall say the state of the game with time lag d, at any instant of time , belongs to the fixed subset G of $C°[0,d]$, where

$$C°[0,d] \triangleq \left\{ x \triangleq \hat{x}(.) : [0,d] \longrightarrow D , \quad x \in C° \right\}$$

Trajectories and Paths

Let $X \triangleq \left\{ x : x \in G, \hat{x}(s) \in Y, s \in [0,d) \right\}$. We shall denote by $S(x^i,p,e,t_j)$, where $x^i \in X$, x^i at time t_i, $t_i < t_j$, the set of functions $\hat{y} : [t_i-d,t_j] \longrightarrow D$, continuous on $[t_i-d,t_j]$, such that

(i) $\hat{y}(t) \in Y$ for all $t \in [t_i-d,t_j-d)$; and

(ii) $\hat{y}(t) = \hat{x}^i(t+d-t_i)$ for all $t \in [t_i-d,t_i]$; and

(iii) $d\hat{y}(t)/dt = f(\hat{y}(t),p(\hat{y}(t-d)),e(\hat{y}(t-d)))$ (2)

for all $t \in [t_i,t_j]$, except possibly on a subset of $[t_i,t_j]$ at most denumerable.

We say that x^i, $x^i \in G$, is __at time__ t_i if the n-th component of $\hat{x}^i(d)$ is t_i.

288

We shall denote by $C(y^i,\hat{u},\tilde{v},t_j)$, $t_i < t_j$, the set of functions $\hat{y} : [t_i,t_j] \longrightarrow D$, continuous on $[t_i,t_j]$, such that $\hat{y}(t_i) = y^i$ and

$$d\hat{y}(t)/dt = f(\hat{y}(t),\hat{u}(t),\tilde{v}(t)) \tag{3}$$

for all $t \in [t_i,t_j]$, except possibly on a subset of $[t_i,t_j]$ at most denumerable, where $\hat{u} : [t_i,t_j] \longrightarrow U$ and $\tilde{v} : [t_i,t_j] \longrightarrow V$.

Trajectories γ^{ij} and ϱ^{ij} are defined by

$$\gamma^{ij} \triangleq \tilde{\gamma}^{ij}(\hat{y}) \triangleq \{\hat{y}(t) : t \in [t_i,t_j]\} , \hat{y} \in S(x^i,p,e,t_j)$$

$$\varrho^{ij} \triangleq \hat{\varrho}^{ij}(\hat{y}) \triangleq \{\hat{y}(t) : t \in [t_i,t_j]\} , \hat{y} \in C(y^i,\tilde{u},\tilde{v},t_j)$$

A terminating trajectory is a trajectory whose end point $\hat{y}(t_j)$ belongs to θ .

Paths π^{ij} are defined by

$$\pi^{ij} \triangleq \tilde{\pi}^{ij}(\hat{y}) \triangleq \{\hat{z}(t,.): t \in [t_i,t_j]\}, \hat{y} \in S'(x^i,p,e,t_j)$$

where $S'(x^i,p,e,t_j)$ is the set of functions $\hat{y}$, $\hat{y} \in S(x^i,p,e,t_j)$, such that $\hat{z}(.,.) : [t_i,t_j] \times [0,d] \longrightarrow D$, defined by $\hat{z}(t,s) = \hat{y}(t+s-d)$ for all $s \in [0,d]$ and for all $t \in [t_i,t_j]$, satisfies $\hat{z}(t,.) \in G$ for all $t \in [t_i,t_j]$.

Let h be the mapping that associates with each path $\pi^{ij} \triangleq \tilde{\pi}^{ij}(\hat{y})$, $\hat{y} \in S'(x^i,p,e,t_j)$, the trajectory $\gamma^{ij} \triangleq \tilde{\gamma}^{ij}(\hat{y})$. We shall say that π^{ij} is a terminating path if $h(\pi^{ij})$ is a terminating trajectory.

We shall say that a strategy pair (p,e) is playable at $x^i \in X$, x^i at time t_i, if there exists a time t_j, $t_j > t_i$, and a function $\hat{y}$, $\hat{y} \in S'(x^i,p,e,t_j)$, such that $\hat{y}(t_j) \in \theta$. We shall let $\mathcal{J}(x^i)$ denote the set of all playable strategy pairs. Likewise we shall say that $(\tilde{u},\tilde{v})$ is playable at point $y \in D$ if there exists a time t_j, $t_j > y_n$, and a function $\tilde{y}$, $\tilde{y} \in C(y,\tilde{u},\tilde{v},t_j)$, such that $\tilde{y}(t_j) \in \theta$. We shall let $J(y)$ denote the set of all such pairs.

Cost, Optimality of a Strategy Pair

For given (x^i,p,e,t_j), x^i in X at time t_i, $t_j = t_i + d$, we shall denote by $g_p(x^i,p)$ and $g_E(x^i,e)$ the functions $\tilde{u}$ and $\hat{v}$, respectively, such that $\tilde{u}(t) = p(\tilde{x}^i(t-t_i))$ and $\tilde{v}(t) = e(\tilde{x}^i(t-t_i))$ for all $t \in [t_i, t_j]$.

For given (x^i,p,e, γ^{ij}), x^i in X at time t_i, $\gamma^{ij} \triangleq \tilde{\gamma}^{ij}(\tilde{y})$, $\tilde{y} \in S(x^i,p,e,t_j)$, we shall denote by $g(x^i,p,e, \gamma^{ij})$ the pair $(\tilde{u},\tilde{v})$, where $\tilde{u}$ and $\tilde{v}$ are such that $\tilde{u}(t) = p(\tilde{y}(t-d))$ and $\tilde{v}(t) = e(\tilde{y}(t-d))$ for all t, $t \in [t_i, t_j]$.

The cost of $\rho^{ij} \triangleq \hat{\rho}^{ij}(\tilde{y})$, $\tilde{y} \in C(y^i,\tilde{u},\tilde{v},t_j)$, is

$$V_\rho(y^i,y^j,\tilde{u},\tilde{v}, \rho^{ij}) \triangleq \int_{t_i}^{t_j} f_o(\tilde{y}(t),\tilde{u}(t),\tilde{v}(t)) \, dt + w(y^j)$$

where $f_o : D \times U \times V \longrightarrow R$, $w : D \longrightarrow R$ such that $w(y^j) = 0$ for all $y^j \notin \theta$, $y^j \triangleq \tilde{y}(t_j)$.

The cost of $\gamma^{ij} \triangleq \tilde{\gamma}^{ij}(\tilde{y})$, $\tilde{y} \in S(x^i,p,e,t_j)$, is

$$V_\gamma(x^i,y^j,p,e, \gamma^{ij}) \triangleq V_\rho(y^i,y^j,\tilde{u},\tilde{v}, \rho^{ij})$$

where $y^i \triangleq \tilde{y}(t_i)$, $y^j \triangleq \tilde{y}(t_j)$, $(\tilde{u},\tilde{v}) = g(x^i,p,e, \gamma^{ij})$, $\rho^{ij} = \gamma^{ij}$.

The cost of $\pi^{ij} \triangleq \tilde{\pi}^{ij}(\tilde{y})$, $\tilde{y} \in S'(x^i,p,e,t_j)$, is

$$V_\pi(x^i,x^j,p,e,\pi^{ij}) \triangleq V_\gamma(x^i,y^j,p,e, \gamma^{ij})$$

where $y^j \triangleq \tilde{y}(t_j)$, $\gamma^{ij} = h(\pi^{ij})$.

If π^{ij} is a terminating path, we shall let

$$V_\pi(x^i, \theta, p, e) \triangleq V_\pi(x^i, x^j, p, e, \pi^{ij})$$

Since, for given x^i and $(p,e) \in \mathcal{J}(x^i)$, terminating path π^{ij} need not be unique, $V_\pi(x^i, \theta, p, e)$ need not be unique.

For all x^i such that $y^i \triangleq \tilde{x}^i(d) \in \theta$, we shall let

$$V_\pi(x^i, \theta, p, e) \triangleq w(y^i).$$

Definition 1. (p^*, e^*) is optimal at $x \in X$ if
(a) $(p^*, e^*) \in \mathcal{J}(x)$; and
(b) $V_\pi(x, \theta, p^*, e^*)$ is defined; and
(c) $V_\pi(x, \theta, p^*, e) \leq V_\pi(x, \theta, p^*, e^*) \leq V_\pi(x, \theta, p, e^*)$
for all $(p^*, e) \in \mathcal{J}(x)$, for all $(p, e^*) \in \mathcal{J}(x)$, and for all terminating paths generated by (p^*, e) and (p, e^*) from state x.

We shall let $\mathcal{J}^*(x)$ denote the set of all strategy pairs optimal at x.

A Restricted Class of DGWITL

From now on we shall consider a restricted class of differential games with time lag (DGWITL); namely, the one for which the following assumptions and definitions hold.

Assumption 2. K_u and K_v are constant sets; that is, they do not depend on y.

Definition 2.
$$Y \triangleq \{ y : t_o \leq y_n \leq t_f \}$$
$$D \triangleq \{ y : t_o \leq y_n \leq t_f + d \}$$
$$\theta \triangleq \{ y : y_n = t_f \}$$
$$G \triangleq C^\circ[0, d]$$
$$S_P \triangleq \{ p : Y \to K_u, \ g_P(x^i, p) \in \mathcal{C}_P \quad \forall \, x^i \in X \}$$
$$S_E \triangleq \{ e : Y \to K_v, \ g_E(x^i, e) \in \mathcal{C}_E \quad \forall \, x^i \in X \}$$

where $\mathcal{C}_P$ and $\mathcal{C}_E$ are the sets of piece-wise continuous functions of time with ranges in K_u and K_v, respectively.

__Assumption 3.__ $\quad \forall x^i, \; x^i$ in X at time t_i, $\; \forall (p,e) \in S_P \times S_E$
$S(x^i, p, e, t_i + d) \neq \emptyset$.

__Assumption 4.__ $\quad \forall \, x \in X \qquad \mathcal{J}^*(x) \neq \emptyset$.

Then we have

__Theorem 1.__ __For the problem stated above, there exists a__ __function__ $V^* : \; Y^* \longrightarrow R$, $\; Y^* \triangleq \{ y : t_o + d \leqslant y_n \leqslant t_f \}$, __such that__ $V_\pi(x, \theta, p^*, e^*) = V^*(\breve{x}(d))$, __for all__ $x \triangleq^f \breve{x}(.) \in X$, __and for all__ $(p^*, e^*) \in \mathcal{J}^*(x)$.

The proof of Theorem 1 is given in (2).

Some Geometric Aspects of a DGWITL

Paths $\prod (\pi^{ij}, C)$ in $R \times G$, $\pi^{ij} \triangleq \tilde{\pi}^{ij}(\tilde{y})$, $\; \tilde{y} \in$ $S'(x^i, p, e, t_j)$, $C \in R$, are defined by

$$\prod (\pi^{ij}, C) \triangleq \{ \xi^k \triangleq (y_o^k, x^k) : \quad y_o^k + V_\pi(x^k, x^j, p, e, \pi^{kj}) = C,$$
$$\pi^{kj} \subseteq \pi^{ij} \}$$

Trajectories $\Gamma (\pi^{ij}, C)$ in $R \times D$ are defined by

$$\Gamma (\pi^{ij}, C) \triangleq \{ \eta^k \triangleq (y_o^k, \tilde{x}^k(d)) : \quad (y_o^k, \tilde{x}^k(.)) \in \prod (\pi^{ij}, C) \}$$

Likewise, for $(p^*, e^*) \in \mathcal{J}^*(x^i)$, let us define in $R \times G$ a P-path $\prod (\pi_P^{ij}, C)$, $\pi_P^{ij} \triangleq \tilde{\pi}^{ij}(\tilde{y}_P)$, $\; \tilde{y}_P \in$ $S'(x^i, p^*, e, t_j)$, an E-path $\prod (\pi_E^{ij}, C)$, $\pi_E^{ij} \triangleq \tilde{\pi}^{ij}(\tilde{y}_E)$, $\tilde{y}_E \in S'(x^i, p, e^*, t_j)$, a PE-path $\prod (\pi_{PE}^{ij}, C)$, $\pi_{PE}^{ij} \triangleq \tilde{\pi}^{ij}(\tilde{y}_{PE})$, $\tilde{y}_{PE} \in S'(x^i, p^*, e^*, t_j)$, by

$$\prod (\pi_P^{ij}, C) \triangleq \{ \xi^k : y_o^k + V_\pi(x^k, x^j, p^*, e, \pi^{kj}) = C,$$
$$\pi^{kj} \subseteq \pi_P^{ij} \}$$

$$\prod (\pi_E^{ij}, C) \triangleq \{ \xi^k : y_o^k + V_\pi(x^k, x^j, p, e^*, \pi^{kj}) = C,$$
$$\pi^{kj} \subseteq \pi_E^{ij} \}$$

$$\Pi(\pi_{PE}^{ij},C) \triangleq \{\xi^k : y_o^k + V_\pi(x^k,x^j,p^*,e^*,\pi^{kj}) = C, \quad \pi^{kj} \leq \pi_{PE}^{ij}\}$$

and a P-trajectory $\Gamma(\pi_P^{ij},C)$, an E-trajectory $\Gamma(\pi_E^{ij},C)$, and a PE-trajectory $\Gamma(\pi_{PE}^{ij},C)$ in $R \times D$, by

$$\Gamma(\pi_P^{ij},C) \triangleq \{\eta^k : (y_o^k,\tilde{x}^k(.)) \in \Pi(\pi_P^{ij},C)\}$$

$$\Gamma(\pi_E^{ij},C) \triangleq \{\eta^k : (y_o^k,\tilde{x}^k(.)) \in \Pi(\pi_E^{ij},C)\}$$

$$\Gamma(\pi_{PE}^{ij},C) \triangleq \{\eta^k : (y_o^k,\tilde{x}^k(.)) \in \Pi(\pi_{PE}^{ij},C)\}$$

PE-paths in $R \times G$ and PE-trajectories in $R \times D$ associated with a terminating path in G are termed optimal paths in $R \times G$ and optimal trajectories in $R \times D$, respectively.

Let us also define trajectories $P(\rho^{ij},C)$ in $R \times D$, $\rho^{ij} \triangleq \tilde{\rho}^{ij}(\tilde{y})$, $\tilde{y} \in C(y^i,\tilde{u},\tilde{v},t_j)$, by

$$P(\rho^{ij},C) \triangleq \{\eta^k \triangleq (y_o^k,y^k) : y_o^k + V_\rho(y^k,y^j,\tilde{u},\tilde{v},\rho^{kj}) = C, \quad \rho^{kj} \leq \rho^{ij}\}$$

For the restricted class of DGWITL defined above, we can define a game surface $\Sigma(C)$, $C \in R$, by

$$\Sigma(C) \triangleq \{\eta \triangleq (y_o,y) : \Phi(\eta) \triangleq y_o + V^*(y) = C\}$$

A given game surface $\Sigma(C)$ separates the set $R \times Y^*$ in two disjoint sets $A/\Sigma(C)$ and $B/\Sigma(C)$; namely

$$A/\Sigma(C) \triangleq \{\eta : \Phi(\eta) > C\}$$

$$B/\Sigma(C) \triangleq \{\eta : \Phi(\eta) < C\}$$

A point $\eta \in A/\Sigma(C)$ will be called an A-point relative to $\Sigma(C)$, and a point $\eta \in B/\Sigma(C)$ a B-point relative to $\Sigma(C)$.

A fundamental property of game surfaces is embodied in

Theorem 2.   (a) <u>No point of a P-trajectory in $R \times D$ emanating from</u> $\eta^i \in \Sigma(C)$ <u>is an A-point relative to</u> $\Sigma(C)$.
(b) <u>No point of an E-trajectory in $R \times D$ emanating from</u> $\eta^i \in \Sigma(C)$ <u>is a B-point relative to</u> $\Sigma(C)$.
(c) <u>All points of a PE-trajectory in $R \times D$ emanating from</u> $\eta^i \in \Sigma(C)$ <u>belong to</u> $\Sigma(C)$.

The proofs of (a) and (b) of Theorem 2 are given in (3). (c) is a direct consequence of (a) and (b).

A Min-Max Principle

For any two functions $m \triangleq \tilde{m}(.)$ and $n \triangleq \tilde{n}(.)$ defined on $[a,b]$ and with ranges in E^n, continuous on $[a,b]$, we shall let

$$d(m,n) \triangleq \sup_{s \in [a,b]} \| \tilde{m}(s) - \tilde{n}(s) \|$$

where $\| . \|$ denotes the euclidean norm in E^n. Indeed, for $x \in X$ we have $[a,b] = [0,d]$.

For $x^k \in X$ and $\alpha > 0$, let $\Delta(x^k, \alpha) \triangleq \{ x : x \in X, d(x,x^k) < \alpha \}$. For $y^k \in D$, let $B(y^k, \alpha) \triangleq \{ y : y \in D, \| y - y^k \| < \alpha \}$, and $T(y^k, \alpha) \triangleq \{ y : y_n = y_n^k, y \in B(y^k, \alpha) \}$
Let $x^i \in X$, x^i at time t_i, $t_i < t_f$. Let $(p^*, e^*) \in \mathcal{J}^*(x^i)$.
Let $\pi^* \triangleq \tilde{\pi}^{if}(y^*)$, $y^* \in S'(x^i, p^*, e^*, t_f)$, and $\rho^* \triangleq h(\pi^*)$.

<u>Assumption 5</u>.   There exists α , $\alpha > 0$ and, for all $x^* \in \pi^*$ there exists $\Delta(x^*, \alpha)$ such that $(p^*, e^*) \in \mathcal{J}^*(x)$ for all $x \in \Delta(x^*, \alpha)$.

<u>Assumption 6</u>.   For all $y^*(t) \in \rho^*$, $t_i \leqslant t < t_f$, there exists a ball $B(y^*(t), \alpha)$ on which p^* and e^* are of class C^1.

<u>Assumption 7</u>.   f is of class C^1 on $D \times U \times V$.

Let $u^* : [t_i, t_f] \longrightarrow U$ and $v^* : [t_i, t_f] \longrightarrow V$ be such that $u^*(t) \triangleq p^*(y^*(t-d))$ and $v^*(t) \triangleq e^*(y^*(t-d))$, for all t, $t \in [t_i, t_f]$. Let $y^i \in D$ such that $y^i_n = t_i$. From the definitions of S_P and S_E and Assumption 3 it follows that $(u^*, v^*) \in J(y^i)$. Then, from Assumptions 6 and 7 it follows that there exists β , $\beta > 0$, such that

$$y^i \in T(y^*(t_i), \beta) \Longrightarrow \tilde{y}(t) \in T(y^*(t), \alpha), \ \forall \ t \in [t_i, t_f]$$

where $\tilde{y} \in C(y^i, u^*, v^*, t_f)$.

Let ρ^{if}_{PE} be the terminating trajectory generated by (u^*, v^*) from $y^i \in T(y^*(t_i), \beta)$, and let

$$P(\ \rho^{if}_{PE}, C') \triangleq \Big\{ \eta^k \triangleq (y^k_o, y^k) : y^k_o + V_\rho (y^k, y^f, u^*, v^*, \rho^{kf}) = C' \\ \rho^{kf} \le \rho^{if}_{PE} \Big\}$$

be the corresponding trajectory in $R \times D$, emanating from $\eta^i \in \Sigma(C)$.

From Assumption 5 one can deduce easily the following

__Theorem 3.__ __All points of__ $P(\ \rho^{if}_{PE}, C')$ __belong to__ $\Sigma(C)$.

Accordingly we have $C' = C$ and, for all $y^k \in \rho^{if}_{PE}$ we have

$$V_\rho (y^k, y^f, u^*, v^*, \rho^{kf}) = V^*(y^k), \qquad \rho^{kf} \le \rho^{if}_{PE} \qquad (4)$$

__Assumption 8.__ w is of class C^1 on E^{n-1}.

__Assumption 9.__ x^i is of class C^1.

From Assumptions 6 and 9 it follows that u^* and v^* are of class C^o on $[t_i, t_f]$ and C^1 on $[t_i, t_i+d]$. Furthermore if $t_i+d < t_f$, u^* and v^* are C^1 on $[t_i+d, t_f]$. Then, it follows from relation (4), and from the fact that there emanates a terminating trajectory, generated by (u^*, v^*), from each point of the ball $T(y^*(t_i), \beta)$,that for all t, $t \in (t_i, t_i+d)$, and $t \in (t_i+d, t_f)$ if $t_i+d < t_f$, there exists a ball $B(y^*(t), \delta)$ on which V^* is twice differentiable. This can be shown by direct computation as for the case of games without retardation. It follows that

$$\text{grad } \Phi \triangleq (1, \partial v^*/\partial y_1, \ \ldots \ \partial v^*/\partial y_n)$$

is defined and continuous on $B(y^*(t), \delta)$.

Finally, by the same arguments as in (1) for games without retardation, one deduces from Theorem 2

Theorem 4. <u>If there exists a strategy pair</u> (p^*,e^*), $(p^*,e^*) \in \mathcal{J}^*(x^i)$, $x^i \in X$, x^i <u>at time</u> t_i, $t_i < t_f$, <u>and if Assumptions 1-9 are satisfied, then, for the problem stated above, there exists a nonzero vector function</u> $\widetilde{\lambda} : [t_i,t_f] \rightarrow E^n$, <u>continuous on</u> $[t_i,t_i+d]$, <u>and on</u> $[t_i+d,t_f]$ <u>if</u> $t_i+d < t_f$, <u>which is a solution of the equation</u>

$$d\lambda /dt = -(\partial f/\partial y)^T \lambda$$

<u>such that</u>

a) $\min\limits_{u \in K_u} H(\widetilde{\lambda}(t), y^*(t), u, e^*(y^*(t-d))) =$

$\qquad \max\limits_{v \in K_v} H(\widetilde{\lambda}(t), y^*(t), p^*(y^*(t-d)), v) =$

$\qquad H(\widetilde{\lambda}(t), y^*(t), p^*(y^*(t-d)), e^*(y^*(t-d)))$

b) $\qquad H(\widetilde{\lambda}(t), y^*(t), p^*(y^*(t-d)), e^*(y^*(t-d)))=0$

<u>for all</u> t, $t \in (t_i,t_i+d)$, <u>and</u> $t \in (t_i+d,t_f)$ <u>if</u> $t_i+d < t_f$; <u>where</u> $H(\lambda ,y,u,v) \triangleq f_o(y,u,v) + \lambda \cdot f(y,u,v)$;

c) $(\widetilde{\lambda}_1(t_f), \ \ldots \ \widetilde{\lambda}_{n-1}(t_f)) = \text{grad } w(y^*(t_f))$

<u>where</u> $\text{grad } w \triangleq (\partial w/\partial y_1, \ \ldots \ \partial w/\partial y_{n-1})$

If $t_i+d < t_f$, the jump condition at point t_i+d can be obtained by invoking the same arguments as in (1), for games without retardation.

References

1. A. Blaquière, F. Gérard and G. Leitmann, "Quantitative and Qualitative Games", Academic Press Inc. (1969).
2. A. Blaquière and P. Caussin, "Jeux différentiels avec retard, démonstration du théorème de base", C. R. Acad. Sc. Paris (to appear).
3. A. Blaquière and P. Caussin, "Jeux différentiels avec retard, propriétés globales d'une surface du jeu", C. R. Acad. Sc. Paris, t. 273, p. 326-328, Série A, (1971).
4. A. Blaquière and G. Leitmann, "Jeux quantitatifs", Mémorial des Sciences Mathématiques, Fasc. 168, Gauthier-Villars, (1969).
5. M. D. Ciletti, "On a Class of Deterministic Differential Games with Imperfect Information", Technical Report No. EE-679, University of Notre Dame, Notre Dame, Indiana, (1967).

STOCHASTIC DIFFERENTIAL GAMES

Avner Friedman

Northwestern University
Evanston, Ill.

1. N-person games with perfect observation

Consider a system of m stochastic differential
equations

$$d\xi = f(t,\xi,y_1,\ldots,y_N)dt + \sigma(t,\xi)dw \qquad (1)$$

for $s \leq t \leq T$, where $w(t)$ is m-dimensional Brownian motion.
The initial condition is

$$\xi(s) = x_0 \qquad (2)$$

where x_0 is a random variable. Let Ω be a bounded domain
in R^m and let $Q = [s,T]\times\Omega$. Denote by τ the exit time of
$\xi(t)$ from Q. (It is assumed that $x_0 \in \Omega$ almost surely.)
The cost functional for the player y_i is

$$J_i(y_1,\ldots,y_N)$$
$$= E\left\{\int_s^\tau h_i(t,\xi,y_1,\ldots,y_N)dt + g_i(\tau,\xi(\tau))\right\}. \qquad (3)$$

We are given a control set Y_i for each player y_i; Y_i is a compact subset of a euclidean space R^{p_i}. Let each player y_i
choose a <u>pure</u> <u>strategy</u>, i.e., a measurable function
$y_i(t,x)$ from $[s,T]\times R^m$ into Y_i. Then (1) becomes

$$d\xi = f(t,\xi,y_1(t,\xi),\ldots,y_N(t,\xi))dt + \sigma(t,\xi)dw. \qquad (4)$$

We assume:

(C) σ, f, h are continuous and bounded and the matrix
$(a_{ij}) = \sigma\sigma^*/2$ is uniformly positive definite; $\partial\Omega$ is in C^2

and g has an extension $\hat{g}$ into $\bar{Q}$ with continuous derivatives

---

This work is partially supported by National Science
Foundation Grant NSF GP-5558.

$D_t \hat{g}$, $D_x \hat{g}$, $D_x^2 \hat{g}$.

According to [13], there exists a unique solution of (4),(2) in a certain sense and (3) is then well defined.

The above scheme is called a game of __perfect observation__. The crucial assumption here is that the players have complete information at each time t regarding the position of $\xi(t)$.

An __equilibrium__ __point__ for such a game is a vector $y^*(t,x) = (y_1^*(t,x),\ldots,y_N^*(t,x))$ of pure strategies such that

$$J_k(y_1^*,\ldots,y_{k-1}^*,y_k,y_{k+1}^*,\ldots,y_N^*) \geq J_k(y_1^*,\ldots,y_N^*) \qquad (5)$$

for any pure strategy, y_k, $1 \leq k \leq N$.

Let p_k be a variable point in R^m and consider the kth __Hamiltonian__ __function__

$$H_k(t,x,y_1,\ldots,y_N,p_k)$$

$$= f(t,x,y_1,\ldots,y_N) \cdot p_k + h_k(t,x,y_1,\ldots,y_N). \qquad (6)$$

Assume the __generalized__ __minimax__ __condition__:

(D) There exist functions $y_1^*(t,x,p),\ldots,y_N^*(t,x,p)$, where $p = (p_1,\ldots,p_N)$, such that:

(i) $y_j^*(t,x,p)$ is measurable in $(t,x) \in Q$ for every p and is continuous in p uniformly with respect to $(t,x) \in Q$;

(ii) $y_j^*(t,x,p) \in Y_j$, and

(iii) for all $(t,x) \in Q$ and for all p,

$$\min_{y_k \in Y_k} H_k(t,x,y_1^*(t,x,p),\ldots,y_{k-1}^*(t,x,p),y_k,y_{k+1}^*(t,x,p),$$

$$\ldots,y_N^*(t,x,p),p_k) = H_k(t,x,y_1^*(t,x,p),\ldots,y_N^*(t,x,p),p_k).$$

__Theorem 1.__ __Assume__ __that__ (C),(D) __hold__ __and__ __that__ $D_x a_{ij}$ __are__ __continuous__ __in__ $\bar{Q}$. __Then__ __there__ __exists__ __an__ __equilibrium__ __point__ __in__ __pure__ __strategies,__ __given__ __by__

$$y_j^*(t,x) = y_j^*(t,x,D_x\phi^*(t,x))$$

__where__ $\phi^* = (\phi_1^*,\ldots,\phi_N^*)$ __is__ __the__ __solution__ __of__ __the__ __semilinear__ __parabolic__ __system__

$$\frac{\partial \phi_k}{\partial t} + \Sigma \, a_{ij} \frac{\partial^2 \phi_k}{\partial x_i \partial x_j} + f(t,x,y^*(t,x,D_x\phi)) \cdot D_x \phi_k \qquad (7)$$

$$+ h_k(t,x,y^*(t,x,D_x\phi)) = 0 \quad \text{in } Q,$$

$$\phi_k = g_k \text{ if } x \in \partial\Omega, \; s < t \leq T \quad \text{or if } x \in \Omega, \; t = T. \tag{8}$$

If $N = 2$ and $J_1 + J_2 = 0$ then (D) can be replaced by the weaker <u>minimax</u> <u>condition</u>:

(D') For any $(t,x) \in Q$ and for every $p \in R^m$,

$$\min_{y_1 \in Y_1} \max_{y_2 \in Y_2} H_1(t,x,y_1,y_2,p) = \max_{y_2 \in Y_2} \min_{y_1 \in Y_1} H_1(t,x,y_1,y_2,p).$$

Now ϕ^* is the solution of

$$\frac{\partial\phi}{\partial t} + \Sigma \, a_{ij} \frac{\partial^2\phi}{\partial x_i \partial x_j} + H(t,x,D_x\phi) = 0, \tag{9}$$

where

$$H(t,x,p) = \min_{y_1 \in Y_1} \max_{y_2 \in Y_2} H_1(t,x,y_1,y_2,p),$$

instead of (7).

The proofs of Theorem 1 and of the last statement are given in Friedman [7]; they employ existence theorems and a priori bounds for parabolic systems obtained in [10]-[12]. The special case of Theorem 1 where $N = 1$ was first treated in Fleming [4].

For $N = 2$, $J_1 + J_2 = 0$, Duncan and Varaiya [3] have considered games with dynamics based on stochastic functional-differential equations

$$dx_t^1 = f^1(t,x^1,y) + dw^1, \quad dx_t^2 = f^2(t,x^2,z) + dw^2 \tag{10}$$

and payoff

$$E \int_{t_0}^{T} c(t,x,y,z)\,dt \quad (x = (x^1,x^2)); \tag{11}$$

here (w^1,w^2) is n-dimensional Brownian motion,

$$f^1 : [t_0,T] \times C^1 \times Y \to R^{n_1},$$

$$f^2 : [t_0,T] \times C^2 \times Z \to R^{n_2},$$

$$c : [t_0,T] \times C^3 \times Y \times Z \to R^1$$

are non-anticipative functionals, C^i is the space of continuous n_i-vector functions, and $n_3 = n = n_1 + n_2$. Defining strategies for y,z as non-anticipative functionals

$$y = y(t,x) : [t_0,T] \times C^3 \to Y,$$

$$z = z(t,x) : [t_0,T] \times C^3 \to Z$$

they proved the existence of a saddle point under the assumption that the sets $f^1(t,x^1,Y)$, $f^2(t,x^2,Z)$ are convex and closed. The functionals $f^1(t,x,y)$, $f^2(t,x^2,y)$ are assumed to be continuous, but neither linear growth in x nor Lipschitz condition in x are assumed. The case N = 1 is also treated in [3]; this case was previously treated by Beneš [1].

Davis and Varaiya [2] have recently derived a necessary and sufficient condition for the saddle point (in [3]) which is a non-Markovian analog of (9).

2. 2-person zero sum games with partial observation

Consider a system of m equations

$$d\xi = f(t,\xi,y,z)dt + \sigma(t,\xi)dw. \tag{12}$$

The payoff is

$$P(y,z) = E\{\int_s^\tau h(t,\xi,y,z)dt + g(\tau,\xi(\tau))\}. \tag{13}$$

y tries to maximize the payoff and z tries to minimize it. The control sets are Y and Z respectively. We now suppose that y and z, at time t, can only observe a quantity $\eta(t)$ which is related to $\xi(t)$ by means of a given dynamics:

$$d\eta = \tilde{f}(t,\xi,\eta,y,z)dt + \tilde{\sigma}(t,\xi,\eta)d\tilde{w}$$

where $\tilde{w}$ is a Brownian motion independent of w. We can then consider the pair (ξ,η) as defining a diffusion process governed by stochastic differential equations and, with respect to these equations, y and z observe precisely the components of η. Thus, simplifying the notation, we may assume at the outset that in the original system (12) y and z observe only

$$\hat{\xi} = (\xi_1,\ldots,\xi_\ell)$$

for some fixed ℓ, $1 \le \ell < m$. A <u>pure strategy</u> for y (z) is now defined as a measurable function $y(t,\hat{\xi})$ $(z(t,\hat{\xi}))$ with values in Y (Z).

It seems doubtful that a saddle point in pure strategies exists for the present game with partial observation. We therefore proceed to introduce a more general concept of strategy.

Let n be any positive integer, $\delta = (T - s)/n$, and $I_j = (t_{j-1} < t \le t_j]$ where $t_j = s + j\delta$. Denote by Y_j the set of all measurable functions $y_j(t,\hat{x})$ from $I_j \times R^m$ into Y. Similarly define Z_j. An <u>upper</u> δ-<u>strategy</u> Γ^δ for y is a vector

$$\Gamma^\delta = (\Gamma^{\delta,1},\ldots,\Gamma^{\delta,n})$$

where $\Gamma^{\delta,j}$ is a map from $Z_1 \times Y_1 \times \cdots \times Z_{j-1} \times Y_{j-1} \times Z_j$ into Y_j. A lower δ-strategy Δ_δ for z is a vector

$$\Delta_\delta = (\Delta_{\delta,1}, \ldots, \Delta_{\delta,n})$$

where $\Delta_{\delta,1}$ is any element of Z_1 and $\Delta_{\delta,j}$ $(j \geq 2)$ is any map from $Z_1 \times Y_1 \times \cdots \times Z_{j-1} \times Y_{j-1}$ into Z_j.

A pair $(\Delta_\delta, \Gamma^\delta)$ determines an outcome $(y^\delta(t,\hat{x}), z_\delta(t,\hat{x}))$ and the corresponding payoff

$$P(y^\delta, z_\delta) \equiv P[\Delta_\delta, \Gamma^\delta] \equiv P[\Delta_{\delta,1}, \Gamma^{\delta,1}, \ldots, \Delta_{\delta,n}, \Gamma^{\delta,n}].$$

The upper δ-value V^δ is defined by

$$V^\delta = \inf_{\Delta_{\delta,1}} \sup_{\Gamma^{\delta,1}} \cdots \inf_{\Delta_{\delta,n}} \sup_{\Gamma^{\delta,n}} P[\Delta_{\delta,1}, \Gamma^{\delta,1}, \ldots, \Delta_{\delta,n}, \Gamma^{\delta,n}].$$

If $\lim_{\delta \to 0} V^\delta$ exists then we call it the upper value of the game and denote it by V^+. Similarly we define the lower value V^-. If $V^+ = V^-$ then we say that the game has value V where $V = V^+ = V^-$.

A sequence $\Delta = \{\Delta_\delta\}$ is called a strategy for z. Similarly we define a strategy $\Gamma = \{\Gamma_\delta\}$ for y. A pair of strategies (Δ, Γ) determines outcomes (y_δ, z_δ) for each δ. If for a subsequence $\{\delta'\}$ of $\{\delta\}$,

$$y_{\delta'}(t,\hat{x}) \longrightarrow \bar{y}(t,\hat{x})$$
$$z_{\delta'}(t,\hat{x}) \longrightarrow \bar{z}(t,\hat{x})$$

weakly where $\bar{y}$, $\bar{z}$ are control functions, and if

$$P(y_{\delta'}, z_{\delta'}) \to P(\bar{y}, \bar{z})$$

then we call $(\bar{y}, \bar{z})$ an outcome of (Δ, Γ); the set of numbers $P(\bar{y}, \bar{z})$ of all outcomes $(\bar{y}, \bar{z})$ is denoted by $P[\Delta, \Gamma]$ and is called the payoff set of (Δ, Γ).

Given sets A, B of real numbers, we write $A \leq B$ if $a \leq b$ for all $a \in A$, $b \in B$ or if A or B are empty. A pair of strategies (Δ^*, Γ^*) is called a saddle point if

$$P[\Delta^*, \Gamma] \leq P[\Delta^*, \Gamma^*] = \{V\} \leq P[\Delta, \Gamma^*] \qquad (14)$$

for all Δ, Γ.

We shall need the condition:

$$(E) \qquad f(t,x,y,z) = f^1(t,x,y) + f^2(t,x,z),$$
$$h(t,x,y,z) = h^1(t,x,y) + h^2(t,x,z).$$

Let $0 < \alpha \leq 1$. If $\partial\Omega$ is in C^2 and the second derivatives in the local representation of $\partial\Omega$ are Hölder continuous with exponent α, then we say that $\partial\Omega$ is in $C^{2+\alpha}$. A function g defined on $[\{s \leq t \leq T\}\times\partial\Omega] \cup [\{t = T\}\times\bar\Omega]$ is said to belong to class $C_\alpha^{2,1}$ if there is a function $\tilde{g}$ which extends g into $\bar{Q}$ such that $D_t\tilde{g}$, $D_x\tilde{g}$, $D_x^2\tilde{g}$ are Hölder continuous with exponent α in x and $\alpha/2$ in t.

$\underline{\text{Theorem 2.}}$ $\underline{\text{Let}}$ (C),(E) $\underline{\text{hold and suppose that}}$ $\partial\Omega$ $\underline{\text{is in}}$ $C^{2+\alpha}$, $g \in C_\alpha^{2,1}$ $\underline{\text{and}}$ $D_x a_{ij}$, $D_t a_{ij}$ $\underline{\text{are continuous in}}$ $\bar{Q}$. $\underline{\text{Then}}$ $\underline{\text{the game (of partial observation) has value.}}$

The proof is given in Friedman [7]. It uses a procedure for deterministic differential games given in [6] together with a priori estimates for parabolic equations [5]. The following theorem is also proved in [7]:

$\underline{\text{Theorem 3.}}$ $\underline{\text{If, in addition to the assumptions made in}}$ $\underline{\text{Theorem 2, f}}$ $\underline{\text{and}}$ $\underline{\text{h}}$ $\underline{\text{are linear in}}$ y,z $\underline{\text{and}}$ Y,Z $\underline{\text{are convex,}}$ $\underline{\text{then there exists a saddle point.}}$

3. More general stochastic differential equations

Consider now, instead of (1), a more general system

$$d\xi = a(t,\xi,y_1,\dots,y_N)dt + \sigma(t,\xi)dw + \int c(t,\xi,\lambda)\tilde{\nu}(dt,d\lambda) \qquad (15)$$

where $\tilde{\nu}([t_1,t_2],A)$ is a random Poisson measure homogeneous of degree 1 in the first variable. We shall take the payoff (3) with $\tau = T$, i.e., with $\Omega = R^m$. The results of Sections 1, 2 then extend to the present case. In fact, the proofs of Theorems 1-3 are based on applying Ito's formula so as to pass to an equivalent setting in terms of parabolic equations. In the present case of (15) we obtain (cf. [9]) in each parabolic equation an additional term acting on the solution ϕ; it has the form

$$M\phi = \int M_\lambda\phi(t,x)\Pi(d\lambda)$$

where

$$M_\lambda\phi(t,x) = \phi(t,x + c(t,x,\lambda)) - \phi(t,x) - D_x\phi(t,x)\cdot c(t,x,\lambda)$$

and $\Pi(d\lambda)$ is a non-negative measure which may be infinite near $\lambda = 0$.

Assuming that

$$\int |c(t,x,\lambda)|^{1+\beta} \Pi(d\lambda) \leq K < \infty,$$

$$\int \left| c(t,x,\lambda) - c(t,\bar{x},\lambda) \right| \Pi(d\lambda) \leq K \left| x - y \right|^{\mu}$$

for some constants $K > 0$, $0 < \mu \leq 1$ and for both $\beta = 0$ and some $\beta \in (0,1]$, we can treat $M\phi$ as a regular perturbation in the parabolic equations in question. Notice however that $M\phi$ is a non-local operator. This forces us to consider the Cauchy problem for the perturbed parabolic system instead of the initial-boundary value problem. It is for this reason that we have taken $\Omega = R^m$, $\tau = T$ in the case of the system (15).

4. Games with perfect observation and non-anticipative controls

In the results of the preceding sections it was essential that $\sigma\sigma^*$ be positive definite. In fact, if this is not the case then, in general, a pair of control functions do not define a payoff $P(y,z)$.

In this section we consider the case where $\sigma\sigma^*$ is not necessarily positive definite.

We shall introduce a model of a zero sum 2-players game with payoff

$$P(y,z) = E\{g(\xi(T)) + \int_s^T h(t,\xi,y,z)dt\} \tag{16}$$

where g and h are random functions $g(x,\omega)$ and $h(t,x,y,z,\omega)$ respectively. Each player makes perfect observation and is free to choose control function ($y(t)$ or $z(t)$) which is non-anticipative in the following sense:

Let $\mathcal{F}_t$ be a given increasing family of σ-fields such that ξ_0 and $w(t)$ are $\mathcal{F}_t$ measurable and for each $s \leq \lambda \leq T$, $w(t+\lambda) - w(\lambda)$ $(0 \leq t \leq T - \lambda)$ is independent of $\mathcal{F}_\lambda$. By a non-anticipative control function $y(t)$ for y we mean a measurable stochastic process $y(t)$ such that, for each t, $y(t)$ is $\mathcal{F}_t$ measurable and $y(t) \in Y$ almost surely. Similarly we define a non-anticipative control function $z(t)$ for z.

We next define spaces Y_j, Z_j analogously to Section 2, but with the elements y_j, z_j being non-anticipative control functions on Ij. We then define upper and lower δ-strategies and δ-value of a game as in Section 2.

We assume that

$$\left| f(t,x,y,z) \right| + \left| \sigma(t,x) \right| \leq K, \tag{17}$$

$$\left| f(t,x,y,t) - f(t,\bar{x},y,z) \right| \leq K \left| x - \bar{x} \right|, \} \tag{18}$$

AVNER FRIEDMAN

$$\left|\sigma(t,x) - \sigma(t,\bar{x})\right| \le K\left|x - \bar{x}\right|.$$

Then (see [9]) for any non-anticipative controls $y(t)$, $z(t)$, the system

$$d\xi = f(t,\xi,y(t),z(t))dt + \sigma(t,\xi)dw(t),$$

$$\xi(s) = \xi_0$$

has a unique solution $\xi(t)$. Using (18) one can show that

$$E\left|\xi(t+h) - \xi(t)\right|^2 \le Ch \quad \text{for any } 0 < h < T - t, \quad (19)$$

where C is a constant independent of the controls.

Assume:

$$\left.\begin{array}{l} h(t,x,y,z) \text{ is uniformly continuous in } t, \\ \left|h(t,x,y,z) \quad h(t,\bar{x},y,z)\right| \le K\left|x-\bar{x}\right|^\alpha \text{ almost surely,} \\ \left|g(x) - g(\bar{x})\right| \le K\left|x - \bar{x}\right|^\alpha \text{ almost surely,} \end{array}\right\} \quad (20)$$

when K, α are positive constants and $\alpha \le 1$.

We can now state:

Theorem 4. <u>Let the conditions (17),(18),(20) and (E)
hold. Then the differential game (with perfect observation
and non-anticipative controls) has value.</u>

The proof is similar to the proof of the existence of value for deterministic differential games of fixed duration. The estimate (19) is used in this proof.

Theorem 4 extends to the case where, for each (t,x,y,z), $f(t,x,y,z)$ is a random variable provided $f(t,x,y,z)$ is measurable and, for each t, $\mathcal{F}_t$ measurable.

Theorem 4 extends to equations of the form (15).

For some measure spaces (for example, for a region of R^n provided with the Lebesgue measure and having surface area zero) Theorem 3 extends to the present game. Thus, if in addition to the assumptions made in Theorem 4, $f(t,x,y,z)$ is linear in x,y,z and Y, Z are convex, then there exists a saddle point.

Finally we mention that if $\tau = T$ and if $y(t)$, $z(t)$ are taken to be non-random control functions then, under some smoothness assumptions on σ,f,h,g, the value is equal to $\lim_{\varepsilon \to 0} W_\varepsilon(s,x_0)$ where W_ε is the solution of the Cauchy problem:

$$\frac{\partial W}{\partial t} + \Sigma(b_{ij} + \varepsilon\delta_{ij})W_{x_i x_j}$$

$$+ \min_{z \in Z} \max_{y \in Y}\{f(t,x,y,z)\cdot\nabla_x W + h(t,x,y,z)\} = 0$$

306

in $s \leq t < T$,

$$W(T,x) = g(x) \text{ in } R^m.$$

The assertion is true even if (E) is replaced by the mini-max condition. The proof for $\sigma \equiv 0$ is given in Friedman [8]; the proof for general σ is similar.

References

[1] V. E. Beneš, SIAM J. Control 9, 446 (1971).

[2] M. H. A. Davis and P. P. Varaiya, to appear.

[3] T. E. Duncan and P. P. Varaiya, SIAM J. Control 9, 354 (1971).

[4] W. H. Fleming, J. Math. Mech. 12, 131 (1963).

[5] A. Friedman, Partial Differential Equations of Parabolic Type, Prentice-Hall, 1964, Englewood Cliffs, New Jersey.

[6] A. Friedman, Differential Games, Wiley-Interscience, 1971, New York.

[7] A. Friedman, J. Diff. Equations, to appear.

[8] A. Friedman, to appear.

[9] I. I. Gikhman and A. V. Skorokhod, Stochastic Differential Equations, Kiev, 1968.

[10] O. A. Ladyzhenskaja, V. A. Solonnikov and N. N. Ural'ceva, Linear and Quasilinear Equations of Parabolic Type, Translations Amer. Math. Soc., vol. 23, 1968, Providence, R.I.

[11] V. A. Solonnikov, Trudy Mat. Inst. Steklova 10, 133 (1964); Amer. Math. Soc. Translations (2), 65, 51 (1967).

[12] V. A. Solonnikov, Trudy Mat. Inst. Steklova 83, 3 (1965).

[13] D. W. Stroock and S. R. S. Varadhan, Comm. Pure Appl. Math. 22, 345, 749 (1969).

STOCHASTIC CONTROL

$$\text{STOCHASTIC MODELS}^{*}$$

E. J. McShane[+]

A dynamical system with state-vector x, affected at time t by noises $\dot{z}^{\rho}(t)$ ($\rho = 1, \ldots, r$), will often evolve in accordance with an equation

$$(1) \quad x(t) = x(a) + \int_a^t f(s,x(s))ds + \sum_{\rho=1}^r \int_a^t g_\rho(s,x(s))dz^\rho(s) \; ;$$

here, as always, x, f and g_ρ denote n-vectors, and each z^ρ is real-valued. But the scientific theory that gives us equation (1) is untrustworthy except for physically realizable z^ρ, and these are at least restricted to be Lipschitzian. On the other hand, in investigating the statistics of solutions of (1) with random noises, engineers have usually preferred to idealize (1) by assuming "white noises", that is Brownian-motion z^ρ. Equation (1) can then no longer be interpreted as usual, with Riemann-Stieltjes integrals. If we interpret the last r integrals in (1) as Itô integrals, meaning is restored, but not necessarily credibility, for the sample functions of the z^ρ are nowhere differentiable w.p.1, and so are not physically realizable. But except in the very simple case in which the g_ρ are independent of x, two major difficulties arise. If we change from x-coordinates to new y-coordinates by a smooth transformation, equation (1) with Lipschitzian z^ρ transforms in a familiar way, forced on us by the requirement that the solutions of the transformed equation be the transforms of the solutions of the original equations. But if we now interpret all the integrals with respect to dz^ρ as Itô integrals, the solutions of the transformed equation are not the transforms of the solutions of the original equation. The procedure of modeling by retaining the form of (1) and

------

*Research supported by Army Research Grant ARO-D-1004.

+Department of Mathematics, University of Virginia, Charlottesville, Virginia.

interpreting the dz^ρ-integrals as Itô integrals fails, because the purported solution depends on the coordinate system we choose to use. Besides this, the solution is unstable, in the sense that there exist sequences z_1^ρ, z_2^ρ, ... ($\rho = 1$, ..., r) with z_j^ρ converging uniformly to z^ρ for each ω, but the solutions x_1, x_2, ... corresponding to them do not converge to the solution of (1). In addition to these major defects, we have the inconvenience that the theory lacks unity; there is one definition of the integral when z is Lipschitzian, another when z is the kind of martingale for which the Itô integral is defined, and none at all for some integral-symbols that have been used and called "Itô integrals" in spite of the fact that the z is not even a martingale.

In several papers ([1],[2]) and in a book now in process of publication ([3]) we have developed a stochastic calculus free of the last-mentioned inconvenience. The family of processes z that we allow, and call "admissible disturbances", are those that satisfy the following conditions.

(2) There is a probability triple (Ω,A,P) and a time-interval [a,b] such that z is real-valued on the set $\{a \leqq t \leqq b, \ \omega \in \Omega\}$.

(3) There is a family $\{F[t]: -\infty < t \leq b\}$ of σ-subalgebras of A such that if $s \leqq t \leqq b$ then $F[s] \subset F[t]$; and z is adapted to $F[\cdot]$.

(4) z is sample-continuous w.p.1; i.e., with probability 1, $z(\cdot,\omega)$ is continuous on [a,b].

(5) There exist a $K > 0$ and a function ψ on $[0,\infty)$, with $\psi(r) \to 0$ as $r \to 0$, such that when $a \leqq s \leqq t \leqq b$ we have w.p.1

$$|E(z(t)-z(s)|F[s])| \leqq K(t-s)$$

$$E([z(t)-z(s)]^2|F[s]) \leqq K(t-s),$$

$$E([z(t)-z(s)]^6|F[s]) \leqq \psi(t-s).$$

All Brownian motions, and all processes with uniformly Lipschitzian sample functions, are admissible disturbances.

In our modeling of dynamical systems we shall need stochastic integrals of first and second order. Their definitions differ from that of the Riemann integral in a small but vital respect. Let f be a process defined on a set D_f that contains $[a,b]$. By a "belated partition" Π of $[a,b]$ we mean a finite sequence of points $(t_1, t_2, \ldots, t_{k+1}, \tau_1, \tau_2, \ldots, \tau_k)$ such that $a = t_1 < t_2 < \ldots < t_{k+1} = b$, and all τ_j are in D_f, and $\tau_j \leq t_j$ $(j = 1, \ldots, k)$. If $z^1, \ldots, z^q$ are processes on $[a,b]$, for each Π we form the "Riemann sums"

$$S(\Pi; f, z^1, \ldots, z^q) = \sum_{i=1}^{k} f(\tau_i)[z^1(t_{i+1}) - z^2(t_i)] \ldots$$

$$[z^q(t_{i+1}) - z^q(t_i)] .$$

If these converge in probability to a random variable J as $\max_i(t_{i+1} - \tau_i)$ tends to 0, J is called (a version of) the q-th order integral

$$\int_a^b f(t) dz^1(t) \ldots dz^q(t) .$$

It is well known that the Riemann definition (with τ_i free to vary in $[t_i, t_{i+1}]$) furnishes nothing useful. In contrast, for f adapted to $F[\cdot]$, the belated integral exists under several sets of hypotheses all resembling the "bounded almost-everywhere-continuous" condition for ordinary Riemann integrability. For instance, it is enough to assume that there is a random variable B such that w.p.1, $|f(\tau, \omega)| \leq B(\omega)$ for all τ in D_f, and that f is continuous in probability at almost points of $[a,b]$. In particular, for such f the q-th order integral is 0 if $q \geq 3$; and if $z^1(\cdot, \omega)$ is Lipschitzian w.p.1, $\int f dz^1 dz^2$ is 0.

For these integrals we can develop an integral calculus which is a direct extension of ordinary calculus, with computation rules that reduce to those of ordinary calculus when f and z^1 are deterministic and z^1 Lipschitzian. This includes a theory of equations

$$(6) \quad x(t) = x(a) + \int_a^t f(s,x(s))ds + \sum_{\rho=1}^r \int_a^t g_\rho(s,x(s))dz^\rho(s)$$

$$+ \sum_{\rho,\sigma} \int_a^t h_{\rho\sigma}(s,x(s))dz^\rho dz^\sigma$$

(which we shall call "differential equations"), wherein the
f, g_ρ and $h_{\rho\sigma}$ are as usual assumed to be continuous in all
variables and Lipschitzian in x on $[a,b]$ R^n. For $F[a]$-
measurable initial values $x(a)$, equations (6) have sample-
continuous solutions that coincide with the solutions of
(1) whenever $z(\cdot,\omega)$ happens to be Lipschitzian. Henceforth,
when we speak of a solution of (6), we shall always mean one
with those properties.

If (1) is a satisfactory mathematical model for a dynamical system when the z^ρ are Lipschitzian, so is (6),
since the second-order integral is then 0. For the extension of the model to all admissible disturbances, (6)
offers us infinitely many possibilities, all consistent with
(1) for the physically realizable Lipschitzian disturbances. The traditional choice $h_{\rho\sigma} = 0$ offers typographical
simplicity along with the major defects already described.
We have proposed instead the choice

$$(7) \quad h_{\rho\sigma}^i(t,x) = \sum_{j=1}^n g_{\rho,x^j}^i(t,x)g_\sigma^j(t,x)/2 \; .$$

With these $h_{\rho\sigma}$, (6) is said to be in "selected form".

If we change from x-coordinates to new y-coordinates,
the new first-order integrals have a form forced on us by
ordinary calculus, applied in the case of Lipschitzian z^ρ.
The selection rule (7) then prescribes the form of the
second-order integrals. Our stochastic calculus contains
a generalization of Itô's differentiation lemma. Using it,
we readily show that the transforms of the solutions of the
selected-form equations in x-coordinates are solutions of
the selected-form equations in y-coordinates. The first of
the major defects has been removed.

The principal purpose of this paper is to show that
when there is only one disturbance z^1, the second defect is
removed in a satisfactory way; changes in z^1 that are small
in the uniform metric produce changes in x that are small

in that metric. This generalizes a theorem of Wong and
Zakai ([5]), in which n = 1 and z is Brownian motion. Such
strong stability is not in general present if there are two
or more disturbances, even when we restrict ourselves to
processes with continuously differentiable sample functions.

We shall henceforth assume

(8) G is the intersection of the strip $\{(t,x): a \leqq t \leqq b,$
   $x \in R^n\}$ with an open set in R^{n+1}; and f^i, g^i (i = 1,
   ..., n) are defined on G and are continuous together
   with their partial derivatives of first and second
   order.

Since ρ and σ can have only the single value 1, we
omit them. In place of (1) we have

$$(9) \quad x(t) = x(a) + \int_a^t f(s,x(s))ds + \int_a^t g(s,x(s))dz(s),$$

and for the corresponding equation in selected form we have

$$(10) \quad x^i(t) = x^i(a) + \int_a^t f^i(s,x(s))ds + \int_a^t g^i(s,x(s))dz(s)$$
$$+ (1/2)\int_a^b [\sum_{j=1}^n g^i_{x^j}(s,x(s))g^j(s,x(s))](dz(s))^2 .$$

An (n+1)-tuple (z,x) of real functions on [a,b] will
be called a "Lipschitzian solution-set" for (9) if z is
Lipschitzian, and (t,x(t)) is in G for all t in [a,b], and
(9) is satisfied. An (n+1)-tuple (z,x) of real functions
on [a,b] is called a "limit-set (for (9))" if (t,x(t)) is
in G for all t in [a,b], and there is a sequence (z_j,x_j) of
Lipschitzian solution-sets converging uniformly to (z,x).

THEOREM I. <u>Let</u> f^i <u>and</u> g^i (i = 1, ..., n) <u>satisfy</u> (8),
<u>and let</u> $(\bar{z},\bar{x})$ <u>be a limit-set for</u> (9) <u>such that</u> $|g(t,\bar{x}(t))|$
> 0 (a $\leqq$ t $\leqq$ b). <u>Then there exist positive numbers</u> δ,L
<u>such that</u>

(i) <u>If</u> $\bar{x}_1$ <u>is an n-vector and</u> z_1 <u>is continuous on</u> [a,b] <u>and</u>
   $|\bar{x}_1 - \bar{x}(a)| < \delta$ <u>and</u> $||z_1 - \bar{z}||_\infty < \delta$, <u>there is a unique</u>
   $x_1(\cdot)$ <u>such that</u> (z_1,x_1) <u>is a limit-set with</u> x(a) = $\bar{x}_1$;

(ii) <u>If</u> $\bar{x}_1,\bar{x}_2$ <u>are n-vectors and</u> z_1,z_2 <u>are continuous on</u>

[a,b] __and__

$$|\bar{x}_\alpha - \tilde{x}(a)| < \delta, \quad ||z_\alpha - \tilde{z}||_\infty < \delta \quad (\alpha = 1,2),$$

the corresponding limit-sets (z_1, x_1), (z_2, x_2) __with__
$x_\alpha(a) = \bar{x}_\alpha \; (\alpha = 1,2)$ __satisfy__

(11) $$||x_1 - x_2||_\infty \leqq L[\, |\bar{x}_1 - \bar{x}_2| + ||z_1 - z_2||_\infty].$$

We first prove this under additional hypotheses, which we later remove:

(12) G is the whole strip $\{(t,x): a \leqq t \leqq b, \; x \in R^n\}$, and f and g are Lipschitzian on G.

(13) There is a positive number β and a number k in $\{1, \ldots, n\}$ such that

$$g^k(t,x) \geqq \beta \quad ((t,x) \text{ in } G) .$$

(14) The z_1 and z_2 in conclusions (i) and (ii) are Lipschitzian.

To simplify notation we assume that $k = 1$ in (13). For each t in [a,b] and each $y^2, \ldots, y^n$ in R there are unique functions

(15) $$(\phi^i(y^1, y^2, \ldots, y^n; t) : -\infty < y^1 < \infty)$$

that satisfy the equations

(16) $$d\phi^i/dy^1 = g^i(t, \; \phi(y^1, y^2, \ldots, y^n; t))$$

and have initial values

(17) $$\phi^1(0, y^2, \ldots, y^n; t) = 0 ,$$

$$\phi^j(0, y^2, \ldots, y^n; t) = y^j \quad (j = 2, \ldots, n) .$$

(Where the range of i is unspecified, as in (15) and (16), it is 1, ..., n.) For each x in R^n there is a single solution of (16) that runs through x. By (13) there is a unique point on this solution-curve with first coordinate 0; we change parameter by an additive constant so that this point has parameter 0. The other coordinates of the point with first coordinate 0 we call $(y^2, \ldots, y^n)$. Then the solution is $\Phi(y^1, y^2, \ldots, y^n; t)$. That is, for each fixed t the map

$$x = \Phi(y;\ t)$$

maps R^n onto itself. There is therefore an inverse map

$$y = \Psi(x;\ t) \quad (x \in R^n) \ .$$

By differentiation with respect to y^j ($j = 1, \ldots, n$) in (16) we see that the Φ_{y^j}, regarded as functions of y^1, satisfy a system of linear differential equations. By (17), the determinant of the $\Phi^i_{y^j}$ at $y^1 = 0$ is

$$\Phi^1_{y^1}(0,y^2,\ldots,y^n;t) = g^1(t,\Phi(0,y^2,\ldots,y^n;t)) \geqq \beta > 0,$$

so the determinant is everywhere non-vanishing. By the implicit functions theorem, the $\Psi(y;t)$ are also twice continuously differentiable.

When (z,x) is any Lipschitzian solution-set or limit-set for (9) we define the corresponding $y(\cdot)$ by

$$(18) \qquad y^i(t) = \Psi^i(x(t);t) \quad (a \leqq t \leqq b),$$

which is equivalent to

$$(19) \qquad x^i(t) = \Phi^i(y(t);t) \quad (a \leqq t \leqq b) \ .$$

If (z,x) is a Lipschitzian solution-set, we find by straightforward computation that the y of (18) satisfies

$$(20) \quad y^i(t) = y^i(a) + \int_a^t F^i(s,y(s))ds + \delta^i_1[z(t)-z(a)],$$

where $\qquad \delta^1_1 = 1,\ \delta^i_1 = 0$ if $i > 1$, and

$$(21) \qquad F^i(t,y) = \Psi^i_t(\Phi(y;t);t)$$

$$+ \sum_{k=1}^{n} \Psi^i_{x^k}(\Phi(y;t);t) f^k(t,\Phi(y;t)).$$

Fix on some $\gamma > 0$, and define

$$(22) \qquad S_\gamma = \{(t,y): a \leq t \leq b, \ |y-\tilde{y}(t)| < \gamma \}.$$

Then F is Lipschitzian on $\bar{S_\gamma}$, say with constant L_γ; and $\psi(x;t)$ is Lipschitzian on the image $\Phi(\bar{S_\gamma})$, say with constant L'_γ. If (z_1,x_1) and (z_2,x_2) are Lipschitzian solution-sets with the graphs of y_1 and y_2 in S_γ, by (20)

$$(23) \qquad |y_1(t)-y_2(t)| \leq |y_1(a)-y_2(a)|$$

$$+ \int_a^t |F(s,y_1(s))-F(s,y_2(s))|\,ds + |z_1(t)-z_2(t)|$$

$$\leq L'_\gamma|x_1(a)-x_2(a)| + ||z_1-z_2||_\infty$$

$$+ L_\gamma \int_a^t |y_1(s)-y_2(s)|\,ds \ .$$

By a form of Gronwall's lemma,

$$(24) \qquad |y_1(t)-y_2(t) \leq [L'_\gamma|x_1(a)-x_2(a)|$$

$$+ ||z_1-z_2||_\infty] \exp L_\gamma(t-a) \ .$$

We can and do choose a positive δ such that

$$(L'_\gamma+1)\delta \exp L_\gamma(b-a) < \gamma \ .$$

If $\bar{x}_1$ and z_1 are as in conclusion (i), (9) has a solution (z_1,x_1) with $x_1(a) = \bar{x}_1$. Let c be the upper bound of numbers c' in $[a,b]$ such that $(t,y_1(t))$ (defined by (18)) is in S_γ when $a \leq t \leq c'$. Let (z_j,x_j) be a sequence of Lipschitzian solution-sets converging uniformly to $(\tilde{z},\tilde{x})$. For all large j the graph of $y_j(t)$ is in S_γ, and by (24) for all $c' < c$

$$|y_1(t)-y_j(t)| \leq (L'_\gamma+1) \exp L_\gamma(b-a) \quad (a \leq t \leq c).$$

Since y_j converges uniformly to $\tilde{y}$, this holds with $\tilde{y}$ in place of y_j. By continuity

$$|y_1(t)-\tilde{y}(t)| \leq (L'_\gamma+1)\delta \exp L_\gamma(b-a) < \gamma \quad (a \leq t \leq c).$$

This contradicts the definition of c unless c = b. So (i)
is established. Now (ii) follows from (24). The proof is
complete under the supplementary hypotheses (12), (13) and
(14); and we have incidentally shown that (11) holds if
(z_1, x_1) is a Lipschitzian solution-set and $(z_2, x_2) = (\tilde{z}, \tilde{x})$.

The removal of the supplementary hypotheses is a rou-
tine matter. We first abandon (12) and replace (13) by:
(25)   There is a k in $\{1, \ldots, n\}$ such that

$$g^k(t, x(t)) > 0 \quad (a \leqq t \leqq b).$$

If (25) holds, the left member has a lower bound $3\beta > 0$.
So there is a positive γ such that $g^k(t, x) \geqq 2\beta$ on the set
$E_\gamma = \{(t, x): a \leqq t \leqq b, |x - \tilde{x}(t)) \leqq \gamma\}$. There exist twice-
continuously-differentiable functions f_0, g_0, Lipschitzian
on the whole strip $\{(t, x): a \leqq t \leqq b\}$, having $g_0^k(t, x) \geqq \beta$
on that strip, and with $f_0 = f$ and $g_0 = g$ on E_γ. We apply
the result already proved to equation (9) with f_0, g_0 in
place of f, g. If $|x_1 - \tilde{x}(a)|$ and $||z_1 - \tilde{z}||_\infty$ are small enough,
(z_1, x_1) will lie in E_γ, and therefore satisfy (9).

A similar result holds, by change of coordinates, if
the inequality in (25) is replaced by $g^k(t, \tilde{x}(t)) < 0$. Under the hypotheses of Theorem I, $[a, b]$ can be subdivided by
points $t_0 = a < t_1 < \ldots < t_h = b$ into subintervals such
that on each $[t_{j-1}, t_j]$ some one of the functions $g^k(t, \tilde{x}(t))$
is non-vanishing. So on each $[t_{j-1}, t_j]$ the conclusions of
the theorem hold with some $\delta_j > 0$ and $L_j \geqq 1$. There is no
difficulty in piecing these conclusions together to obtain
the conclusions of Theorem I.

We still must remove the supplementary hypothesis (14).
If $\bar{x}_1$ and $\bar{x}_2$ satisfy $|\bar{x}_\alpha - \tilde{x}(a)| < \delta$, and z_1 and z_2 are con-
tinuous on $[a, b]$ and satisfy $||z_\alpha - \tilde{z}||_\infty < \delta$, we can approxi-
mate z_1 uniformly by a sequence $z_3, z_5, z_7, \ldots$ of Lipschitzian functions, and similarly approximate z_2 by $z_4, z_6,$
$z_8, \ldots$ For all large j we can apply the conclusions already established to $\bar{x}_1, \bar{x}_2, z_{2j+1}$ and z_{2j+2} to obtain
Lipschitzian pairs (z_j, x_j), for which inequality (11) holds.
In particular, the sequence $x_3, x_5, x_7, \ldots$ satisfies the
condition for uniform convergence; we denote its limit by
x_1. Likewise x_{2j} converges uniformly to a limit x_2. Again
by (11) we have

$$||x_{2j+1}-x_{2j+2}||_\infty \leq L[|\bar{x}_1-\bar{x}_2| + ||z_{2j+1}-z_{2j+2}||_\infty],$$

whence (11) (as written) follows. The uniqueness of the solution follows from (11), and the proof is complete.

The application of Theorem I to stochastic differential equations is by way of a theorem, proved in ([3]), and closely related to one proved by Wong and Zakai ([5]). Let the coefficients in (6) satisfy the requirements listed just after that equation, and let z^1, ..., z^r be admissible disturbances. For each partition Π of $[a,b]$ let $z_\Pi^\rho(\cdot,\omega)$ be the piecewise linear function that coincides with $z^\rho(\cdot,\omega)$ at the division-points of Π, and let x_Π be the solution of (6) (that is, of (1)) with z_Π^ρ in place of z^ρ. Then if (6) has selected form, the $h_{\rho\sigma}^i$ satisfying (7), from any sequence of partitions with maximum interval tending to 0 we can select a subsequence $\Pi(j)$ ($j = 1,2,...$) such that the $x_{\Pi(j)}(\cdot,\omega)$ converge uniformly to $x(\cdot,\omega)$ w.p.1. That is, w.p.1 the set $(z(\cdot,\omega),x(\cdot,\omega))$ furnished by (6) is a limit-set for (1). So from Theorem I we obtain two corollaries.

Corollary 1. If (6) has selected form, then w.p.1, if $\bar{x}_j \to x(a,\omega)$ and $z_j(\cdot)$ is Lipschitzian and converges uniformly to $z(\cdot,\omega)$, the solution of

$$x_j(t) = \bar{x}_j + \int_a^t f(s,x_j(s))ds + \int_a^t g(s,x_j(s))dz_j(s)$$

converges to the sample-function $(z(\cdot,\omega),x(\cdot,\omega))$ of the solution of (6), uniformly on $[a,b]$.

Corollary 2. If (6) has selected form, there is a subset Ω_0 of Ω with $P(\Omega_0) = 1$ such that if ω_0, ω_1, ω_2, ... are in Ω_0, and $x(a,\omega_j) \to x(a,\omega_0)$, and $z(\cdot,\omega_j)$ converges uniformly to $z(\cdot,\omega_0)$, then $x(\cdot,\omega_j)$ converges uniformly to $x(\cdot,\omega_0)$.

As another example we consider one of the simplest of the estimations studied by W. M. Wonham in ([6]). Here $\dot{x}$ is a random variable assuming values $+1$ and -1, each with probability 1/2. One observes

$$y(t) = xt + \int_a^t \beta(s)dz(s) \quad (t \geq 0),$$

where β is continously differentiable and $\beta(s) \geqq \beta_0 > 0$ on $[0,\infty)$. Wonham shows that when z is Brownian motion, the difference $q(t)$ between the conditional probabilities of $x = +1$ and $x = -1$, given $(y(s): 0 \leqq s \leqq t)$, satisfies

$$(26) \qquad q(t,\omega) = \tanh \int_o^t \beta(s)^{-2} \, dy \ .$$

(The last integral is not in fact an Itô integral, since y is not a martingale; but it can be taken to be a belated integral.) We seek a differential equation that generates $q(\cdot)$ as $y(\cdot)$ becomes known. By the Itô differentiation lemma, this is

$$(27) \qquad q(t) = - \int_o^t q(s)(1-q(s)^2)\beta(s)^{-2} ds$$

$$+ \int_o^t (1-q(s))^2 \beta(s)^{-2} dy(s).$$

This is not in selected form. But if we use a substitution theorem proved in ([3]), together with a well-known property of Brownian motion expressed by $(dz)^2 = dt$, we obtain

$$(1/2) \int_a^t [\partial(1-q^2)\beta^{-2}/\partial q](1-q^2)\beta^{-2}(dy^2)$$

$$= - \int_o^t q(1-q^2)\beta^{-2} ds.$$

So (27) can be written as

$$q(t) = \int_o^t (1-q(s)^2)\beta(s)^{-2} dy(s)$$

$$+ \int_o^t h(s)(dy(s))^2 \ ,$$

with h given by (7). If for any b we take $G = \{(t,q): 0 \leqq t \leqq b, -1 < q < +1\}$, the hypotheses of Theorem I hold. Since by (26) we always have $|q| < 1$, all sample functions of the solution of (27) have their graphs in G, and so are uniformly closely approximated by the solutions of

$$\tilde{q}(t) = \int_o^t (1-\tilde{q}(s)^2)\beta(s)^{-2} d\tilde{y}(s)$$

with $\tilde{y}(\cdot)$ Lipschitzian and uniformly close to $y(\cdot)$.

Theorem I does not apply to linear stochastic differential equations in which the coefficients as well as the disturbances are random functions. To include this case also, we state and sketch the proof of an extension of Theorem I in which the f and g, as well as the $x(a)$ and $z(\cdot)$, are permitted to change. Suppose that $\tilde{f}^i$ and $\tilde{g}^i$ satisfy (8), and that $(\tilde{z},\tilde{x})$ is a limit-set for (9) with $\tilde{f},\tilde{g},\tilde{x}(a),\tilde{z}$ in place of $f,g,x(a),z$. Let γ be a positive number (or $+\infty$) such that the closure of the set

$$E(\gamma) = \{(t,x): a \leqq t \leqq b, \ |x-\tilde{x}(t)| < \gamma\}$$

is contained in G. If x is an n-vector and z is continuous and f,g as in (6), we define $||(x,z,f,g)||_\gamma^\sim$ to be the greatest of the 2n+6 numbers $|x|$, $||z||_\infty$, and the suprema of the absolute values of f, g and their first partial derivatives on $E(\gamma)$.

THEOREM II. <u>Let</u> $\tilde{f}^i$, $\tilde{g}^i$ <u>satisfy</u> (8), <u>and let</u> $(\tilde{z},\tilde{x})$ <u>be a limit-set for</u> (9). <u>Then there are positive numbers</u> γ, δ <u>and L such that</u>

(i) <u>if</u> x_1, z_1, f_1, g_1 <u>are such that</u> $||(\bar{x}_1-\tilde{x}(a),z_1-\tilde{z},f_1-\tilde{f},g_1-\tilde{g})||_\gamma^\sim < \delta$, <u>there is a unique</u> $x_1(\cdot)$ <u>such that</u> (z_1,x_1) <u>is a limit-set for</u> (9) <u>with</u> x_1, z_1, f_1, g_1 <u>in place of</u> $x(a)$, z, f, g; <u>and the graph of</u> $x_1(\cdot)$ <u>is in</u> $E(\gamma)$;

(ii) <u>if</u> $(\bar{x}_\alpha,z_\alpha,f_\alpha,g_\alpha)$ $(\alpha=1,2)$ <u>both satisfy</u> $||(\bar{x}_\alpha-\tilde{x}(a),z_\alpha-\tilde{z},f_\alpha-\tilde{f},g_\alpha-\tilde{g})||_\gamma^\sim < \delta$, <u>then the corresponding limit-sets</u> (z_α,x_α) <u>satisfy</u>

$$||x_1-x_2||_\infty \leqq L||(\bar{x}_1-\bar{x}_2,z_1-z_2,f_1-f_2,g_1-g_2)||_\gamma^\sim .$$

We again begin by making the supplementary hypotheses (12), (13) and (14) for $\tilde{f}$ and $\tilde{g}$, and we take $\gamma = \infty$. If f_1, g_1 satisfy (12) and the inequality in (i) holds with $\delta = \beta/2$, (13) (with $\beta/2$) holds for f_1, g_1 also. The mapping Φ defined by (16) with $\tilde{g}$ in place of g we denote by $\tilde{\Phi}$, and we define Φ_1 analogously; their inverses are $\tilde{\Psi}$ and $\tilde{\Psi}_1$. If x_1 is the solution of (9) with $x_1(a)$, z_1, f_1, g_1 in place of $x(a)$, z, f, g, and

$$y_1(t) = \tilde{\Psi}_1(x_1(t);t) \quad (a \leqq t \leqq b),$$

then as in proving (20) we show that

$$y_1^i(t) = y^i(a) + \int_a^t F_1^i(s,y(s))ds + \delta_1^i[z_1(t)-z_1(a)],$$

where F_1^i is defined by (21) with Φ_1, Ψ_1 in placed Φ, Ψ. We define and treat y_2 similarly. If we define S_γ as in (22), as long as $(s,y_1(s))$ and $(s,y_2(s))$ are in S_γ for $a \leqq s \leqq t$ we have

$$|y_1(t)-y_2(t)| \leqq |y_1(a)-y_2(a)| + |z_1(t)-z_2(t)|$$

$$+ \int_a^t |F_1(s,y_1(s)) - F_1(s,y_2(s))|\,ds$$

$$+ \int_a^t |F_1(s,y_2(s)) - F_2(s,y_2(s))|\,ds.$$

But $\Phi_1-\Phi_2$ and its derivates do not exceed a constant multiple of $||(\bar{x}_1-\bar{x}_2,z_1-z_2,f_1-f_2,g_1-g_2)||_\infty^\sim$ on S_γ, and similarly for $\psi_1-\psi_2$ on $\Phi(S_\gamma)$, so $|F_1-F_2|$ behaves similarly on S_γ. So by Gronwall's lemma, for some constant L_γ' and L_γ

$$|y_1(t)-y_2(t)| \leqq [L_\gamma'||(x_1-x_2,z_1-z_2,f_1-f_2,g_1-g_2)||_\infty^\sim].$$

$$\exp L_\gamma(t-a) \ .$$

This replaces (24), and the rest of the proof closely resembles that of Theorem I.

References

1. E. J. McShane, _Toward a stochastic calculus_, parts I and II; Proc. Nat. Acad. Sci. vol. 63 (1969), pp. 275-280 and 1084-1087.

2. E. J. McShane, _Stochastic Differential Equations and Models of Random Processes_, to appear in Proc. VI Berkeley Symposium on Mathematical Statistics and Probability.

3. E. J. McShane, _Stochastic Calculus and Stochastic Models_, Holt, Rinehart and Winston, to appear in 1972.

4.   E. Wong and M. Zakai, <u>On the relation between ordinary and stochastic differential equations</u>, Int. Jour. Engng. Sci. vol. 3 (1965), pp. 213-219.

5.   E. Wong and M. Zakai, <u>On the convergence of ordinary integrals to stochastic integrals</u>, Annals of Math. Stat. vol. 36 (1965), pp. 1560-1564.

6.   W. M. Wonham, <u>Some applications of stochastic differential equations to optimal non-linear filtering</u>, J. SIAM Control, ser. A vol. 2 (1965), pp. 347-369.

DYNAMICAL SYSTEMS WITH SMALL STOCHASTIC TERMS

Wendell H. Fleming[*]

Brown University
Providence, Rhode Island

1. Introduction

Consider a dynamical system in n-dimensional R^n, governed by the ordinary differential equations

$$d\xi^o = f[\xi^o(t)]dt. \qquad (1.1)$$

The vector $\xi^o(t) = (\xi_1^o(t), \ldots, \xi_n^o(t))$ is called the state at time t. We are concerned with stochastic perturbations of (1.1), obtained by adding a white noise term multiplied by a coefficient depending on a small positive parameter ϵ. Specifically, we consider Markov diffusion processes ξ^ϵ which satisfy the stochastic differential equations

$$d\xi^\epsilon = f[\xi^\epsilon(t)]dt + (2\epsilon)^{\frac{1}{2}}\sigma[\xi^\epsilon(t)]dw, \qquad (1.2)$$

where w is a brownian motion of some finite dimension. For both (1.1) and (1.2) a given vector x is prescribed as initial data:

$$\xi^\epsilon(0) = \xi^o(0) = x. \qquad (1.3)$$

Many interesting properties of the process ξ^ϵ are determined by expectations $E_x\Phi[\xi^\epsilon(t)]$ for various choices of Φ {the notation E_x is used rather than (say) ξ_x^ϵ to indicate dependence of ξ^ϵ on the initial data (1.3)}. For instance, if $\Phi(x) = x_i^\ell$, then $E_x\Phi[\xi^\epsilon(t)]$ is the ℓth

[*] This research was supported in parts by the Air Force Office of Scientific Research under grant AF-AFOSR 71-2078, and National Science Foundation under grant GP 15132.

moment of the component $\xi_i^\epsilon(t)$ of the vector $\xi^\epsilon(t)$, $i = 1,\ldots,n$. Depending on the structure of the unperturbed system (1.1), various questions are of interest. For fixed (finite) t, there is an expansion (2.2) of $E_x\Phi[\xi^\epsilon(t)]$ in powers of ϵ. Successive coefficients in the expansion are computed as integrals along the trajectories of (1.1).

If Φ is constant on the trajectories of (1.1), then another question of interest is to estimate the size of $E_x\Phi[\xi^\epsilon(t^\epsilon)]$ when $t^\epsilon = O(\epsilon^{-1})$ as $\epsilon \to 0$. See §3. This question was treated, for some special 2-dimensional systems, by Carrier [2] using different methods. If the process ξ^ϵ is ergodic, then one can ask for an estimate of the steady-state expectation $\lambda^\epsilon = E_{\mu^\epsilon}\Phi$ where μ^ϵ is the equilibrium measure. This is discussed briefly in §4 for the special case when (1.1) is linear.

Our method is based on the Ito stochastic differential rule. See [5, Chap. 8]. If $\psi(t,x)$ has continuous partial derivatives ψ_t, ψ_{x_i}, $\psi_{x_ix_j}$, $i, j = 1,\ldots,n$, then this rule implies that

$$d\psi(t-s,\xi^\epsilon(s)) = -\psi_t ds + \psi_x d\xi^\epsilon + \epsilon\ \mathrm{tr}\ \alpha\ \psi_{xx}ds, \qquad (1.4)$$

where $\psi_x = (\psi_{x_1},\ldots,\psi_{x_n})$ is the gradient in the variables x,

$$\alpha = \frac{1}{2}\sigma\sigma^T, \qquad \mathrm{tr}\ \alpha\ \psi_{xx} = \sum_{i,j=1}^n \alpha_{ij}\psi_{x_ix_j}.$$

If we introduce the notation

$$\mathcal{L}^\epsilon\psi = \epsilon\ \mathrm{tr}\ \alpha\ \psi_{xx} + \psi_x \cdot f, \qquad (1.5)$$

then (1.4) becomes

$$d\psi(t-s,\xi^\epsilon(s)) = (-\psi_t+\mathcal{L}^\epsilon\psi)ds + (2\epsilon)^{\frac{1}{2}}\psi_x\sigma dw.$$

The functions on the right side are evaluated at $(t-s,\xi^\epsilon(s))$. Take $E_x \int_0^t$ on both sides. If $E_x\int_0^t|\psi_x\sigma|^2 ds < \infty$, then $E_x\int_0^t \psi_x\sigma dw = 0$. We get using (1.3)

$$E_x\psi(0,\xi^\epsilon(t)) - \psi(t,x) = E\int_0^t(-\psi_t+\mathcal{L}\ \psi)ds. \qquad (1.6)$$

Assumptions

Let us assume that f, σ, Φ are of class C^∞. Moreover, it is assumed that a positive constant M and a positive C^∞ function V exist such that:

$$|\sigma(x)| \leq M(1+|x|); \tag{1.7}$$

$$\mathcal{L}^\epsilon V \leq M(1+V), \qquad 0 \leq \epsilon \leq 1; \tag{1.8a}$$

$$(1+|x|)|V_x| \leq M(1+V); \tag{1.8b}$$

$$V(x) \to \infty \quad \text{as} \quad |x| \to \infty; \tag{1.8c}$$

$$|\Phi| \leq M(1+V)^\ell \quad \text{for some} \quad \ell > 0. \tag{1.9}$$

By methods of [6, Chap. III] it can be shown that (1.7) and (1.8) guarantee that the process ξ^ϵ in (1.2) is defined for all $t > 0$ and $EV[\xi^\epsilon(t)] < \infty$. See also [4]. For any $\ell = 1,2,\ldots,V^\ell$ also satisfies (1.8) with M replaced by a suitable constant M_ℓ. By (1.9) the expectation $E\Phi[\xi^\epsilon(t)]$ which we wish to estimate exists.

2. Results for Fixed t (Finite)

$$\varphi^0(t,x) = \Phi[\xi^0(t)], \qquad \varphi^\epsilon(t,x) = E_x\Phi[\xi^\epsilon(t)], \tag{2.1}$$

where ξ^0, ξ^ϵ have initial data (1.3). We wish to expand φ^ϵ in powers of ϵ:

$$\varphi^\epsilon = \varphi^0 + \epsilon\theta_1 + \epsilon^2\theta_2 + \cdots + \epsilon^k\theta_k + o(\epsilon^k), \tag{2.2}$$

where k is finite but arbitrary. From the fact that $\varphi^0(t-s,\xi^0(s))$ is constant $(= \Phi(x))$ for $0 < s < t$, we find that $\varphi^0(t,x)$ satisfies the first order partial differential equation

$$-\varphi^0_t + \varphi^0_x \cdot f = 0, \qquad t \geq 0. \tag{2.3}$$

It can be shown that for $\epsilon > 0$, φ^ϵ satisfies the second order equation

$$-\varphi^\epsilon_t + \mathcal{L}^\epsilon\varphi^\epsilon = 0, \qquad t \geq 0. \tag{2.4}$$

See [4], and under more restrictive assumptions [1], [5, Chap. 8]. Both (2.3) and (2.4) have the Cauchy data

$$\varphi^0(0,x) = \varphi^\epsilon(0,x) = \Phi(x). \qquad (2.5)$$

By formally differentiating (2.4) repeatedly with respect to ϵ and setting $\epsilon = 0$, we get

$$-(\theta_j)_t + (\theta_j)_x \cdot f + \text{tr } \alpha(\theta_{j-1})_{xx} = 0, \quad j = 1,2,\ldots \qquad (2.6)$$

with Cauchy data $\theta_j(0,x) = 0$, and with $\theta_o = \varphi^0$ in (2.6) when $j = 1$. By the method of characteristics

$$\theta_j(t,x) = \int_0^t \text{tr } \alpha(\theta_{j-1})_{xx} ds, \quad j = 1,2,\ldots, \qquad (2.7)$$

where the integrand is evaluated at $(t-s,\xi^0(s))$. These are the equations which the coefficients in (2.2) must satisfy.

In outline, our proof of (2.2) proceeds as follows. Let

$$\theta_k^\epsilon = \epsilon^{-k}(\varphi^\epsilon - \varphi^0 - \epsilon\theta_1 - \cdots - \epsilon^{k-1}\theta_{k-1}).$$

By direct calculation one finds that

$$-(\theta_k^\epsilon)_t + \mathscr{L}^\epsilon \theta_k^\epsilon = - \text{tr } \alpha(\theta_{k-1})_{xx}.$$

We use (1.6) with $\psi = \theta_k^\epsilon$, together with the convergence of ξ^ϵ to ξ^0 uniformly on $[0,t]$ with probability 1 as $\epsilon \to 0$. This gives $\theta_k^\epsilon(t,x) \to \theta_k(t,x)$ as $\epsilon \to 0$, as required. In making this argument precise some technical steps are involved; for details see [4].

Note that in (2.2) the square of the coefficient $(2\epsilon)^{\frac{1}{2}}$ in (1.2) occurs, not the coefficient itself. Thus the expansion is in powers of a noise covariance coefficient.

Expansions like (2.2) of solutions to the linear second order equation (2.4) have been proved by nonprobabilistic methods. However, our method works when (2.4) is replaced by a nonlinear equation which occurs in the dynamic programming approach to optimal stochastic control [3], as well as for other kinds of nonlinear equations which arise for instance in population models involving both growth and diffusion [4].

3. Conservative Systems

In this section we assume that

$$\Phi_x \cdot f = 0. \tag{3.1}$$

By (1.1), $\Phi[\xi^o(t)]$ is constant. If we regard $\Phi(x)$ as an "energy" when the state is x, then energy is conserved by the unperturbed system. We are interested in the expected energy $E_x \Phi[\xi^\epsilon(t)]$ when $t = t^\epsilon$ is of order ϵ^{-1}. In addition to (1.9), we assume that for some constant M

$$(1+|x|)|\Phi_x| \le M(1+\Phi); \tag{3.2a}$$

$$|\text{tr } \alpha \, \Phi_{xx}| \le M(1+\Phi). \tag{3.2b}$$

In many instances, we may take $\Phi = V$. Then (3.2) is a consequence of (1.8) and (3.1). In (1.6) we now take $\psi(t,x) = \Phi(x)$, and obtain

$$E_x \Phi[\xi^\epsilon(t)] = \Phi(x) + \epsilon E_x \int_0^t \text{tr } \alpha \, \Phi_{xx} ds. \tag{3.3}$$

This is an exact expression, which replaces the approximation (2.2). However, the integrand is now evaluated at $\xi^\epsilon(s)$ rather than along the unperturbed trajectory as in (2.7).

Let us put $g(t) = E_x \Phi[\xi^\epsilon(t)]$. Here, we regard the initial state x as fixed. By (3.2b) and (3.3),

$$g(t) \le g(0) + M\epsilon \int_0^t (1+g(s))ds.$$

By Gronwall's inequality and $\Phi(x) = g(0)$,

$$g(t) \le (\Phi(x) + M\epsilon t)\exp(M\epsilon t). \tag{3.4}$$

Thus, the expected energy $g(t)$ remains bounded so long as ϵt is bounded. However, the bound is rather poor because of the exponential on the right side of (3.4). In a number of cases, a better bound can be found by exploiting special features of the system (1.2).

As an illustration, let us consider the following two dimensional system:

$$d\xi_1^\epsilon = \xi_2^\epsilon dt$$
$$d\xi_2^\epsilon = -(\xi_1^\epsilon + (\xi_1^\epsilon)^3)dt + (2\epsilon)^{\frac{1}{2}}\xi_1^\epsilon dw.$$

(3.5)

This is equivalent to the randomly perturbed nonlinear oscillator

$$\ddot{\eta} + \eta + \eta^3 = (2\epsilon)^{\frac{1}{2}}\eta\dot{w}$$

(3.5')

studied by Carrier [2], with $\eta = \xi_1^\epsilon$ and $\dot{w}$ a white-noise. In [2], (3.5') came from modelling the lateral oscillations of an elastic string whose length is subjected to random changes. For (3.5) the energy function

$$\Phi(x_1,x_2) = \frac{1}{2}(x_1^2+x_2^2) + \frac{1}{4}x_1^4$$

(3.6)

is conserved when $\epsilon = 0$. Moreover,

$$\sigma(x) = \begin{pmatrix} 0 \\ x_1 \end{pmatrix}, \quad \alpha(x) = \begin{pmatrix} 0 & 0 \\ 0 & x_1^2 \end{pmatrix},$$

and $\operatorname{tr} \alpha \Phi_{xx} = x_1^2 \Phi_{x_2 x_2}$. Then (3.3) becomes

$$E_x\Phi[\xi^\epsilon(t)] = \Phi(x) + \epsilon E_x \int_0^t [\xi_1^\epsilon(s)]^2 ds.$$

(3.7)

By Cauchy-Schwarz and (3.6)

$$E_x[\xi_1^\epsilon(s)]^2 \le (E_x[\xi_1^\epsilon(s)]^4)^{\frac{1}{2}} \le 2g(s)^{\frac{1}{2}},$$

(3.8)

$$g(t) \le g(0) + 2\epsilon \int_0^t g(s)^{\frac{1}{2}}ds;$$

from which

$$g(t) \le \left(\sqrt{g(0)} + \epsilon t\right)^2$$

(3.9)

In this case, the expected energy grows quadratically in ϵt, rather than exponentially. In [2], Carrier considered expected mean square displacement $E_x[\xi_1^\epsilon(t)^2]$ rather than expected energy. By (3.8) and (3.9),

$$E_x[\xi_1^\epsilon(t)^2] \le 2\left[\sqrt{g(0)} + \epsilon t\right].\qquad (3.10)$$

This is the same (linear) growth rate in ϵt found in [2]. The method of [2] uses in a more sophisticated way the specific form of (3.5') and thereby gets a better constant on the right side of (3.10) ($\frac{1}{2}$ rather than 2).

4. Linear Systems

Let us now suppose that

$$f(x) = Ax, \qquad \sigma(x) = \sigma_o x + \sigma_1.\qquad (4.1)$$

It is often of interest to take quadratic $\Phi(x) = \sum b_{ij} x_i x_j$. If $\sigma_o = 0$, then $\operatorname{tr} \alpha \Phi_{xx}$ is a constant γ. In (2.2), it turns out that $\theta_1 = \gamma t$, $\theta_2 = \theta_3 = \cdots = 0$. If (3.1) holds, then $\varphi^o(t,x) = \Phi(x)$. The expected energy at time t is $\Phi(x) + \gamma \epsilon t$.

If $\sigma_o \ne 0$, then the expected energy may grow exponentially with ϵt as in (3.4). For instance, this occurs in the randomly perturbed harmonic oscillator obtained by omitting the η^3 term in (3.5'). The energy function $\Phi(x_1, x_2) = \frac{1}{2}(x_1^2 + x_2^2)$ is conserved. One gets again (3.7). However, Cauchy-Schwarz is no longer available. From the obvious estimate $E[\xi_1^\epsilon(s)^2] \le 2g(s)$, (3.7), and Gronwall's inequality one gets for the expected energy $g(t) \le g(0)\exp(2\epsilon t)$. It can be shown that the growth of $g(t)$ is, in fact, exponential [2].

5. Concluding Remarks

In many problems the process ξ^ϵ lives only until the time τ^ϵ when $\xi^\epsilon(t)$ reaches the boundary ∂B of a given region B. Suppose that one wishes to find

$$\varphi^\epsilon(x) = E_x \Phi[\xi^\epsilon(\tau^\epsilon)].$$

Under various assumptions, φ^ϵ satisfies for $\epsilon > 0$ the second order equation $\mathcal{L}^\epsilon \varphi^\epsilon = 0$ in B, with $\varphi^\epsilon = \Phi$ on ∂B. For $\epsilon = 0$ the equation drops to first order, and boundary data are not used for those $x \in \partial B$ where $f(x) \cdot \nu(x) < 0$, $\nu(x)$ the exterior normal to B. This

creates a "boundary layer" effect. It would be interesting
to treat probabilistically the boundary layer expansions
which occur in the theory of singular perturbations [7].
In [4] we obtain probabilistically a "regular" expansion
for $\varphi^\epsilon(x)$ similar to (2.2), valid in portions of B from
which the trajectories of (1.1) lead to ∂B and intersect
∂B nontangentially.

If the Markov process ξ^ϵ in (1.2) is ergodic, then
one may wish to find $\lambda^\epsilon = E_{\mu^\epsilon}\Phi$ where μ^ϵ is the equilibrium probability measure. In particular, it would be of
interest to show under reasonable assumptions on f, σ, Φ
that $\lambda^\epsilon = \epsilon a_1 + \epsilon^2 a_2 + \cdots + \epsilon^k a_k + O(\epsilon^k)$. An example in
[8] suggests that such a result could not be extended to
steady-state optimal stochastic control problems. In that
example, λ^ϵ appears to be of order a fractional power of
ϵ, in contrast to the results of [3] for control on a
finite time interval.

It would also be of interest to extend the results in
§'s 2, 3 to systems governed by white-noise driven partial
differential equations of the type

$$u_t^\epsilon = Au^\epsilon + (2\epsilon)^{\frac{1}{2}}\sigma(u^\epsilon)w_t, \qquad (5.1)$$

where A is a (possibly nonlinear) elliptic differential
operator and w a Wiener process in the appropriate
Hilbert space. For the case when A is a monotone operator and σ is constant, a theorem about existence and
uniqueness of solutions to (5.1) was proved in [9].

References

1. Yu. N. Blagovescenskii and M. I. Freidlin, Doklady Akad
    Nauk SSSR 138 (1962), 508-511. Translation Soviet
    Math. Doklady 2, 633-636.
2. G. F. Carrier, J. Fluid Dynamics 44 (1970), 249-264.
3. W. H. Fleming, SIAM J. Control 9 (1971), 473-517.
4. W. H. Fleming, to appear.
5. I. I. Gikhman and A. V. Skorokhod, Introd. Theory
    Random Proc., W. B. Saunders, 1969.
6. H. Kushner, Stochastic Stability and Control, Academic
    Press, 1969.
7. W. Wasow, Asymptotic Expansions for Ordinary Differ-
    ential Equations, Interscience, 1965.

8. W. M. Wonham and W. F. Cashman, Inter. J. Control $\underline{10}$ (1969), 77-98.
9. A. Bensoussan and R. Temam, Equations aux derivees partielles stochastiques nonlineaires, to appear.

OPTIMALITY OF CONTROLS FOR
SYSTEMS WITH JUMP MARKOV DISTURBANCES

R. W. Rishel

Bell Telephone Laboratories, Incorporated
Whippany, New Jersey

Introduction

Let $r(t)$ denote a finite state Markov process. Consider the system of controlled stochastic processes which are solutions of the stochastic differential equation

$$\dot{x}(t) = f[t,x(t),r(t),u(t,x(t),r(t))] \tag{1}$$

with initial condition $x(t_O) = x_O$. Suppose terminal conditions are given by specifying a finite number of functions at the terminal time. That is, let

$$M = \{(t,x): \phi_i(t,x) = 0 \quad i = 1,\ldots,\ell\} \tag{2}$$

be the set of possible terminal conditions. Let $\mathcal{U}$ be a set of controls for which (1) has a solution which satisfies the terminal conditions with probability one. That is for each solution $x(t)$ of (1) there is a random variable τ such that

$$P\{x(\tau) \in M\} = 1. \tag{3}$$

Suppose a performance index of the form

$$E\left\{\int_{t_O}^{\tau} f_O[s,x(s),r(s),u(s,x(s),r(s))]ds\right\} \tag{4}$$

is given. Consider the problem of finding the control in the class U which makes (4) a minimum.

In deterministic optimization problems similar to the stochastic optimization problem just described examples show that optimal controls may be discontinuous and that the partial differential equation of Dynamic Programming may not have a smooth solution. A similar situation is to be expected in these stochastic problems.

This paper shows if the system of partial differential equations of Dynamic Programming has a generalized solution, and if there is some control whose performance function is equal to this solution that this control is optimal. A generalized solution of the system of partial differential equations is defined to be one which is locally Lipschitzian and such that the system of partial differential equations is satisfied almost everywhere with respect to Lebesgue measure.

Problems similar to the one described above were considered in a sequence of papers by Krassovskii and Lidskii. References [3] and [6] are typical of these. They were also discussed by Florentine in [2] and by Kushner in [4]. In [5] Kushner defined a Stochastic Pontryagin Principle. In [9] Sworder discussed necessary conditions for optimality for similar problems through use of a Stochastic Pontryagin Principle.

The ideas of the current paper are very similar to those of [3] and [6] and could be considered as a rigorous treatment of results which were obtained formally in [3] and [6]. In the deterministic case similar results to the present ones were given by Boltyanskii in [1]. In a companion paper [8] a Stochastic Pontryagin Principle and Transversality Conditions are defined for the optimization problem. The Stochastic Pontryagin Principle defined in [8] differs from those of [5] and [9]. It is shown in [8] under some technical conditions that the performance function of a control which satisfies the Stochastic Pontryagin Principle and the Transversality Condition is a generalized solution of the system of partial differential equations of Dynamic Programming. Hence by the results of this paper such a control must be optimal.

Preliminaries

Let us begin by stating more precisely the conditions to be assumed on the quantities involved in the optimization problem. Let R be the finite set $R = \{1,\ldots,k\}$. Let Ω, β, P be a triple of probability space Ω, Borel field β, and probability measure P. The Markov process $r(t) = r(t,\omega)$ will be understood to be defined on this probability space with state space R. The process $r(t)$ will be assumed to have stationary transition probabilities. Such a Markov process is characterized by numbers λ_{ij} which satisfy

$$\lambda_{ij} \geq 0 \qquad\qquad \lambda_{ii} = -\sum_{j \neq 1} \lambda_{ij}$$

$$P\{r(t+h) = j \mid r(t) = i\} = \lambda_{ij}h + o(h) \qquad \text{if} \quad j \neq i$$

$$P\{r(t+h) = i \mid r(t) = i\} = 1 + \lambda_{ii}h + o(h)$$

An open subset G of E^{n+1} will be considered to be the space of time state pairs (t,x) in which the actions of the controlled system can take place. The values of the controls will be assumed to be restricted to a compact subset U of E^n. Let $f(t,x,i,u)$ and $f_o(t,x,i,u)$

$$f: G{\times}R{\times}U \to E^n \qquad f_o: G{\times}R{\times}U \to E^1$$

be continuous functions continuously differentiable in x and u for fixed t,i. $\phi_i(t,x)$, $i = 1,\ldots,\ell$ be continuously differentiable real valued functions defined on G.

Controls and Corresponding Processes

The controls shall be considered to be functions of the current time, state, and value of the random process. Examples show that optimal controls may be discontinuous, consequently we shall wish to include discontinuous controls in our class of controls. When the control is discontinuous the concept of a corresponding solution of (1) needs to be clarified. This will be done by reducing the problem to one for deterministic systems. To achieve this let us require the controls of U to have the following property.[1]

A. For each control $u(t,x,i) \in U$, $(t,x) \in G$, and $i \in R$ there is a unique solution of the deterministic ordinary differential equation

$$\dot{x}^i(t) = f[t,x^i(t),i,u(t,x^i(t),i)] \tag{5}$$

with initial condition $x^i(t) = x$ on an interval $[t,t^i]$. During time $[t,t_i)$ the solution lies in $G - M$.

At time t^i, $x^i(t^i) \in M$.

A sample function interpretation of the stochastic differential equation (1) will be given by the following argument. Let a decomposition $\{s_m\}$ of the interval $[t,\infty)$

$$t = s_o < \ldots < s_m < \ldots$$

be given by a divergent sequence. Let a sequence of states $\{j_m\} \in R$ be given. Assumption A implies that by piecing together appropriate solutions of (5) that a continuous deterministic function $x(s)$ can be contructed on an interval $[t,\tau]$ which has the following properties:

a) $x(t) = x$.

b) If $s_{m-1} < s < s_m \leq \tau$ or $s_{m-1} < s < \tau < s_m$, $x(s)$ is a solution of

$$\dot{x}(s) = f[s,x(s),j_m,u(s,x(s),j_m)] \tag{6}$$

on either the interval

$$s_{m-1} < s < s_m \quad \text{or} \quad s_{m-1} < s < \tau.$$

c) Either $\tau = \infty$ or $x(\tau) \in M$.

---

[1] Conditions could be given on the controls $u(t,x,i)$ which would imply A held, for instance see [1] pp 341-342, however, we shall prefer to assume A directly as a condition on the controls of U.

Let $S(t,x)$ denote the set of functions $x(s)$ satisfying a, b, and c.

To give a sample function interpretation of (1) recall that the sample functions of $r(t)$ are step functions with probability one. Hence with probability one the sample function $r(t,\omega)$ will determine a divergent sequence $\{s_m\}$ and a sequence of states $\{j_m\}$ by the condition that $r(t) = j_m$ if $s_{m-1} < s < s_m$. For this sample function Equation (1) is identical to Equation (6). Let us define the solution of (1) on a random interval $[t,\tau(\omega)]$ to be such that with probability one the sample function $x(t,\omega)$ is the element of $S(t,x)$ which satisfies Equation (6) for the sequences induced by the corresponding sample function $r(t,\omega)$.

The following is a concept concerning the set $S(t,x)$ that will be important for later use. Define the <u>domain of influence</u> of a point (t,x) under the control $u(t,x,i)$ to be the set of points.

$$\{(s,y): \exists \, x(\cdot) \in S(t,x) \ni x(s) = y\}. \tag{7}$$

Further assumptions will be made about the nature of the discontinuities of the controls and several other properties of the controls. A set will be said to be a smooth polyhedron if it is a smooth homeomorphic image of a polyhedron. A set will be said to be piecewise smooth if it is a countable union of smooth polyhedrons such that only a finite number intersect any compact set. The following further conditions will be assumed for each control $u(t,x,i) \in \mathcal{U}$.

B. For each $i \in R$ there are decompositions of G by piecewise smooth sets S_m^i

$$G = S_{n+1}^i \supset S_n^i \ldots \supset S_o^i$$

such that

$$\text{dimension} \left(S_m^i \right) = m.$$

Each component of $S_m^i - S_{m-1}^i$ is a smooth m-dimensional polyhedron. Call these components cells. In each cell there is a continuously differentiable function

$\tilde{u}(t,x,i)$ defined on a neighborhood of the closure of the cell which agrees with $u(t,x,i)$ on the cell.

C. Each trajectory of (5) with initial condttions in a given cell passes through the same finite set of cells in the same order to reach the terminal manifold.

D. The domain of influence of (t,x) has compact closure in G.

E. If $u \in U$ and $u(s,x,i) \in U$ and if

$$u^*(s,x,i) = \begin{cases} u & \text{if} \quad t \le s \le t+h \\ u(s,x,i) & \text{otherwise} \end{cases}$$

then $u^*(s,x,i) \in U$.

Notice that assumption D implies that with probability one all the sample functions of the stochastic process $x(t)$ which is the solution of (1) reach the terminal manifold and do so before a fixed time.

Dynamic Programming

Consider the value of the optimization problem as a function of the initial conditions of Equation (1) and of the initial value of the stochastic process $r(t)$. That is the "value function" $V(t,x,i)$ of the optimization problem is defined on $G \times R$ by

$$V(t,x,i) = \inf_{u(t,x,i) \in U} E\left\{ \int_t^\tau f_o ds \,\middle|\, x(t) = x, \; r(t) = i \right\} \tag{8}$$

A control will be called optimal if for that control

$$V(t,x,i) = E\left\{ \int_t^\tau f_o ds \,\middle|\, x(t) = x, \; r(t) = i \right\} \tag{9}$$

The following theorem can be obtained by a dynamic programming argument.

$\underline{\text{Theorem 1}}$: Let $u(t,x,i)$ be an optimal control. At each point at which $V(t,x,i)$ has a tangent plane and $u(t,\mathbf{x},i)$ is continuous, the equation

$$\underset{u \in U}{\text{Min}} \left\{ V_t(t,x,i) + V_x(t,x,i)f[t,x,i,u] \right.$$

$$\left. + f_o(t,x,i,u) + \sum_j \lambda_{ij} V(t,x,j) \right\} \qquad (10)$$

$$= V_t(t,x,i) + V_x(t,x,i)f[t,x,i,u(t,x,i)]$$

$$+ f_o[t,x,i,u(t,x,i)] + \sum_j \lambda_{ij} V(t,x,j) = 0$$

is satisfied.

Notice if it was known a priori that each $V(t,x,i)$ was locally Lipschitzian on G, Radamacher's theorem would imply that $V(t,x,i)$ had a tangent plane almost everywhere. Motivated by this and by concepts from the Theory of Distributions let us say that a function $V(t,x,i)$ defined on $G \times R$ is a generalized solution of the system of partial differential equations of dynamic programming.

$$\underset{u \in U}{\text{Min}} \left\{ V_t(t,x,i) + V_x(t,x,i)f(t,x,i,u) \right.$$

$$\left. + f_o(t,x,i,u) + \sum_j V(t,x,j) \right\} = 0, \qquad (11)$$

if the functions $V(t,x,i)$ for each $i \in R$ are locally Lipschitzian on G and the system of Equations (11) is satisfied almost everywhere on G with respect $(n+1)$-dimensional Lebsegue measure.

Theorem 2 is the basic result of the paper mentioned in the introduction.

Theorem 2: If $W(t,x,i)$ is a generalized solution of the system of partial differential equations of dynamic programming such that $W(t,x,i) = 0$ for $(t,x) \in M$ and $u(t,x,i)$ is a control in $\mathcal{U}$ such that for the corresponding performance function

$$W(t,x,i) = E\left\{\int_t^\tau f_0 ds \,\Big|\, x(t) = x, \; r(t) = i\right\}, \tag{12}$$

then $u(t,x,i)$ is an optimal control.

Theorem 2 will be proven through the use of two Lemmas and some technical machinery similar to that of [7].

Lemma 1: Let $W(t,x,i)$ be a generalized solution of the partial differential equation of dynamic programming. Let $u(t,x,i)$ be an arbitrary control in $\mathcal{U}$. For $t,x \in G$ let $x^i(s)$ be the solution of

$$\dot{x}^i(s) = f[(s,x^i(s),i,u(s,x^i(s),i))]$$

with initial conditinn $x^i(t) = x$. Then there is a $h_0 > 0$ such that for $0 \le h \le h_0$

$$W(t,x,i) - W\big(t+h,x^i(t+h),i\big) \tag{13}$$

$$\le \int_t^{t+h} \left[f_0\big(s,x^i(s),i,u(s,x^i(s),s)\big) + \sum_j \lambda_{ij} W\big(s,x^i(s),j\big) \right] ds$$

Proof: Assumptions B and C on the controls of $\mathcal{U}$ imply there is some interval $[t,t+h_0]$ such that the curve $(s,x^i(s))$ for $t \le s \le t+h_0$ lies in a single cell and there is a continuously differentiable function $\tilde{u}(t,x,i)$ on the cell. Consider the deterministic ordinary differential equation

$$\dot{x} = f[s,x,i,\tilde{u}(s,x,i)]. \tag{14}$$

Theorems on continuous dependence of solutions imply that there is closed spherical neighborhood N of x such that for

$y \in N$, (14) has a solution with initial condition $x(t) = y$ which lies in G and is defined on the entire interval $[t, t+h_o]$. Denote this solution by $x^i(s;y)$.

Consider the mapping

$$T: [t, t+h_o] \times N \to G$$

defined by

$$T(s,y) = \left(s, x^i(s;y)\right).$$

This mapping is continuously differentiable and has a non-singular Jacobian matrix given by

$$J = \begin{pmatrix} 1 & f[s,x,i,\tilde{u}(s,x,i)] \\ 0 & \delta x^i(s) \end{pmatrix}$$

where $\delta x^i(s)$ is the solution of

$$\delta x^i = f_x \delta x^i + f_u \tilde{u}_x \delta x^i \tag{15}$$

with initial condition $\delta x^i(t) = $ identity matrix.

Since $[t, t+h_o] \times N$ is a compact set its image under the mapping T is compact. Since $W(t,x,i)$ is locally Lipschitzian, Radamacher's theorem implies that it has a tangent plane at all points of G except for a set of $(n+1)$-dimensional Lebesgue measure zero. Let H denote the subset of G at which either some $W(t,x,i)$ does not have a tangent plane or the system of partial differential equations of dynamic programming do not hold. Hence $\Lambda(H) = 0$ where Λ denotes n+1-dimensional Lebesgue measure. Uniqueness theorems for differential equations imply that T is a one-one mapping. Since J is nonsingular

$$\Lambda\left(T^{-1}(H)\right) = \int_{H \cap T([t, t+h_o] \times N)} |J|^{-1} dt dx \tag{16}$$

Hence $T^{-1}(H)$ has $(n+1)$-dimensional Lebesgue measure zero.

Thus by Fubini's theorem for almost every $y \in N$
$\{s \in [t,t+h_o]: (y,s) \in T^{-1}(H)\}$ has zero one-dimensional
Lebesgue measure. For such a y, for almost every
$s \in [t,t+h_o]$, equation (11) holds. Hence

$$\frac{d}{dt} W(s,x^i(s;y),i) = W_t + W_x f \geq -f_o - \sum_j \lambda_{ij} W(s,x^i(s;y),j)$$

(17)

Since $W(t,x,i)$ is locally Lipschitzian and $T([t,t+h_o]\times N)$ is
a compact set $W(t,x,i)$ is Lipschitzian on $T([t,t+h_o]\times N)$.
This implies $W(s,x^i(s;y),i)$ is a Lipschitzian function of s
on $[t,t+h_o]$. Hence it is the integral of its derivative.
Integrating both sides of (17) from t to t+h gives

$$W(t,y,i) - W(t+h,x^i(t+h;y),i)$$
$$\leq \int_t^{t+h} \left[f_o + \sum_j \lambda_{ij} W(s,x^i(s,y),j) \right] ds.$$

(18)

Now let y_n be a sequence of N approaching x for which (18)
holds. The functions $x^i(s;y_n)$ converge uniformly to
$x^i(s;x)$ on $[t,t+h_o]$. Hence taking limits of both sides of
(18) gives

$$W(t,x,i) - W(t+h,x^i(t+h),i) \leq \int_t^{t+h} \left[f_o + \sum \lambda_{ij} W(x,x^i(s),j) \right] ds$$

(19)

which is the conclusion of the lemma.

Theorem 2 will be proven using Lemma 1 and some techniques similar to those of [7]. The following concepts are
nearly identical to corresponding ones in [7]. Consider a
fixed $(t_o,x_o,j_o) \in G \times R$ and the solution of $x(t)$ of (1)
with initial conditions $x(t_o) = x_o$ and $r(t_o) = j_o$. Recall that the stochastic process $r(t)$ and consequently $x(t)$
are defined on the probability space Ω, with Borel field β,
and probability measure P. Let $\beta_t = \mathfrak{F}\{x(s),r(s); t_o \leq s < t\}$

be the sigma field generated by the past of the processes $x(t)$, $r(t)$.

For each control $u(t,x,i) \in \mathcal{U}$, define a space $L_1(u)$ of functionals on the processes $x(t)$, $r(t)$ corresponding to this control by defining $L_1(u)$ to be the collection of functions

$$L_1(u) = \{\phi(t,\omega)\}$$

such that

    a) $\phi(t,\omega)$ is jointly measurable in (t,ω)

    b)   for fixed t, $\phi(t,\omega)$ is β_t measurable as a function of ω

    c) $E\left\{ \int_{t_o}^{\tau} |\phi(t,\omega)| ds \right\} < \infty.$

Let Λ denote one dimensional Lebesgue measure. In $L_1(u)$ let equality be interpreted by equality except on a set of $\Lambda \times P$ measure zero. The following statements can be demonstrated by modifications of proofs of similar statements in [7] pp 561-564. $L_1(u)$ is a Banach space. The Dual space of $L_1(u)$ is $L_\infty(u)$ which is defined by:

$$L_\infty(u) = \{\Psi(t,\omega)\}$$

such that

    a) $\Psi(t,\omega)$ is jointly measurable in (t,ω)

    b)   for fixed t, $\Psi(t,\omega)$ is β_t measurable as a function of ω

    c) $\underset{\Lambda \times P}{\text{ess sup}} |\Psi(t,\omega)| < \infty.$

Each linear functional on $L_1(u)$ has the form

$$L(\phi) = E\left\{ \int_{t_o}^{\tau} \phi(s,\omega)\Psi(s,\omega)ds \right\} \tag{20}$$

for some element Ψ of $L_\infty(u)$. The notation (ϕ, Ψ) will be used to denote the right hand side of (20).

Define operators R_h on $L_\infty(u)$ by

$$R_h[\phi](t,\omega) = \begin{cases} \phi(t-h,) & \text{if} \quad t \geq h \\ 0 & \text{if} \quad t < h \end{cases}.$$

Then R_h is a normdecreasing semigroup of operators on $L_\infty(u)$ satisfying

$$\lim_{h \downarrow 0} \| R_h \phi - \phi \| = 0.$$

Let $\chi(t,\omega)$ be defined by

$$\chi(t,\omega) = \begin{cases} 1 & \text{if} \quad t \leq \tau(\omega) \\ 0 & \text{if} \quad t > \tau(\omega). \end{cases}$$

Then the adjoint semigroup of operators on $L_\infty(u)$ has the representation

$$T_h[\Psi](t,\omega) = E\{\chi(t+h)\Psi(t+h)|\beta_t\}.$$

By Theorem 3, p 564 of [7] an element Ψ of $L_\infty(x)$ is in the domain D_A of the weak infinitesimal generator A_u of T_h if and only if $\exists\ K \ni$

$$\| T_h \Psi - \Psi \| \leq Kh. \tag{21}$$

The following comparison theorem can be proven eaactly as theorem [9] of [7].

<u>Theorem 3</u> Suppose for each $u(t,x,i) \in U$ there is a stochastic process $W_u(t,\omega)$ and a constant W independent of u such that

$$\lim_{t \downarrow t_o} E\{W_u(t,\omega)\} = W$$

$$W_u \in L_\infty(u) \cap D_A$$

and

$$-A_u W_u \leq E\{\chi_u(t) f_o\big(t,x(t),r(t),u(t,x(t),r(t))\big)|\beta_t\}$$

$\Lambda \times P$ almost everywhere, then

$$W \leq E\left\{\int_{t_o}^{\tau} f_o\big(t,x(t),r(t),u(t,x(t),r(t))\big)ds\right\} .$$

Having collected these results of [7] let us show that a generalized solution of the Dynamic Programming partial differential equations "evaluated along the process $(x(t),r(t))$" satisfies the hypothesis of the comparison theorem.

Lemma 2: Let $W(t,x,i)$ be a generalized solution of the partial differential equations of dynamic programming such that $W(t,x,i) = 0$ if $(t,x) \in M$. Let

$$W(t,\omega) = \chi(t,\omega)W\big(t,x(t,\omega),r(t,\omega)\big).$$

Then

$$W(t,\omega) \in L_\infty(u) \cap D_A .$$

and

$$-A_u[W](t,\omega) \leq \chi(t,\omega)f_o\big(t,x(t),r(t),u(t,x(t),r(t))\big).$$

Proof: Let $I(t_o,x_o)$ denote the closure of the domain of influence of (t_o,x_o) under the control $u(t,x,i)$. By assumption D this is a compact subset of G. Hence both

$$\|W\| = \sup_{(t,x,i)\in I(t_o,x_o)\times R} |W(t,x,i)|$$

and

$$\|f\| = \sup_{(t,x,i,u)\in I(t_o,x_o)\times R\times U} |f(t,x,i,u)|$$

are finite. In addition if (t,x) and $(t',x') \in I(t_o,x_o)$
there is a constant K such that

$$|W(t,x,i) - W(t',x',i)| \leq K[|t-t'| + |x-x'|].$$

Since the values $t,x(t,\omega)$ for $t_o \leq t \leq \tau(\omega)$ must lie
with probability one in the domain of influence of (t_o,x_o)
the function

$$W(t,\omega) = \chi(t,\omega)W(t,x(t,\omega),r(t,\omega))$$

is bounded with probability one. This and the definition
of $W(t,\omega)$ imply that $W(t,\omega)$ satisfies the properties A, B,
and C required of a function in $L_\infty(u)$.

To show that $W(t,\omega) \in D_A$ we must show for some K that

$$\|T_h[W] - W\| \leq Kh.$$

Let α denote the time of the first jump of r after time t.
Let γ_h denote the characteristic function of the set

$$\{\omega\colon \alpha(\omega) > t+h,\ \tau(\omega) > t+h\}$$

and δ_h the characteristic function of the set

$$\{\omega\colon t \leq \alpha(\omega) \leq t+h,\ \tau(\omega) > t+h\}.$$

Now

$$T_h[W] - W$$

$$= E\{\gamma_h[W(t+h,x(t+h),r(t+h)) - W(t,x(t),r(t))]|\beta_t\}$$

$$+ E\{\delta_h[W(t+h,x(t+h),r(t+h)) - W(t,x(t),r(t))]|\beta_t\}$$

$$- W(t,x(t),r(t))P\{t < \tau \leq t+h|\beta_t\} \tag{22}$$

Let $x^i(s)$ be the solution of (5) with initial condition
$x^i(t) = x(t)$. From the definition of γ_h

$$E\{\gamma_h[W(t+h,x(t+h,r(t+h)) - W(t,x(t),r(t))]|\beta_t\}$$

$$\leq |W(t+h,x^i(t+h),i) - W(t,x^i(t),i)| \leq K[1 + \|f\|]h. \tag{23}$$

Using the formula, [10] p. 350, for the distribution of α, and the definition of δ_h

$$E\{\delta_h[W(t+h,x(t+h),r(t+h)) - W(t,x(t),r(t))]\,|\,\beta_t\}$$

$$\leq 2\|W\|P\{t \leq \alpha \leq t+h\,|\,\beta_t\} \leq 2\|W\|\,|\lambda_{ii}|h$$

Whenever $P\{t < \tau < t+h\,|\,\beta_t\} > 0$ there must be some curve in $S(t,x(t))$ such that this curve goes from $t,x(t)$ to M in time less than h. Consequently, there is a $\tilde{t},\tilde{x}$ such that $t \leq \tilde{t} < t+h$, $\tilde{x} \in M$, $\{|x(t) - \tilde{x}| \leq \|f\|h$ and $W(\tilde{t},\tilde{x},i) = 0$. Therefore whenever $P\{t < \tau < t+h\,|\,\beta_t\} > 0$

$$|W(t,x(t),r(t))|$$

$$= |W(t,x(t),r(t)) - W(\tilde{t},\tilde{x},r(t))| \leq K[1 + \|f\|]h \qquad (24)$$

Combining the estimates above

$$\|T_h W - W\| \leq 2[K(1 + \|f\|) + |\lambda_{ii}|\,\|W\|]h. \qquad (25)$$

Hence $W \in D_A$.

To compare $A_u W$ and f_0 notice that (22), Lemma 1 and formulas for the distributions of the first jumps of $r(t)$ imply

$$T_h W - W \geq - \left[\int_t^{t+h} f_0 + \sum_j \lambda_{ij} W(s,x^i(s),j)ds\right] P[\alpha>t+h,\tau>t+h\,|\,\beta_t]$$

$$+ \sum_{j \neq i} \int_t^{t+h} \lambda_{ij} e^{\lambda_{ii}(\alpha-t)} E\{\chi(t+h)W(t+h,x(t+h),r(t+h))\,|\,x(\alpha)=x^i(\alpha),r(\alpha)=j\}d\alpha$$

$$- W(t,x,r(t))P\{\tau>t+h,t<\alpha<t+h\,|\,\beta_t\} - W(t,x,r(t))P\{t<\tau\leq t+h\,|\,\beta_t\}.$$

$$(26)$$

Notice that $P\{t < \tau \leq t+h\,|\,\beta_t\} = 0$, if $\chi(t) = 0$ or if distance $(x(t),M) \geq \|f\|h$. Hence

$$\lim_{h \downarrow 0} \frac{P\{t < \tau \leq t+h \mid \beta_t\}}{h} = 0$$

From the form of the distribution of α

$$\lim_{h \downarrow 0} \frac{P\{\tau > t+h, t < \alpha < t+h \mid \beta_t\}}{h} = -\lambda_{ii} \chi(t).$$

The continuity of the paths of $x(t)$ and the continuity of $W(t,x,i)$ in (t,x) imply that

$$\lim_{h \downarrow 0} \left| \int_t^{t+h} \frac{\lambda_{ij} e^{\lambda_{ii}(\alpha-t)} E\{\chi(t+h)W(t+h,x(t+h),r(t+h)) \mid x(\alpha)=x^i(\alpha), r(\alpha)=j\} d\alpha}{h} \right.$$

$$= \lambda_{ij} \chi(t) W\left(t, x^i(t), j\right).$$

Hence the right side of (26) divided by h converges to

$$-\chi(t) f_0[t, x(t), r(t), u(t, x(t), r(t))].$$

The right side of (26) is bounded as a function of h, hence Lebesgue's dominated convergence theorem implies that for each $\phi \in L^1(x)$ such that $\phi \geq 0$ that

$$\lim_{h \downarrow 0} - \frac{(T_h W - W, \phi)}{h} \leq \left(\chi(t) f_0, \phi\right)$$

or that

$$-A_u W \leq \chi(t) f_0\left(t, x(t), r(t), u(t, x(t), r(t))\right)$$

$\Lambda \times P$ almost everywhere which is the conclusion of Lemma 2.

Theorem 2 now follows readily from Theorem 3 and Lemma 2.

Proof of Theorem 2: Since (1) has initial conditions $x(t_0) = x_0$, the process $x(t)$ has continuous paths, $r(t)$ is continuous in probability, and $W(t,x,i)$ is continuous in (t,x)

$$\lim_{t \downarrow t_o} E\{\chi(t,\omega)\, W(t,x(t),r(t))\} = W(t_o,x_o,j_o).$$

Lemma 2 and Theorem 3 imply that

$$W(t_o,x_o,j_o) \tag{27}$$

$$\le E\left\{\int_{t_o}^{\tau} f_o\big(s,x(s),r(s),u(s,x(s),r(s))\big)ds \,\big|\, x(t_o)= x_o, r(t_o)= j_o\right\}$$

for any control $u \in \mathcal{U}$. The argument establishing (27) is valid for an arbitrary point $(t,x,j) \in G{\times}R$ hence

$$W(t,x,j) \le E\left\{\int_{t_o}^{\tau} f_o\,ds \,\big|\, x(t) = x, r(t) = j\right\} \tag{28}$$

Therefore any control for which equality holds in (28) must be optimal.

References

[1] V. G. Boltyanskii, "Sufficient Conditions for Optimality and the Justification of the Dynamic Programming Method," Izv. Akad. Nauk SSSR Ser. Mat. 28(1969) pp. 481-514; English translation SIAM Journal on Control 4 (1966) pp. 326-361.

[2] J. J. Florentin, "Optimal Control of Systems with Generalized Poisson Inputs," Journal of Basic Engineering (1963) pp. 217-221.

[3] N. N. Krassovskii and E. A. Lidskii, "Analytical Design of Controllers in Stochastic Systems with Velocity Limited Controlling Action," Applied Math. and Mech. 25 (1961) pp. 627-643.

[4] H. J. Kushner, "Optimal Stochastic Control," IRE Transactions on Automatic Control, Vol. AC-7 (1962)

[5]   H. J. Kushner, "On the Stochastic Maximum Principle:
      Fixed Time of Control," Journal of Mathematical
      Analysis and Applications, 11 (1965) pp. 78-92.

[6]   E. A. Lidskii, "Optimal Control of Systems with
      Random Properties," Applied Math. and Mechanics, 27
      (1963) pp. 33-45.

[7]   R. W. Rishel, "Necessary and Sufficient Dynamic Pro-
      gramming Conditions for Continuous Time Stochastic
      Optimal Control," SIAM Journal on Control 8 (1970)
      pp. 559-571.

[8]   R. W. Rishel, "Dynamic Programming and a Stochastic
      Pontryagin Principle for Systems with Jump Markov
      Disturbances," to appear.

[9]   D. D. Sworder, "Feedback Control of a Class of Linear
      Systems with Jump Parameters," IEEE Transactions on
      Automatic Control, Vol. AC-14 (1969) pp. 9-14.

[10] I. I. Gikhman and A. V. Skorohod, "Introduction to
      the Theory of Random Process," (1969) Saunders,
      Philadelphia.

ECONOMICS

OPTIMAL SYSTEMS FOR
INFORMATION AND DECISION

Jacob Marschak
Western Management Science Institute
University of California
Los Angeles, California

1. INTRODUCTION. "Optimal" means the same as "economical." My talk might be entitled: "Economics of producing, storing, transporting and using knowledge." Optimal ways to produce and use knowledge (sometimes called "optimal design of experiment") have been studied by statisticians such as Neyman and Pearson (1928-32), Wald (1950), Blackwell (1953), De Groot (1962, 1968). The study of economical codes for storing and transporting knowledge was introduced by Shannon (1948), and this optimality aspect of communication theory was emphasized by the Yagloms (1960, esp. Ch. IV, Section 1) and Wolfowitz (1961). The traditional separation between the economics of communication and of statistical decision is, in general, "sub-optimal." A precise estimate based on a large statistical sample may be not worth its cost if it is garbled or outdated by the time it reaches the decision-maker, owing to the low capacity or deficient coding in communication. Using well-established theorems, I shall try to show how the expected benefit and cost of an information-and-decision system depend on the choice of its links.

We shall thus distinguish between the system's expected benefit (its "value in use," i.e., the least upper bound on the price that the organizer, i.e., the buyer or renter of the system should be willing to pay); and the system's expected cost (the greatest lower bound that the sellers of its various components are willing to charge).

Another important contrast is that between the storage and communication costs, which (similar to the cost of storing and transporting commodities) are, by and large, additive and independent of content; and the costs of inquiring and deciding, which do not have these properties.

2. PURPOSIVE SYMBOL-PROCESSING SYSTEMS. Define a finite sequence of sets

$$(X^0, X^1, \ldots, X^n) = X$$

with generic elements $x^i \epsilon X^i$, $i = 0, \ldots, n$. All sets X^i are assumed finite for simplicity, and are interpreted thus:

$$X^i \text{ consists of } \begin{cases} \text{events,} & i = 0 \\ \text{symbols,} & 1 < i < n \\ \text{actions,} & i = n. \end{cases}$$

Define a sequence (chain) of "processors"

$$(P^1, \ldots, P^n) = P \epsilon \wp$$

$$P^i = \langle X^{i-1}, X^i, \varphi^i, d^i, \varkappa^i \rangle.$$

X^{i-1}, X^i are, respectively, the sets of inputs and outputs of P^i.

The Markov matrix φ^i is the transpose of the matrix of conditional probabilities, $[p(x^i | x^{i-1})]$ (say) and is called a stochastic transformation, degenerating in the special, deterministic case (with a unit element in each row) into an ordinary function, $X^{i-1} \to X^i$, also to be denoted by φ^i. $\varkappa^i$ is the "cost" function per time unit, from X^{i-1} onto non-negative reals (in "cost units" such as dollars per time unit). As in the usual accounting practice, it consists of a constant component (depreciation and maintenance of equipment, other long-term commitments) and a random variable component depending on the input processed. d^i, a non-negative real (in time units) is the delay between inputs and outputs of P^i.

$\wp$ denotes the sets of conceivable (not necessarily feasible) sequences P of processors. We write similarly

$$(\varphi^1, \ldots, \varphi^n) = \varphi \epsilon \varPhi$$

$$(\varkappa^1, \ldots, \varkappa^n) = \varkappa \epsilon \mathcal{K}$$

$$(d^1, \ldots, d^n) = d \epsilon \mathcal{D}.$$

Hence $\wp$ is the cartesian product

$$\wp = \Phi \times \mathcal{K} \times \mathcal{D}.$$

<u>Given</u> to the organizer (buyer) of a system is, to begin <u>with</u>, the <u>feasible</u> subset

$$\wp_f \subset \wp;$$

for example, he is aware that he has to trade off high speed (a low d^i) for high cost ($\varkappa^i$), or for low "precision" (a property of φ^i, also called "informativeness": see Section 4).

Also <u>given</u> to the organizer are
π, a <u>probability</u> function on X^0, represented for convenience by a diagonal matrix (non-negative, with trace = 1); and

β, the <u>benefit</u> function from $X^0 \times X^n$ onto the reals (in dollars <u>per action</u>), conveniently represented by a matrix, $\beta = [\beta_{x^n x^0}]$. There is no loss of generality in eliminating repeated rows and columns of β, thus defining X^0 and X^n as the sets of "benefit-relevant" events and actions, respectively [Marschak (1963)]; and also eliminating dominated rows of β (corresponding to "inadmissible actions"). Clearly, the <u>expected benefit</u> earned <u>per action</u> is

$$B = \text{trace } \pi\varphi^1 \ldots \varphi^n \beta \equiv B_{\pi\beta}(\varphi), \tag{1}$$

with the givens in the subscript. On the other hand, <u>expected cost</u> spent <u>per time unit</u> is

$$K = \Sigma E[\varkappa^i(x^{i-1})] \equiv K_\pi(\varphi,\varkappa). \tag{2}$$

If for simplicity we disregard the queueing problem by assuming [in the language of network theory: e.g., Iri (1969)] that "the flows do not stagnate," so that

$$d^i = \bar{d}, \text{ all } i,$$

then we can define <u>expected utility per time unit</u> as

$$B_{\pi\beta}(\varphi)/\bar{d} - K_\pi(\varphi,\varkappa) \equiv U_{\pi\beta}(\varphi,\varkappa,d), \tag{3}$$

conceiving the utility function of benefit and cost simply
as the difference between them (admittedly a special
case). [In case the delay $\bar{d}$ is long, or the equipment is
very durable, one must introduce discounting for time.
Note also the effect of delays on B (not K) whenever they
cause loss in informativeness: see Section 4.] Under our
assumptions, the problem is then, by the definition of
expected utility as the maximand for a consistent choice-
maker [see, e.g., Savage (1954)]:

$$\text{Max}_{(\varphi,\varkappa,d)\,\epsilon\rho_f}\,U_{\pi B}(\varphi,\varkappa,d). \qquad (4)$$

The notations of this Section may be reinterpreted, and
some assumptions relaxed, to allow greater generality.
More general than a chain is a network of processors.
This would permit to describe, for example, an organization with a well-defined objective (benefit) function, the
incentives to a given member being represented by a cost
function. A network can be reduced to a chain by defining the states (of the outside world and of the system) at
time-points i-1 and i as the inputs x^{i-1} and outputs x^i of
the i-th processor. This would take care of feedbacks.
It is also possible to reinterpret an event, an action,
and any intermediate output as a time-sequence and also to
include among possible actions that of "getting new information;" sequential sampling is a special case of such
"learning while earning." For events to remain independent of actions, X^0 can be also interpreted as a set of
unknown and rival "laws of nature": an interval for a
physical constant, a statistical parameter space. Another
generalization is to permit the collection (say, $\mathcal{X}$) of the
in- and output sets of all processors to vary: for example, the chooser of a system may be able to choose between
business enterprises belonging to different industries!

   Among important assumptions that need to be relaxed
is the finiteness of the in- and output; and the neglecting of the queueing problem. Yet, the following Sections
will have to introduce less rather than more generality.

3. EXPECTED UTILITY OF INQUIRING. In this and the subsequent Sections, and on the Diagram at the end of the
paper, the notations of Section 2 will be replaced by more
specific ones, using the following correspondence:

$$x^0 \in X^0 \rightarrow z \in Z \; : \; \text{events;}$$
$$\omega^1 \rightarrow \eta \qquad : \; \text{inquiring;}$$
$$\varkappa^1 \rightarrow \varkappa_\eta \qquad : \; \text{cost of inquiring;}$$
$$x^1 \in X^1 \rightarrow y \in Y \; : \; \text{observed data (or,}$$
$$\text{possibly, estimates,}$$
$$\text{diagnoses);}$$
$$\cdot \cdot \cdot \cdot \cdot \cdot \cdot \cdot \cdot$$
$$\omega^n \rightarrow \alpha \qquad : \; \text{deciding;}$$
$$\varkappa^n \rightarrow \varkappa_\alpha \qquad : \; \text{cost of deciding;}$$
$$x^n \in X^n \rightarrow a \in A \; : \; \text{actions.}$$

(See the Diagram for further notations.)

Essentially, statistical decision theory considers only two processors: inquiring and deciding. This can be stated formally in two alternative ways:

either (a) $\varphi^1 = \eta$, $\omega^i =$ identity matrix, $1 < i < n$;
or (b) $\omega^1 \omega^2 \ldots \omega^{n-1} = \eta$.

In the first case, communication is neglected and the statistician concentrates on the comparative advantages of alternative methods of sampling and of producing estimates on which decisions are based. In the second case, action is taken in response to a "message received," the terminal output of a chain or network of processors--observing, storing, encoding, transmitting, decoding--jointly characterized by a stochastic transformation η and by total cost and delay. Although possibly neglecting sampling details, the second case, too, can be said to capture much of the contents of the statistical decision theory. In any case, this theory (though not all statistical practice) makes two further assumptions, making the cost of deciding, and the delays, constant:

$$\varkappa_\alpha = 0 \text{ (or a constant)}; \quad \overline{d} = 1 \text{ (or a constant} > 0\text{)}.$$

With the decision cost neglected, all functions from Y onto A ("pure strategies") as well as their mixtures become available. The set of optimal decision rules includes then at least one pure strategy. Hence the set of transformations α can be reduced to deterministic ones. The expected utility is by (3)

$$U_{\pi\beta}(\eta,\alpha) = B_{\pi\beta}(\eta,\alpha) - K_\pi(\eta,\varkappa_\eta),$$

to be maximized, by (4), with respect to η, $\varkappa_\eta$, α subject to the feasibility of $(\eta, \varkappa_\eta)$ (and with no constraint on α, any function $Y \to A$). Define the <u>value of information</u> and use (1) [see also McGuire (1971)]:

$$V_{\pi\beta}(\eta) \equiv V(\eta) \equiv \underset{\alpha}{\text{Max}}\, B_{\pi\beta}(\eta,\alpha) \equiv \underset{\alpha}{\text{Max}}\, \text{trace } \pi\eta\beta \qquad (5)$$

$$\underset{\eta,\varkappa_\eta,\alpha}{\text{Max}}\, U_{\pi\beta}(\eta,\alpha) = V(\eta) - \underset{\eta,\varkappa_\eta}{\text{Min}}\, K_\pi(\eta,\varkappa_\eta). \qquad (6)$$

If Z consists of m events, all the considered matrices η have m rows. If these are all identical ($\eta_{zy} = \lambda_y$, all z), η is called <u>null-information</u>. If η is an identity matrix, it is called <u>complete information</u>. All null-information matrices can be shown to have the same value, viz.,

$$\underset{a \in A}{\text{max}}\, \Sigma_z \beta_{az}\pi_{zz} \equiv V^{\min} \leq V(\eta) \leq \Sigma_z(\text{max}_z \beta_{az})\pi_{zz} \equiv V(I), \qquad (7)$$

for all η with the same number of rows. The difference

$$G(\eta) \equiv V(\eta) - V^{\min} \geq 0, \qquad (8)$$

will be called "gain due to information η." The following example illustrates the <u>non-additivity</u> of information gains, <u>contrasted with the additivity</u> of the transmission rate (mutual information), a parameter of the joint distribution $\pi\eta$ of Z and Y which we shall denote by

$$R_\pi(\eta) \equiv H(Z) + H(Y) - H(Z \times Y), \qquad (9)$$

where, given the probabilities p_t, $t \in T$, $H(T) \equiv -\Sigma p_t \log p_t$, also conveniently denoted by $H(p)$. Define the Markov matrices $\eta^1, \eta^2, \eta^{12}$ by their typical entries

$$\eta^1_{z,y1} = p(y^1|z), \quad \eta^2_{z,y2} = p(y^2|z), \quad \eta^{12}_{z,y1y2} = p(y^1,y^2|z)$$

and let y^1, y^2 be independent. Given the distribution π of z, a real valued function $F_\pi(\eta)$ is additive only if

$$F_\pi(\eta^{12}) = F_\pi(\eta^1) + F_\pi(\eta^2). \qquad (10)$$

Thus $R_\pi(\eta)$ is known to be additive. But the gain $G_{\pi\beta}(\eta)$ (for β_π fixed) need not be. For example,[*] let event

---
[*] If our y and z are parameters describing, respectively, the estimated and the actual distribution of some variable, our example would contradict the additivity desideratum imposed by I.J. Good (1969) on the "utility of asserting" an estimated distribution, different from the actual one.

$z = (z_1, z_2)$ be a pair of independent random variables with

$$\Pr(z_i=1) = 1-\Pr(z_i=0) = \pi_1 = 1-\pi_0 < \tfrac{1}{2}; \quad i = 1,2,$$

where $z_i = 1$ (0) means fire (no fire) in suburb i. The possible fire danger is 1 money (or, better, utility) unit. To fix ideas let the expense of alerting 1 fire brigade be $= \pi_1$ (i.e., adjusted to the risk). Denote the number of alerted brigades (the <u>action</u>) by $a = 0,1,2$ and neglect the expense of putting out the fire, once alerted. Then the matrix of losses (the negative of the <u>benefits</u>), $[-\beta_{az}]$, has the following entries, row by row:

$$-\beta_{0z} = z_1+z_2; \quad -\beta_{1z} = \pi_1+z_1z_2; \quad -\beta_{2z} = 2\pi_1;$$

the probabilities of the events (1,1), (1,0), (0,1), (0,0) are, respectively: π_1^2, $\pi_1\pi_0$, $\pi_1\pi_0$, π_0^2. Assuming negligible decision costs, only pure strategies α need be considered. Now define four information systems, each with a different (and deterministic) transformation η; and compute for each η

say, its value $V_{\pi\beta}(\eta) = \max_\alpha \mathcal{E}_{zy}(\beta_{\alpha(y)}, z) = \mathcal{E}_{zy}(\beta_{\alpha*(y)}, z)$,

and also its gain $G_{\pi\beta}(\eta) = V_{\pi\beta}(\eta) - V_{\pi\beta}^{min}$ and the transmission rate $R_\pi(\eta)$:

Information system η:	"Complete", η^{12}	"Partial", η^1	"Partial", η^2	"Null", η^0
Message received, $y = \eta(z)$	(z_1, z_2)	z_1	z_2	constant
Optimal strategy, $a = \alpha*(y)$	z_1+z_2	1	1	1
Value, $V_{\pi\beta}(\eta)$	$-2\pi_1^2$	$-\pi_1\pi_0$	$-\pi_1\pi_0$	$-\pi_1\pi_0$
Gain, $G_{\pi\beta}(\eta)$	$\pi_1\pi_0$	0	0	0
Transmission rate $R_\pi(\eta)$	$2H(\pi)$	$H(\pi)$	$H(\pi)$	0

Note that η^1 and η^2 are "useless," calling as they do for the same optimal constant action $(a = 1)$ as does η^0. They yield no gain, although η^1 (or η^2) adds information on one variable, z^1 (or z^2), to the information at η^0. With η^{12}, information on still another variable is added. The function $G_{\pi\beta}$ (or G for brevity) is not additive: $G(\eta^{12}) >$

$G(\eta^1) + G(\eta^2)$. Super-additivity (and also sub-additivity) of the values in use (demand prices) occurs in other economic fields and is called complementarity (possibly negative): the value of a cow with a barn is larger than the value of a cow without a barn (in a city apartment, for example!) <u>plus</u> the value of a barn without a cow. A bag of fertilizer and 10 bushels of seeds have greater value together than the sum of their separate values.

On the other hand, the cost of transporting that bag and those bushels together is indeed (by and large) equal to the sum of the costs of transporting them separately. Similarly, the economic significance of the additivity of transmission rates is due to the fact that $R_\pi(\eta)$ measures (in the limit of long sequences of messages) the expected number of required code symbols, and this is, by and large, proportional to the cost of communication; and the cost of communicating two independent sequences would be equal to the sum of communicating them separately.

Unlike the costs of transportation, costs of production are, in general, non-additive. In our example, the cost of observing (as distinct from that of communicating about) fires in several suburbs may or may not equal the sum over separate observations. In statistical decision theory, sampling cost may be proportional to sample size but the choice of optimization procedures will depend on whether or not the expected utility $U_{\pi\beta}(\eta)$ is a concave function of sample size (case of "decreasing returns"): see Marschak (1971).

Another property of transportation costs is their (approximate) independence on contents: it costs about the same to transport over a given distance 100 pounds of seeds or of fertilizer. Similarly, communication costs do not depend on whether the message belongs to one or another equivalence class of messages with the same "meaning;" nor (economically more important!) on whether it belongs to a given (more inclusive) class of messages calling for the same optimal action. [This corresponds presumably to the distinction between "semantic" and "pragmatic" theories, as formulated by Cherry (1957) and others. Optimization is, then, "pragmatic."] Thus $R_\pi(\eta)$ does not depend on β. But $V_{\pi\beta}(\eta)$ does. Detailed and accurate information on trivial matters may be, on the average, less valuable than rough information on matters of life and death. This was clearly stated, e.g., in the very introductions to the main works of Shannon (1949) and Brillouin (1960). [Still, misunderstandings do occur; see, e.g., Charkewitch (1960)]. To illustrate, let $a = (a_1, a_2)$, $z = (z_1, z_2)$, where

$$z_1 = {1 \atop 0} \text{ if operation } {\text{succeeds} \atop \text{doesn't}}; a_1 = {1 \atop 0} \text{ means: } {\text{operate} \atop \text{don't}}; \eta^1(z) = y_1 = z_1$$

$$z_2 = {1 \atop 0} \text{ if stock } {\text{rises} \atop \text{doesn't}}; a_2 = {1 \atop 0} \text{ means: } {\text{buy} \atop \text{don't}}; \eta^2(z) = y_2 = z_2.$$

Let $\Pr(z_1=1) \equiv \pi_1 > P(z_2=1) = \frac{1}{2}$. Then $R_\pi(\eta^1) < R_\pi(\eta^2)$; but β may be such that $V_{\pi\beta}(\eta^1) > V_{\pi\beta}(\eta^2)$. Let $\beta_{az} = \sum_1^2 k_i \cdot \delta_{a_i z_i}$ (Kronecker delta), with $k_1 > k_2 > 0$. Then the optimal response to message y_i ($i = 1,2$) is: $a_i = y_i$, $a_j = 1$ ($j \neq i$). Hence

$$V_{\pi\beta}(\eta^1) = k_1 + \tfrac{1}{2}k_2 > V_{\pi\beta}(\eta^2) = \pi_1 k_1 + k_2 \text{ iff } \frac{k_1}{k_2} > \frac{1}{2(1-\pi_1)}.$$

Thus, although under our assumptions less information bits are conveyed by correct prediction of the outcome of surgical rather than stock market action, the former prediction is more valuable, provided surgical success has sufficiently high relative importance and sufficiently low probability.

4. PARTIAL ORDERING BY INFORMATIVENESS. Let the typical entries of matrices η, η' be $\eta_{zy} = p(y|z)$ and $\eta'_{zy'} = p(y'|z)$, respectively. Define the following conditions on any pair η, η' with the same number of rows:

(g) η g η' (η is "garbled," or "cascaded" into η') means:
$$p(y'|y,z) = p(y'|y).$$

(q) η q η' (η is "quasi-garbled" into η') means:[*]
$$\exists \text{ a Markov matrix q with } \eta' = \eta q.$$

(>) $\eta > \eta'$ (η is "more informative" than η') means:
$$V_{\pi\beta}(\eta) \geq V_{\pi\beta}(\eta'), \text{ all } \pi, \beta.$$

(R) η R η' (η has no lower transmission rate than η) means:
$$R_\pi(\eta) \geq R_\pi(\eta'), \text{ all } \pi.$$

---

[*] This terminology seems preferable to that used in earlier publications.

It has been shown [Blackwell (1953)] that (q) and (>) are
equivalent. Moreover,

$$(g) \mapsto (q) \Leftrightarrow (>) \mapsto (R), \tag{11}$$

where $(a) \mapsto (b)$ means: (b) if, but only if, (a). Whenever (q) is not satisfied there exist (π, β) and (π', β')
such that

$$V_{\pi\beta}(\eta) > V_{\pi\beta}(\eta') \text{ and } V_{\pi'\beta'}(\eta') > V_{\pi\beta}(\eta),$$

as already illustrated. Thus (>) orders the information
systems only partially. When the matrix q is deterministic
the relation $\eta q \eta'$ can be read "η is coarsened (contracted)
into η'." It follows immediately from (11) that lengthening
memory (or any addition of detail) cannot decrease informativeness. And it can be shown that increasing obso-
lescence (increasing delay) of complete information
about event-sequences cannot decrease informativeness, provided the sequences are Markovian. [See Marschak and
Radner (1972).]

5. CASE WHEN DECISION COST IS NOT NEGLIGIBLE. If $\varkappa_\alpha$ is
not negligible, the optimization problem cannot be reduced
to that stated in (5), (6) but becomes, more generally, as
in (4),

$$\text{Max}_{(\eta, \varkappa_\eta, \alpha, \varkappa_\alpha) \in \mathcal{P}_f} [B_{\pi\beta}(\eta, \alpha) - K_\pi(\eta, \varkappa_\eta, \varkappa_\alpha)],$$

where $\mathcal{P}_f$ is some <u>feasible set</u> of quadruples $(\eta, \varkappa_\eta, \alpha, \varkappa_\alpha)$.
In particular, some α can occur only in combination with
$\varkappa_\alpha$ infinite; such strategies themselves are then often
called non-feasible. It is also possible that a mixed
strategy is cheaper than some of its components and cannot
therefore be disregarded (as was done in Section 3). It
may be implemented by an error-making decider, whether man
or machine (not to be confused with the "organizer," our
"rational" chooser of an optimal information system in $\mathcal{P}_f$,
not of a single action in A) to be chosen by the organizer
in preference to an expensive error-free one.

It is therefore important to note the results obtained
by Perez (1964) who concerned himself with the "average
risk," i.e., the negative of our "expected benefit"
$B_{\pi\beta}(\eta, \alpha)$, with α, generally mixed--and not only with the
"Bayes risk" which is the negative of our "information
value," $V_{\pi\beta}(\eta)$. An upper bound on the difference between

the expected benefits of (η,α) and (η',α') [a lower bound is obtained by interchanging the two pairs] is given by

$$[B_{\pi\beta}(\eta,\alpha) - B_{\pi\beta}(\eta',\alpha')]$$

$$\leq [2B_{\beta^2\pi}(\eta',\alpha')\cdot H(\pi\eta\alpha\beta|\pi\eta'\alpha'\beta)]^{\frac{1}{2}},$$

using our notations and adjusting to our context (Perez considers varying π, and not only η, α); where $\beta^2(a,z) = (\beta(a,z))^2$, and $H(p|p')$ stands for the "equivocation" $\equiv H(p) - R_p(p')$. Note that the latter expression depends not only on the compared joint probability distributions $\pi\eta$, $\pi\eta'$ but also on $\alpha\beta$, $\alpha'\beta$ characterizing decisions and benefits.

What do we know about decision cost $\varkappa_\alpha$? While the inquiry cost, $\varkappa_\eta$, is reasonably transparent at least in the case of statistical sampling (including quality control), I wish I knew of some systematic work on the cost of deciding and its "technology," as exemplified by the number of required computer operations and by psycho-physiological observations of problem-solving behavior. Bongard (1963) may be mentioned here as one who tried to justify the entropy formula as a measure of the average "degree of difficulty of a problem." He appeals to an "intuitive feeling" that the "difficulty" is roughly proportional to the expected logarithm of the number of search acts. This psychological hypothesis reminds one of Daniel Bernoulli's (1738) logarithmic utility of monetary wealth which inspired G.Th. Fechner (1859) to postulate subjective "sensations" as logarithms of the intensity of a physical stimulus: mathematically attractive but without supportive evidence.

6. EXPECTED BENEFIT AND COST OF COMMUNICATION. Essentially, communication theory is isolated out of a comprehensive theory of information and decision by a device analogous to (a) of Section 3, page 5. Using the notations of our Diagram, assume

$\eta = \alpha$ = identity matrix; or, alternatively, if   (12)
    messages are decoded directly into the
    language (set) of actions,

$\eta = \delta$ = identity matrix.   (12a)

In much of the literature, a restriction is also imposed upon the benefit function by assuming

$$\beta_{az} = \begin{matrix} 0 \\ -1 \end{matrix} \ \text{if} \ a \overset{=}{\neq} z, \tag{13}$$

so that, by (12) or (12a), expected benefit is the negative
of the probability of communication error. It is more
interesting in our context, not to use restriction (13)
but to follow the more general and recent approach of
Shannon (1960) anticipated in his earlier work (1949) for
continuous message sets only. He has called the negative
of our β_{az} the "distortion measure." His example: to re-
ceive message "all's well!" when the message sent is "emergency!" is a much worse distortion than the converse. In
our terminology, to optimize a communication system is to
maximize expected utility, i.e., the expected difference
between benefit and cost, per time unit, as in (3).
As on our Diagram, let the transmission channel have a
Markov matrix τ, of order m × n, say. Using R for transmission rate as before, define its capacity

$$C(\tau) \equiv \max_{\lambda \, \in \Lambda} R_\lambda (\tau) \ \text{bits per letter}, \tag{14}$$

where Λ is the set of all probability vectors of order m.
It would be more conforming with the usual economic terminology to call capacity the quantity

$$vC(\tau) \ \text{bits per time unit}, \tag{15}$$

where v is the channel speed, i.e., the number of letters
received per time unit. Presumably, the higher (15) the
higher the cost, constant per time unit, of maintenance
and depreciation. [It seems, in fact, that, for fixed v,
channels with equal $C(\tau)$ are assumed to be equally expensive, even in the case like

$$C \begin{pmatrix} .89 & .11 \\ .11 & .89 \end{pmatrix} \cong .32 \cong C \begin{pmatrix} 1 & 0 \\ \tfrac{1}{2} & \tfrac{1}{2} \end{pmatrix}.]$$

On the other hand, the expected benefit per letter, $B_\lambda(\tau,\gamma)$,
will depend on the matrix τ and the code $\gamma \equiv (\sigma,\delta)$ (see
Diagram); the expected benefit per time unit is decreased,
as in (3), by delays; and these depend on the number of
letters, w (say) sent per time unit. No expected benefit
for letter can exceed that for τ noiseless (an identity
matrix), B^{max}, say, analogous to the value V^{max} of complete
information in (7). For a preassigned expected benefit B*,
and a given code γ, define

$$Q_{\lambda\beta}(B^*,\gamma) \text{ bits per letter} \equiv \underset{\tau}{\text{Max }} R_\lambda(\tau)$$
$$\text{taken over all (m$\times$n)-Markov matrices,} \tag{16}$$
$$\text{subject to } B^* \leq B_{\lambda\beta}(\tau,\gamma) \leq B^{max}.$$

Shannon proved that for any positive ϵ there exists a number $N(\epsilon)$ and a sequence of codes $\gamma_1,\ldots,\gamma_{M(\epsilon)},\ldots$ such that, if $M(\epsilon) > N(\epsilon)$ then

$$B^* - B_{\lambda\beta}(\tau,\gamma_{M(\epsilon)}) \leq \epsilon, \tag{17}$$

provided the channel capacity and speed is sufficiently large in the sense that

$$w\cdot Q_{\lambda\beta}(B^*,\gamma_{M(\epsilon)}) \leq v\cdot C(\tau). \tag{18}$$

[It can be verified that when the benefit function is that of (13), then the left side of (17) becomes the probability of error and that of (18) becomes $w\cdot H(\lambda)$ as in the "Fundamental Theorem for Discrete Channel with Noise" of Shannon's original paper (1949).]

However, large capacity entails high constant cost per time unit; and the required code $\gamma_{M(\epsilon)}$ may cause long delays in coding, as well as in the storage preliminary to encoding large "blocks" of messages. Thus, while increasing the benefit per letter, $\gamma_{M(\epsilon)}$ may result in a decreased expected utility.

7. CONCLUDING REMARK. The "economic" interpretation of the theory of statistics and of information is not the only possible one. In the context of statistical inference, Sir Ronald Fisher (1955) was unhappy about the modern Russians' "ideal that research in pure science can and should be geared to technological performance" and disliked the American fashion "to confuse the process appropriate for drawing conclusions with those aimed at ... saving money." And a leader in information mathematics, A. Renyi (1965) stated the difference between the "pragmatic" interest of J. Wolfowitz in optimal coding, and his own "axiomatic" approach that seeks "to find all the expressions which possess (those) postulated properties which a reasonable measure of information should satisfy." The latter approach is also that of S. Kullback's (1959) book on our subject. But in the present Colloquium on Optimization Techniques no apologies are needed for the approach used in this paper.

The support given by the Western Management Science

Institute and the National Science Foundation and the
Office of Naval Research is gratefully acknowledged.

References

D. Bernoulli (1738) Exposition of a new theory on the
measurement of risk. English translation, Econometrica, 22
(1954).

D. Blackwell (1953) Equivalent comparisons of experiments. Ann. Math. Stat, 24.

M. Bongard (1963) On the concept of "useful information."
Problemy Kibernetiki, 9.

L. Brillouin (1960) Science and Information Theory.
Academic Press.

A. Charkewitch (1960) On value of information. Problemy
Kibernetiki, 4.

C. Cherry (1957) On Human Communication. Wiley.

M. De Groot (1962) Uncertainty, information, and sequential experience. Ann. Math. Stat., 33. -- (1970) Optimal
Statistical Decisions. McGraw-Hill.

G. Th. Fechner (1859) Elemente der Psychophysik. Leipzig.

R. Fisher (1955) Statistical methods and scientific
induction. J. Royal Stat. Soc., B, 17.

I.J. Good (1969) What is the use of a distribution?
Multivariate Analysis, Vol. 2. Academic Press.

S. Kullback (1959) Information Theory and Statistics.
Wiley.

J. Marschak (1963) The payoff-relevant description of
states and acts. Econometrica, 31. -- (1971) Economics of
information systems, J. Amer. Stat. Assn., 66.

-- and K. Miyasawa (1968) Economic comparability of information systems. Internat. Econ. Rev., 9.

-- and R. Radner (1972) Economic Theory of Teams. Yale
University Press.

C.B. McGuire (1971) Comparisons of information structures. Decision and Organization, R. Radner and C.B.
McGuire, eds. North-Holland Publishing Co.

J. Neyman and E. Pearson (1928-38) Joint Statistical
Papers. Republished by University of California Press,
1967.

A. Perez (1964) Information theory methods in reducing
complex decision problems. Trans. 4. Prague Conf. Inform.
Theory. Prague.

A. Renyi (1965) On the foundations of information
theory. Bul. Internat. Stat. Inst. 33.

L.J. Savage (1954) The Foundations of Statistics. Wiley.

C. Shannon (1949) The Mathematical Theory of Communication. Illinois Univ. Press (with W. Weaver). -- (1960) Coding theorems for a discrete source with a fidelity criterion. Information and Decision Processes, R.E. Machol, ed. McGraw-Hill.

A. Wald (1950) Statistical Decision Functions. Wiley.

J. Wolfowitz (1961) Coding Theorems of Information Theory. Springer.

A. and I. Yaglom (1960) Probability and Information, 2nd Ed. (in Russian). Fizmat.

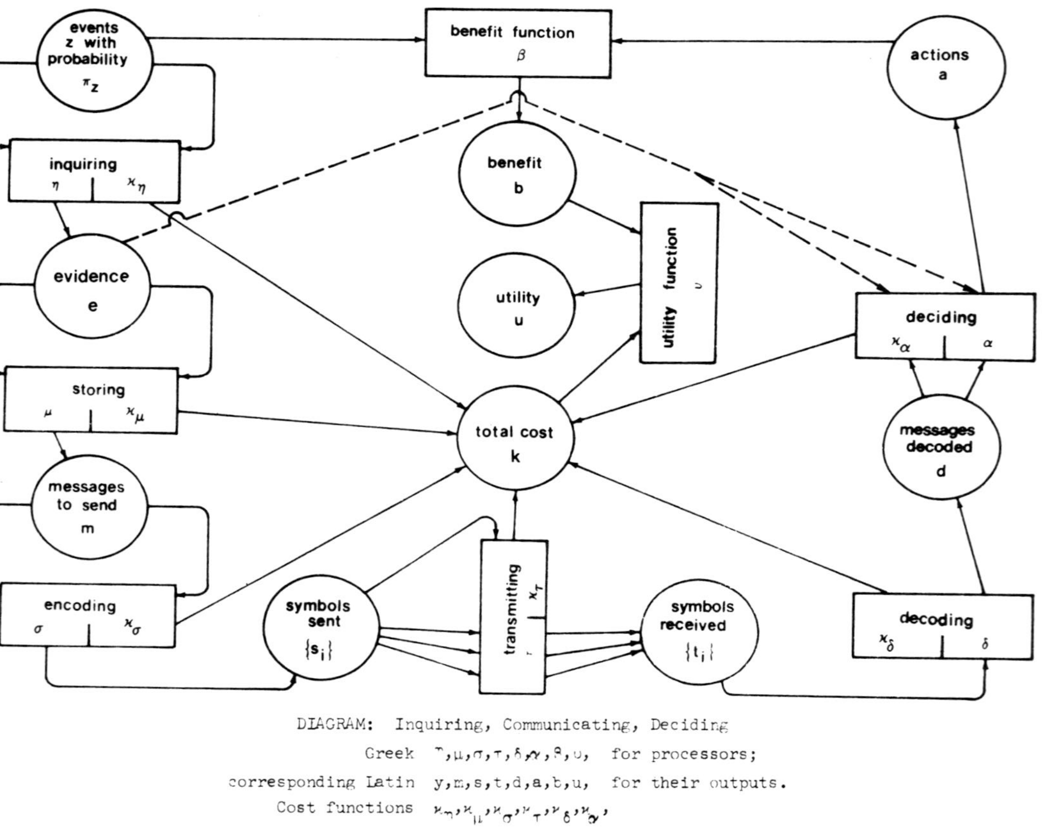

DIAGRAM: Inquiring, Communicating, Deciding

Greek $\eta, \mu, \sigma, \tau, \delta, \alpha, \beta, \upsilon$, for processors;

corresponding Latin y,m,s,t,d,a,b,u, for their outputs.

Cost functions $\kappa_\eta, \kappa_\mu, \kappa_\sigma, \kappa_\tau, \kappa_\delta, \kappa_\alpha$,

with k as joint output.

SHADOW PRICES FOR INFINITE GROWTH PROGRAMS:
THE FUNCTIONAL ANALYSIS APPROACH*

M. Majumdar and R. Radner

Stanford University and
University of California, Berkeley

Abstract

This paper summarizes some recent results on price
systems characterizing efficient or optimal programs in
economic models for which the commodity space can be
represented as a subset of L_∞. In general, a price system
is represented as a continuous linear functional on L_∞.
An efficient program is called regular if there is an
associated price system of the form $p(y) = \int gy$, where g is
in L_1. Various conditions are given under which efficient
programs are regular, and also under which efficient
programs can be approximated by regular programs. The
approach is illustrated by examples of economic optimization problems involving time and uncertainty.

I. Introduction and Summary

Some of the most intensively studied economic problems
center around the characterization of efficient programs of
resource allocation by a price system, and the use of a
price mechanism to attain such an allocation in economies
in which decision-making is decentralized. A well-known
proposition is that, under an appropriate convexity
assumption on technology, an efficient production program
maximizes the value of net output if value is calculated
using a proper system of prices. The extension of the
standard theory to programs involving infinite numbers of
"commodities" poses certain mathematical difficulties,

---

*Research supported by National Science Foundation
grants GS-30377 and GS-2078.

which in turn raise deeper conceptual problems concerning the proper definition of a "price system" when the commodity space is infinite dimensional. Alternative approaches have been suggested by different writers (see Debreu, 1954; Hurwicz, 1958; Malinvaud, 1953; and Peleg and Yaari, 1970); and in the present paper we survey some recent results obtained with the tools of functional analysis.

From the point of view of mathematical programming or control theory, prices can be identified with "Lagrangean multipliers" or "dual variables." In addition to suggesting how economic decisions can be decentralized, the study of price systems associated with efficient or optimal programs has been used to provide qualitative information about such programs, e.g., about existence and asymptotic properties (see, e.g., the collection of articles on efficient and optimal economic growth in the <u>Review of Economic Studies</u>, Vol. 34, 1967).

Recall that with a finite dimensional commodity space, say R^ℓ, a price system (p_i) is simply an ℓ-vector such that if $y = (y_i)$ is any consumption program involving these commodities, the <u>value</u> of y relative to p is given by

$$(1.1) \quad p(y) = \sum_{i=1}^{\ell} p_i y_i .$$

From the expression (1.1) the following useful properties are clear:
- (a) the value is well-defined for any y in R^ℓ and any p in R^ℓ;
- (b) for fixed p it is a homogeneous linear function of y, and vice versa;
- (c) it is continuous in p and y.

By virtue of (a), it is possible to compare the values of any two programs; and a price being associated with each commodity it is possible to talk about ratios of prices, which may be used to measure the rates of exchange between commodities.

In Section II we present several examples of situations involving infinitely many commodities, and show in each case that the set of all technologically possible consumption programs (measured in suitable units) can be represented as a proper subset of the linear space $L_\infty(\Omega, \mathcal{F}, \mu)$, where $(\Omega, \mathcal{F}, \mu)$ is some suitably chosen complete, σ-finite measure space, and may be interpreted as the set of "commodities."

By assigning the μ- essential sup-norm to L_∞, and considering a price system or a value functional to be a linear functional continuous in the norm topology, it is clearly possible to conserve (a) and (b) and continuity in y; if, further, the space of all continuous linear functionals (L_∞^*) is endowed with the weak* topology, one gets joint continuity. It can be shown that, if Y is a convex set representing all possible consumption programs and y* is an efficient point of Y (i.e., maximal in the vector ordering), then y* is indeed <u>value-maximizing</u> relative to some non-zero p*, i.e., there exists a non-zero p* in L_∞^* with $p^*(y^*) \geq p^*(y)$ for all y in Y.

In most applications to economic theory, if p* is to have a useful economic interpretation as a price system, then it must be representable in the form

$$(1.2) \qquad p^*(y) = \int_\Omega g^* y d\mu \; ,$$

where g* is a μ-integrable function on the set of commodities. Unfortunately, this representation is not possible in general. A theorem of Yosida and Hewitt (1952) shows that a general representation of p* is

$$(1.3) \qquad p^*(y) = \int_\Omega g^* y d\mu + \int_\Omega y d\nu \; ,$$

where g* is as in (1.2), and ν is a bounded, purely finitely additive measure that vanishes on sets of μ-measure zero (see Section III). In problems of dynamic optimization with an infinite time horizon, the second term on the right side of (1.3) reflects the asymptotic behavior of the program y, whereas in problems of decision in the face of uncertainty, this term can be interpreted as reflecting the behavior of the program y on sets of "arbitrarily small probability." It is, of course, the first part of (1.3) that corresponds most directly to (1.1), since $g^*(\omega)$ can be interpreted as the "price of commodity ω."

Two approaches can be taken to deal with the problems introduced by the second term in (1.3). First, one can give various compactness conditions under which every maximal program in the feasible set can be approximated by maximal programs for which the associated price system is of the form (1.2) (Section IV). Second, one can give conditions (Section V) that guarantee that one can choose a price system of the form (1.2), even though there may be other price systems in which the second part of (1.3) is not zero.

No proofs will be included in the present paper; for the less obvious proofs the reader will be referred to items in the bibliography at the end.

II. Economic Examples

In this section we present several economic models, in order to motivate the more formal and abstract presentation that follows in the subsequent sections, as well as to illustrate the diversity of models whose analysis can be unified by the functional analysis approach.

Example 1. A Discrete Time Infinite Horizon Model

Some of the typical difficulties involved in problems of intertemporal resource allocation appear even in a simple aggregative model with just one producible good that can be used either in consumption or as an input for further production of the same good (or it can be thrown away--"free disposal"). Let q_t denote the output at date t, k_t the capital stock, I_t the (gross) investment, and y_t the consumption (t = 1,2,..., ad. inf.). The initial capital stock, k_0, is given. The output at date t is determined by the capital stock at the previous date:

$$(2.1) \qquad q_t = f(k_{t-1}) \ .$$

The capital stock is assumed to depreciate at a constant rate, δ, which is strictly between 0 and 1; thus, if $\sigma \equiv 1 - \delta$,

$$(2.2) \qquad k_{t+1} = \sigma k_t + I_t \ .$$

By convention, consumption plus investment cannot exceed output, and the capital stock must be nonnegative:

$$(2.3) \qquad y_t + I_t \leq q_t \ , \qquad k_t \geq 0 \ .$$

The nonnegative function f is called the production function. A negative investment may be interpreted as "eating up" or "running down" the capital stock. A negative consumption implies a need to borrow or obtain "aid" from outside the system.

Given the initial capital stock, k_0, the set Y of feasible consumption sequences, (y_t), is determined by (2.1)-(2.3) for $t \geq 0$. A consumption sequence $y^* = (y_t^*)$ in Y is <u>efficient</u> or <u>maximal</u> in Y if there is no other sequence y in Y such that for all t, $y_t \geq y_t^*$, and for some t, $y_t > y_t^*$. If U is a real-valued function on Y, then $\hat{y}$ is called <u>optimal</u> in Y with respect to U if it maximizes $U(y)$ in Y.

Let p_t denote the price of the (single) commodity at date t. The total value of consumption is (assuming no disposal)

$$(2.4) \quad \sum p_t y_t = \sum p_t [f(k_{t-1}) - k_t + \sigma k_{t-1}]$$

$$= p_1 [f(k_0) + \sigma k_0] + \sum_{t \geq 2} (p_{t+1}[f(k_t) + \sigma k_t] - p_t k_t) .$$

A consumption sequence $y^* = (y_t^*)$ maximizes the value (2.4) in the set Y if, and only if, for each $t \geq 1$, the corresponding capital stock, k_t^*, maximizes

$$(2.5) \quad p_{t+1}[f(k_t) + \sigma k_t] - p_t k_t ,$$

or, if $p_t > 0$, the expression

$$(2.6) \quad [1/(1 + r_t)][f(k_t) + \sigma k_t] - k_t ,$$

where the ratio $(1 + r_t) = (p_t/p_{t+1})$ is called the <u>interest factor</u> at date t, and r_t is the corresponding <u>interest rate</u>.

Since k_t is the input of capital at date t, and $[f(k_t) + \sigma k_t]$ is the resulting new output plus the surviving capital, the expression (2.6) may be interpreted as the "intertemporal profit" at date t, where the value of new output plus surviving capital stock is "discounted" according to the interest rate r_t. Alternatively, expression (2.5) may be interpreted as "intertemporal profit at date t discounted from date 0."

Thus, if an efficient program can be characterized as maximizing the value (2.4) for some sequence (p_t), then it can equivalently be characterized as maximizing intertemporal profit at each date, provided the sum (2.4) converges. Notice that if the sequences in Y are uniformly bounded,

and (p_t) is absolutely summable, then the sum (2.4)
converges for every program in Y. However, discounted
intertemporal profit (2.5) is well defined at each date for
any program (y_t) and any sequence (p_t), whether or not the
total value (2.4) converges.

Example 2. A Continuous Time Model

Even with a finite horizon, the treatment of time as a
continuous variable leads to a continuum of commodities.
As an illustration, we present now a version of the
previous example, with continuous time and a finite
horizon. We also introduce the idea of a primary resource,
which is necessary for production but is not produced and
is supplied exogenously.

Consider the time interval $[0,T]$, where 0 is the
present instant, and $T > 0$ is the "horizon." At time t,
let output, capital stock, labor input, and investment be
denoted by $Q(t)$, $K(t)$, $L(t)$, and $I(t)$, respectively;
output, labor and investment are flows, whereas capital is
a stock. Corresponding to (2.1) and (2.2) of Example 1, we
have

$$(2.7) \quad Q(t) = F[K(t),L(t)],$$

$$(2.8) \quad \dot{K}(t) = -\delta K(t) + I(t), \quad K(t) \geq 0.$$

Assume that the labor input is determined exogenously, by

$$(2.9) \quad L(t) = L(0)e^{\lambda t}, \quad \text{and } L(0) > 0,$$

and that the production function, F, is homogeneous of
degree one ("constant returns to scale"). With these two
assumptions we can reduce the variables of the system to
per capita terms. Define

$$(2.10) \quad q(t) = Q(t)/L(t), \quad k(t) = K(t)/L(t),$$
$$y(t) = [Q(t)-I(t)]/L(t), \quad f(x) \quad F(x,1).$$

Then, corresponding to (2.3), (2.7), and (2.8), one can
verify that

$$(2.11) \quad q(t) = f[k(t)],$$

$$(2.12) \quad y(t) + \dot{k}(t) \leq q(t) - (\delta + \lambda)k(t), \quad k(t) \geq 0.$$

If we further impose the condition that per capita consumption, $y(t)$, be nonnegative, then (2.11) and (2.12) reduce
to

(2.13a) $\dot{k}(t) \leq f[k(t)] - (\delta + \lambda)k(t), \quad k(t) \geq 0,$

(2.13b) $y(t) = f[k(t)] - (\delta + \lambda)k(t) - \dot{k}(t).$

We also impose initial and terminal conditions on the capital stock:

(2.13c) $k(0) = k^*, \quad k(T) = k_*,$

where k^* is the given initial ratio of capital stock to labor input, and k_* is the terminal value of $k(t)$ that one wants to attain.

Define the set π of <u>feasible capital accumulation programs</u> as the set of nonnegative absolutely continuous functions k on $[0,T]$ satisfying (2.13a-c), and let Y be the set of corresponding <u>per capita consumption programs,</u> defined by (2.13b).

<u>Example 3. A Multisector Model</u>

A natural extension of the one-producible-good model discussed in the first two examples is one in which there is a finite number of goods (say ℓ) at each date. Again, taking time as a discrete variable, assume that at each date $t(= 1,2,\ldots)$, for every nonnegative <u>input</u> vector a in R^ℓ, there is a set $P_t(a)$ of possible <u>output</u> vectors available at date $(t+1)$ for possible consumption or for use as an input in production. The correspondence P_t is the <u>production possibility correspondence</u> at date t. The <u>graph</u> G_t of P_t is the set $\{(a,b): b \text{ is in } P_t(a)\}$ in $R^{2\ell}$.

A <u>production program</u> is a sequence $\{(a_t,b_{t+1}): t = 1,2,\ldots\}$ such that $a_t \geq 0$, $b_{t+1} \geq 0$, and b_{t+1} is in $P_t(a_t)$. For any production program, the <u>net output program</u> is defined by

(2.14) $y_1 = -a_1 ,$

$$y_t = b_t - a_t , \qquad t \geq 2 .$$

The set of all net output programs $y = (y_t)$ corresponding to all possible production programs is denoted by Y.

One may wish to impose an additional constraint, corresponding to the constraint that consumption be nonnegative. Suppose that at each date t there is a nonnegative vector w_t of quantities of commodities exogenously supplied to the economy. Since consumption at t equals $(y_t + w_t)$, we may impose the constraint $(y_t + w_t) \geq 0$; the corresponding subset of Y will be denoted by Y_w.

A particular case of the foregoing is the <u>linear activity analysis</u> model. For each t let A_t and B_t be given nonnegative matrices, called the input- and output-coefficient matrices, respectively, and let the production correspondence P_t be defined by

$$(2.15) \qquad P_t(a_t) = \{b_{t+1}: \text{ for some } x_t \geq 0, \; b_{t+1} \leq B_t x_t,$$

$$a_t \geq A_t x_t \} \; .$$

The vector x_t is called the vector of activity levels at date t.

For a more general discussion of such multisector models, see Radner (1967) and the references given there, especially Malinvaud (1953).

Example 4. <u>Allocation under Uncertainty</u>

Finally, we present an example of dynamic resource allocation under uncertainty. In this example, the technology and resources at each date depend upon the history of the (stochastic) environment up to that date. The model presented here is an extension of the model of Example 3.

Let S_t denote the finite set of all events that can occur at date t, and let S^t denote the set of t-tuples, $s^t = (s_1,\ldots,s_t)$, with s_t in S_t. The t-tuple s^t is a possible history of the environment from date 1 through date t. Suppose that $(s_1, s_2, \ldots, \text{ ad inf.})$ is a stochastic process, i.e., there is a probability measure on $S = \overset{\infty}{\underset{t=1}{X}} S_t$, and that at each date the conditional distribution of future events, given the past events and economic decisions, does not depend on the economic decisions taken. In other words, the environment is stochastic and exogenous (this is by now a standard approach in decision theory and economics; see Savage, 1954, and Ch. 7 of Debreu, 1959).

The production correspondence of Example 3 is generalized as follows: if a_t is the vector of inputs at date t, then $P_t(a_t, s^t)$ is a set of "stochastic outputs," i.e., a set of <u>functions</u> from S^{t+1} to R^ℓ_+ , the nonnegative orthant of R^ℓ. A production program is a sequence (α_t, β_t) such that (i) α_t and β_t are functions from S^t to R_+^ℓ , and (ii) for every s^t in S^t, β_{t+1} is in $P_t[\alpha_t(s^t),s^t]$.   If we write $a_t = \alpha_t(s^t)$ and $b_t = \beta_t(s^t)$, then the (stochastic) <u>net outputs</u>, y_t, are determined by (2.14), and we may write $y_t = \eta_t(s^t)$.   The set Y of <u>net output programs</u> is therefore the set of sequences (η_t) corresponding to the set of production programs.

If the exogenous supplies of resources, w_t (see Example 3), are also stochastic, i.e., determined by corresponding functions ν_t on S^t, and we constrain consumption to be nonegative, then, corresponding to the set Y_w of Example 3 we get the set Y_ν , where $\nu = (\nu_t)$.

Models along the lines of the present example have been used in Radner (1971a, 1971b), Jeanjean (1971), and Stigum (1971).

III.   Programs and Prices

Let $(\Omega, \mathcal{F}, \mu)$ be a complete, σ-finite measure space. In our models, Ω can be identified with the set of all commodities.   Consider $L_\infty(\Omega, \mathcal{F}, \mu)$, the linear space of all real-valued measurable functions that are bounded μ-almost everywhere.   For y in $L_\infty(\Omega, \mathcal{F}, \mu)$ one can define the "μ-essup norm" as

$$(3.1) \qquad \|y\| = \inf_N \; \sup_{\omega \in \Omega \sim N} \; |y(\omega)| \; ,$$

where N ranges over sets of μ-measure zero.   A special example is obtained by taking Ω to be any countable set (in this case, without loss of generality Ω can be chosen to be the set of all positive integers), $\mathcal{F}$ the σ-field of all subsets of Ω and μ the counting measure; in such a case, L_∞ is simply the space of all bounded real sequences (known as ℓ_∞).

At the end of this section we shall indicate how for each of the examples in the previous section the set Y of feasible consumption programs can be identified with a convex subset of L_∞ , provided certain assumptions are made and the units of measurement are suitably chosen.   Since a point y in L_∞ specifies the quantity $y(\omega)$ of every commodity ω, we shall call points of L_∞ <u>programs</u>.

379

A program y in L_∞ is nonnegative, written $y \geq 0$, if $y(\omega) \geq 0$ for almost every (a.e.) ω in Ω; y is _positive_, written $y > 0$, if $y \geq 0$ _and_ $\mu\{\omega: y(\omega) > 0\} > 0$. For y_1, y_2 in L_∞, $y_1 \geq y_2$ (resp. $y_1 > y_2$) if $y_1 - y_2 \geq 0$ (resp. $y_1 - y_2 > 0$).

For a subset Y of L_∞, y^* in Y is a _maximal_ or _efficient_ element of Y if there does not exist another y in Y, with $y > y^*$.

A _price system_ or a _value functional_ is simply a linear functional on L_∞ that is continuous in the topology generated by the norm defined in (3.1). The set of all value functionals is denoted by L_∞^*. For p in L_∞^*, define

$$(3.2) \qquad \|p\|_* = \sup_{\|y\| \leq 1} |p(y)| \ .$$

For p in L_∞^*, p is _nonnegative_ (written $p \geq 0$) if $p(y) \geq 0$ for all $y \geq 0$; p is _positive_ ($p > 0$) if $p \geq 0$ and $p \neq 0$; and p is _strictly positive_ ($p >> 0$) if $p \geq 0$ and $p(y) > 0$, for all $y > 0$.

Since the measure space $(\Omega, \mathcal{F}, \mu)$ is assumed to be complete and σ-finite, if p is in L_∞^*, one has the following representation (Yosida and Hewitt, 1952):

$$(3.3) \qquad p(y) = \int_\Omega ygd\mu + \int_\Omega yd\nu_p \ ,$$

where g is an integrable function on Ω (i.e., g belongs to $L_1(\Omega, \mathcal{F}, \mu)$ and ν_p is a bounded, purely finitely additive signed measure that vanishes on all sets of μ-measure zero. g will be called the "L_1-part of p" and ν_p the "purely finitely additive part of p." If $\nu_p \equiv 0$, we shall say that "p is in L_1." When p is in L_1, we have

$$p(y) = \int_\Omega ygd\mu \ ,$$

so that the value of the program y relative to p [i.e., the number $p(y)$] is an integral over the set of commodities; $g(\omega)$ is the unit price of commodity ω, and the situation is analogous to that of the special case in which Ω is finite and value is the scalar product of the price and quantity vectors.

In the case of ℓ_∞, we have, for any $y = (y_\omega)$ in ℓ_∞ and p in ℓ_∞^*

$$(3.4) \quad p(y) = \sum_{\omega=1}^{\infty} g_{\omega} y_{\omega} + \int_{\Omega} y d\nu_p$$

where (g_{ω}) is in ℓ_1 .

The existence of the purely finitely additive part
creates problems of economic interpretation. It should be
mentioned at the outset that even for a nonzero $p \geq 0$, it
is possible for the L_1-part of p to be zero. However, if p
is represented by a purely finitely additive measure, then
only a "small" set of commodities can possibly have nonzero
value, as the following lemma shows:

Lemma 3.1 (Yosida and Hewitt, 1952): If ψ is a
countably additive measure on $(\Omega, \mathcal{F})$ with $\psi(\Omega)$ finite, and
ν is any purely finitely additive measure, then for any
$\epsilon > 0$, there is A_ϵ in $\mathcal{F}$ with $\psi(A_\epsilon) < \epsilon$ and $\nu(\Omega \sim A_\epsilon) = 0$.

Going back to the example of a countable Ω--and thus
without loss of generality Ω may be chosen to be the set of
positive integers ("dates" $\omega = 1,2,\ldots$)--the Yosida-Hewitt
lemma implies that if the ℓ_1-part of p is zero, the value
of any plan $y = (y_{\omega})$ is determined by what happens "at
infinity." Thus if $y = (y_{\omega})$ converges to $\bar{y}$ (as $\omega \to \infty$),
then for any p (with $\|p\|_* = 1$) that is representable by a
purely finitely additive measure, $p(y) = \bar{y}$; more generally
one can show that

$$(3.5) \quad \lim \inf y_{\omega} \leq p(y) \leq \lim \sup y_{\omega}$$

whenever p has a zero ℓ_1-part and $\|p\|_* = 1$ (in Lemma 3.1,
let ψ assign the measure $(1/2^{\omega})$ to the point ω in Ω).

If Y is a subset of L_∞, y^* is a program in Y, p is
a price system in L_∞^* , and

$$(3.6) \quad p(y) \leq p(y^*) \qquad \text{for all } y \text{ in } Y,$$

then we shall say that y^* is value maximizing (in Y) with
respect to p, or equivalently that p supports Y at y^*. The
following theorem relates the concepts of efficiency and
value maximization (see Majumdar, 1971a).

Theorem 3.1: If p in L_∞^* supports a subset Y of L_∞ at y^*
in Y, and $p \gg 0$, then y^* is maximal in Y. On the other
hand, if Y is convex, and y^* is maximal in Y, then there is
a $p > 0$ in L_∞^* that supports Y at y^*.

Examples of programs supported by a purely finitely

additive price system are given in Radner (1967), McFadden (1968), and Peleg-Yaari (1970).

We now reconsider the examples of Section II. In each case we shall show that the set Y of feasible consumption programs has certain compactness properties. The relevance of these compactness properties will become apparent in the next section.

<u>Example 1</u>. Take Ω to be the nonnegative integers (the dates), and μ to be the counting measure. Assume that the production function, f, is nonnegative and concave, that $f(0) = 0$, and that $f'(k)$ approaches 0 as k increases without limit. Under these assumptions, Y is convex, and one can show (see Koopmans, 1965) that for every k_0 there is a constant c such that $y_t \le c$ for every feasible consumption program. Hence the set Y_+ of nonnegative feasible consumption programs is a convex norm-bounded subset of ℓ_∞ . One can also show that Y_+ is closed in the product topology, and therefore in the $w(\ell_\infty, \ell_1)$ topology. Hence Y_+ <u>is a convex</u>, $w(\ell_\infty, \ell_1)$--<u>compact subset of</u> ℓ_∞ (see Dunford and Schwarz, 1957, V.4.3).

<u>Example 2</u>. Take Ω to be the interval $[0,T]$, $\mathcal{F}$ to be the Borel sets of $[0,T]$, and μ to be Lebesgue measure. Assume that $F(0,L) = F(K,0) = 0$ (i.e., that both capital and labor are necessary in production), that F is strictly concave and homogeneous of degree one, and that f is twice differentiable, with $f'(0) > \delta + \lambda$. Koopmans (1965) has shown that under these assumptions, π is a norm-bounded subset of $\mathcal{E}[0,T]$, considered as a subset of L_∞ . From this one can show that π <u>is a norm-compact convex subset of</u> $\mathcal{E}[0,T]$, and that <u>Y is a</u> $w(L_\infty, L_1)$ -<u>compact convex subset of</u> L_∞ (see Majumdar, 1971b).

<u>Example 3</u>. We shall say that a production possibility correspondence P is <u>neoclassical</u> if (i) $P(0) = \{0\}$, (ii) the graph of P is a closed convex cone, and (iii) b in $P(a)$, $b' \le b$, $a' \ge a$, imply b' in $P(a')$ ("free disposal"). One can show (see Radner, 1967) that if at every date t, P_t is neoclassical, then for any given sequence (w_t) of exogenous resources there is a sequence (h_t) of nonnegative numbers such that, for all consumption programs (y_t) in Y_w, $\|y_t\| \le h_t$ for all t (where $\|\ \|$ denotes the Euclidean norm in R^ℓ). It follows from this that there is a change of units at each date such that Y_w can be represented as a

convex, norm-compact subset of ℓ_∞ .

<u>Example 4</u>. Since at each date t there are only finitely many partial histories $s^t = (s_1,\ldots,s_t)$, and since inputs and outputs at each date t are functions of s^t, the analysis of this example is similar to that of Example 3, and will not be given here (see Majumdar, 1970b). The situation is more complicated if we wish to study "stationary" stochastic economies, in which case it is convenient to think of the history of the environment as extending infinitely far back into the past. (We may imagine that we are observing the economy after it has been operating under a given policy for a long time.) In this case one is led to represent Y as a subset of L_∞ , where Ω is the set of doubly infinite sequences (s_t), $\mathcal{F}$ is the sigma-field generated by cylinder sets, and μ is a stationary probability measure on Ω, i.e., a measure invariant under the shift transformation $(Ts)_t = s_{t+1}$ (see Radner, 1971b).

IV. Approximation of Maximal Programs
by Regular Maximal Programs

A maximal program y^* in Y will be called <u>regular</u> if Y is supported at y^* by some nonnegative, nonzero price system in L_1, i.e., if the supporting price system has no purely finitely additive part. In this section we give conditions under which maximal programs can be approximated by regular maximal programs.

Our first theorem is taken from Majumdar (1971a).

<u>Theorem 4.1</u>: <u>Suppose Y is a convex $w(L_\infty,L_1)$-compact subset of L_∞; then the $w(L_\infty,L_1)$-closure of the set of regular maximal points of Y contains the set of maximal points of Y.</u>

Since the weak [or, $w(L_\infty,L_\infty^*)$] topology of L_∞ is stronger than the weak* [or, $w(L_\infty,L_1)$] topology, by using Eberlein's theorem (see Dunford-Schwartz, 1957, V.6.1) and the proof of Theorem 4.1, one can get the following:

<u>Theorem 4.2</u>: <u>Let Y be a convex, $w(L_\infty L_\infty^*)$-compact subset of L_∞ ; then the $w(L_\infty,L_\infty^*)$-sequential closure of the set of regular maximal points of Y contains the set of maximal points of Y.</u>

When the measure space $(\Omega,\mathcal{F},\mu)$ is finite, one can substitute "$w(L_\infty,L_1)$-sequential closure" for "$w(L_\infty,L_1)$-closure" in the statement of Theorem 4.1 (imitate the first

383

part of the proof of Dunford-Schwartz, 1957, V.6.1).

For a subset Y of L_∞, $w(L_\infty, L_1)$-compactness is equivalent to boundedness in norm and closedness in $w(L_\infty, L_1)$-topology (see Dunford-Schwarz, 1957, V.4.3). Relatively mild restrictions (like continuity of the production function) ensure closedness. Boundedness is guaranteed by a suitable choice of units when the exogenous resources impose a restriction on the rate of growth of the economy.

Weak convergence is to be interpreted as <u>convergence in value</u>; thus, $\{y^d: d\epsilon D\}$ converges to y^* in the $w(L_\infty, L_1)$-topology if the values $p(y^d)$ converge to $p(y)$ for every value functional p in L_1 .

A sharpened version of convergence--convergence of programs, prices and values--can be proved if Y is assumed to be convex and compact in the norm topology of L_∞ . The latter undoubtedly is a strong restriction.

In what follows L_∞^* is endowed with the $w(L_\infty^*, L_\infty)$ or weak* topology; any product of topological spaces is, of course, assigned the product topology.

<u>Theorem 4.3</u>: <u>Let Y be a convex, norm-compact subset of L_∞ and y^* be a maximal point of Y; then</u>
  (<u>i</u>) <u>there is $p^* > 0$ in L_∞^* , $\|p^*\|_* = 1$, supporting Y at y^*;</u>
  (<u>ii</u>) (y^*, p^*) <u>is the limit of a net</u> $\{(y^d, p^d): d\epsilon D\}$, <u>such that for each $d\epsilon D$,</u>
    (<u>ii.a</u>) y^d <u>is a maximal point of Y ,</u>
    (<u>ii.b</u>) $p^d > 0$, $\|p^d\|_* = 1$, p^d <u>in L_1, and p^d supports Y at y^d.</u>

Thus, y^* is the norm-limit of a net $\{y^d: d\epsilon D\}$ of regular maximal points: the value functionals p^d associated with the approximating y^d are themselves convergent to some p^* relative to which y^* is value maximizing. Of course, $p^d(y^d)$ converges to $p^*(y^*)$, and $\{y^d: d\epsilon D\}$ converges to y^* in $w(L_\infty, L_\infty^*)$-topology.

Proof of convergence of prices is based on a direct application of Alaoglu's theorem (see Majumdar, 1971a, for details). In this connection it ought to be remarked that the unit ball of ℓ_∞^* is compact, but <u>not</u> sequentially compact in the $w(\ell_\infty^*, \ell_\infty)$-topology (see the counterexample in Köthe, 1969, p. 311).

If Y is a convex subset of ℓ_∞ that is compact in the product topology, any maximal point y^* is the limit of a sequence $y^n = (y_t^n)$ of maximal points such that for each n there is a strictly positive $p^n = (p_t^n)$ $(p_t^n > 0$ for all $t)$, with $\sum\limits_{t-1}^{\infty} p_t^n y_t^n \geq \sum\limits_{t=1}^{\infty} p_t^n y_t^n$ for all y in Y. Hence, if Y is compact in any stronger topology, the price systems associated with the approximating regular maximal points can be chosen to be strictly positive. Also, note that, in ℓ_∞ convergence in the $w(\ell_\infty, \ell_1)$-topology implies pointwise convergence; hence, Theorem 4.1 when applied to ℓ_∞ also yields the result that any maximal point y^* is the limit in the topology of pointwise convergence of a net of regular maximal points (in addition to convergence in value relative to each p in ℓ_1). A detailed discussion of the approximation question for subsets in ℓ_∞ is to be found in Majumdar (1970a) and Peleg (1971).

V.   Conditions under Which a Maximal
Program is Regularly Maximal

The theorems of Section IV indicate that maximal programs are regular, i.e., are value-maximizing with respect to a price system in L_1, except in "limiting" cases. The present section gives two situations in which it can be directly verified that a maximal program is regular.

In the first situation, the maximal program also maximizes a function on L_∞ that has suitable continuity, concavity, and monotonicity properties. Such a function may be interpreted as the "utility" or "objective" function.

A real-valued function U on L_∞ is called <u>quasi-concave</u> at $\bar{y}$ if $\{y$ in $L_\infty: U(y) \geq U(\bar{y})\}$ is convex. We shall say that U is <u>weakly-monotone if</u> $x \geq y$ implies $U(x) \geq U(y)$ and $x \gg y$ implies $U(x) > U(y)$ [where $x \gg y$ means that $x(\omega) > y(\omega)$ a.e.]; further, U is <u>strongly-monotone</u> if $x > y$ implies $U(x) > U(y)$. Denote by $t(L_\infty, L_1)$ the Mackey topology on L_∞ relative to L_1, i.e., the strongest topology on L_∞ such that the points of L_1 are continuous on L_∞ (see Köthe, 1969).

<u>Theorem 5.1</u>: Let y^* <u>be a maximal point of a convex set</u> Y <u>in</u> L_∞ . <u>If there exists a real-valued function</u> U <u>such that</u>
   (<u>a</u>) $U(y^*) \geq U(y)$ <u>for all</u> y <u>in</u> Y,
   (<u>b</u>) U <u>is quasi-concave at</u> y^*,
   (<u>c</u>) U <u>is weakly-montone and is continuous in the</u>
       $t(L_\infty, L_1)$-<u>topology of</u> L_∞ ,

385

then there exists a nonzero p^* in L_1 , $p^* > 0$, such that $p^*(y^*) \geq p^*(y)$ for all y in Y.

If U satisfies the strong-monotonicity condition, then p^* is strictly positive, i.e., $p^* \gg 0$. (For a proof of Theorem 5.1, see Majumdar, 1971a).

Our second theorem is applicable to a situation in which the set Y satisfies a certain "mixture" condition. In this situation, if a program is value-maximizing with respect to a price system, then it is value-maximizing with respect to the L_1-part of the price system (see Section III).

We shall say that a set Y in L_∞ has the mixture property if for every y' and y" in Y, and every set A in $\mathcal{F}$, the program y is also in Y, where y is defined by $y(\omega)= y'(\omega)$ for ω in A and $y(\omega) = y''(\omega)$ for ω not in A.

Theorem 5.2: If Y has the mixture property, and y^* is value-maximizing with respect to p, then y is also value-maximizing with respect to the L_1 part of p.

Corollary: If Y has the mixture property, if y^* is maximal in Y, if p^* supports Y at y^*, and if p^* is not purely finitely additive, then y^* is regular.

(For a proof, see Majumdar, 1971a.) This result, and related techniques, have been applied in Bewley (1970) and Radner (1971b).

VI. Efficiency and Intertemporal Profit Maximization

In this section we reconsider the multi-sector model (Example 3), and briefly indicate how by applying Theorem 4.3, one can derive a system of prices relative to which efficient programs can be characterized in terms of inter-temporal profit maximization. First, for simplicity, let us assume that there are primary resources, and that the units of measurement are, in fact, chosen in such a way that Y_w as well as all the sequences $(a_t),(b_{t+1})$ of inputs and outputs generating the elements of Y_w are in ℓ_∞ (see Radner, 1967). A production program $(\hat{a}_t, \hat{b}_{t+1})$ is intertemporally profit maximizing with respect to a sequence (p_t) of vectors in R^ℓ if for every $t \geq 1$

$$(6.1) \quad p_{t+1}b_{t+1} - p_t a_t \leq p_{t+1}\hat{b}_{t+1} - p_t\hat{a}_t \text{ for all } (a_t,b_{t+1}) \text{ in } G_t.$$

The following theorem relates value maximization to intertemporal profit maximization (for a proof, see Radner, 1967, and Majumdar and Kurz, 1970).

Theorem 6.1: Suppose p^* in ℓ_∞^* supports Y_W in ℓ_∞ at $y^* = (y_t^*)$. If (a_t^*, b_{t+1}^*) is the production program generating $y^* = (y_t^*)$, then (a_t^*, b_{t+1}^*) is intertemporally profit maximizing, and $y^* = (y_t^*)$ is value maximizing relative to the ℓ_1-part of p^*.

Thus, if the ℓ_1-part of the value functional supporting an efficient program in nonzero, one can simply disregard the purely finitely additive part of p^*. From (3.5) it follows that if $y^* = (y_t^*)$ is such that there is another $y = (y_t)$ in Y_W with $\lim \inf y_t > \lim \sup y_t^*$, then the ℓ_1-part cannot be zero. Even when the ℓ_1-part of the supporting functional p^* is zero, provided that the efficient program $y^* = (y_t^*)$ satisfies a particular technological condition, it is possible to obtain a nonzero sequence (p_t) of prices relative to which the corresponding production program is intertemporally profit maximizing. Actually, Theorem 6.1 can be stated in a way such that it is applicable even if there are no primary resources (see Radner, 1967). We shall state the next theorem in full generality, making it directly comparable to the original version of Malinvaud (1953). We first need a few definitions.

A good is nonproducible at date t if the corresponding coordinate of the output vector b is zero for every (a,b) in G_t; otherwise, it is producible. Suppose that the (possibly empty) set of nonproducible goods is the same for each t; inputs, a, and outputs, b, are representable in the form

$$(6.2) \quad a = \begin{pmatrix} a_1 \\ \\ a_2 \end{pmatrix} \quad b = \begin{pmatrix} b_1 \\ \\ 0 \end{pmatrix},$$

where a_1, a_2 are the inputs of producible and nonproducible goods, respectively, and b_1 is the vector of producible goods. A program $(a_t, b_{t+1}; y_t)$ is called tight at date t if there is no (a', b') in G_t such that

(6.3) $a_2' \ll (a_2)_t$,

$\qquad$ $b_1' \gg (b_1)_{t+1}$,

where "$\ll$" means each coordinate of a_2 is strictly greater than the corresponding coordinate of a_2' , etc. Of course the first line of (6.3) applies only if there are nonproducible goods.

The proof of the following theorem is in Radner (1967).

<u>Theorem 6.2</u>:   <u>Suppose</u>

($\underline{i}$)   <u>for each</u> t, G_t <u>is closed, convex and contains</u> (0,0),

($\underline{ii}$) λ_t <u>is a given sequence of positive real numbers converging to zero</u>,

($\underline{iii}$) $\|\hat{y}_t\| \le \lambda_t$, $\|\hat{a}_t\| \le \lambda_t$, $\|\hat{b}_{t+1}\| \le \lambda_{t+1}$, <u>for all</u> $t \ge 1$,

($\underline{iv}$) $(\hat{a}_t, \hat{b}_{t+1} ; \hat{y}_t)$ <u>is a feasible program such that</u> $(\hat{y}_t)$ <u>is efficient in</u> $Y \cap \ell_\infty$,

($\underline{v}$) $(\hat{a}_t, \hat{b}_{t+1}; \hat{y}_t)$ <u>is not tight at any date</u> t ;

<u>then there exists a sequence</u> $(\bar{p}_t)$ <u>of nonnegative vectors in</u> R^ℓ, <u>not all zero, such that</u> $(\hat{a}_t, \hat{b}_{t+1})$ <u>is intertemporally profit maximizing with respect to</u> $(\bar{p}_t)$.

It should be remarked that, under condition (i), if $(\hat{a}_t, \hat{b}_{t+1}; \hat{y}_t)$ is <u>any</u> feasible program, then units of measurement can always be chosen so that condition (iii) is satisfied. In addition, if there are primary resources, units can be chosen so that for <u>all</u> programs in Y_w, (iii) will be satisfied (see Radner, 1967, for details). An example of a tight program such that there is no nonzero system of prices relative to which it is profit maximizing is given by McFadden (1968). Intertemporal profit maximization relative to a strictly positive system of prices is not sufficient for efficiency (see Malinvaud, 1953).

References

1.   T. Bewley (1970), Equilibrium Theory with an Infinite Dimensional Commodity Space, Ph.D. Diss., Dept. of Economics, Univ. of Calif., Berkeley.

2. G. Debreu (1954), "Valuation Equilibrium and Pareto Optimum," _Proc. Nat. Acad. Sci. of the U.S.A._ 40, 588-592.

3. G. Debreu (1959), _Theory of Value_, Wiley, New York.

4. N. Dunford and J. Schwarz (1957), _Linear Operators_ (Part I), Interscience, New York.

5. L. Hurwicz (1958), "Programming in Linear Spaces," in Arrow, Hurwicz and Uzawa (eds.), _Studies in Linear and Non-Linear Programming_, Stanford Univ. Press, Stanford.

6. P. Jeanjean (1971), Optimal Growth in a Multi-Sector Economy with Stochastic Technology, Ph.D. Diss., Dept. of Economics, Univ. of Calif., Berkeley.

7. T. C. Koopmans (1965), "On the Concept of Optimal Economic Growth," in _Study Week on the Econometric Approach to Development Planning_, North Holland, Amsterdam, pp. 225-300.

8. G. Köthe (1969), _Topological Vector Spaces_ (Vol. I), Springer-Verlag, Berlin.

9. M. Majumdar (1970a), "Some Approximation Theorems on Efficiency Prices for Infinite Programs," _J. Econ. Theory_ 2, 399-410.

10. M. Majumdar (1970b), A General Theory of Efficiency Prices in Economies with Infinitely Many Commodities, Ph.D. Diss., Dept. of Economics, Univ. of Calif., Berkeley.

11. M. Majumdar (1971a), "Some General Theorems on Efficiency Prices with an Infinite Dimensional Commodity Space" (to appear in _J. Econ. Theory_).

12. M. Majumdar (1971b), "Compactness Properties of the Set of Feasible Growth Programs with Continuous Time," Working Paper, IMSSS, Stanford Univ., Stanford.

13. M. Majumdar and M. Kurz (1970), "Efficiency Prices in Infinite Dimensional Spaces: A Synthesis" (to appear in _Rev. Econ. Stud._).

14. E. Malinvaud (1953), "Capital Accumulation and Efficient Allocation of Resources," _Econometrica_ 21, 233-268.

15. D. McFadden (1968), "An Example of the Non-Existence of Malinvaud Prices in a Tight Economy," Working Paper No. 134, IBER, Univ. of Calif., Berkeley.

16. B. Peleg (1971), "Efficiency Prices for Optimal Consumption Plans: II," _Israel J. Math._ 9, 222-234.

17. B. Peleg and M. Yaari (1970), "Efficiency Prices in an Infinite Dimensional Space," _J. Econ. Theory_ 2, 41-85.

18. R. Radner (1967), "Efficiency Prices for Infinite Horizon Production Programmes," _Rev. Econ. Stud._ 34, 51-66.

19. R. Radner (1971a), "Balanced Stochastic Growth at the Maximum Rate," in G. Bruckmann and W. Weber (eds.), _Contributions to the Von Neumann Growth Models (Zeitschrift für Nationalökonomie, Suppl. 1)_, Springer-Verlag, Vienna, pp. 39-62.

20. R. Radner (1971b), "Optimal Stationary Consumption with Stochastic Production," MSSB Working Paper No. 17, Dept. of Economics, Univ. of Calif., Berkeley.

21. L. J. Savage (1954), _Foundation of Statistics_, Wiley, New York.

22. B. Stigum (1971), "Competitive Equilibrium with Infinitely Many Commodities (II)," MSSB Working Paper No. 15, Dept. of Economics, Univ. of Calif., Berkeley.

23. K. Yosida and E. Hewitt (1952), "Finitely Additive Measures," _Trans. Amer. Math. Soc._ 72, 46-66.

PRICE   DECENTRALIZATION   IN
THE CASE OF INTERRELATED PAYOFFS

Alain Bensoussan

European Institute for Advanced Studies in Management (Brussels)
and
Institut de Recherche d'Informatique et d'Automatique (Paris)

Motivation of the problem

The problem of price decentralization in the case of independent payoffs has been extensively studied, for instance [1]
in Economics as well as in Management. Loosely speaking, one can
formulate it as follows.

There are m decision makers, whose decision variables are
denoted by $v_1 \ldots v_m$ (v_i being in general a vector). Each of
them has his own payoff $J_i(v_i)$. When decision maker i
chooses v_i, he consumes a quantity $B_i(v_i)$ of a scarce resource.
Denoting by b the total amount of the scarce resource,
one gets the constraint

$$(1.1) \qquad \sum_{i=1}^{m} B_i(v_i) \leq b$$

(in general (1.1) is a vector inequality). Thus, the decision
makers are trying to get a non cooperative equilibrium.

The result known as price decentralization is this : there
exists a price p of the scarce resource (p in general a vector) such that, leaving decision makers buy the amount they
want and optimize their own cost functional, one gets the desired equilibrium.

This result is very interesting because it can be implemented easily in practical applications. Indeed, the payoffs $J_i(v_i)$
are changed into $J_i(v_i) + (p, B_i(v_i))$ and decision makers
can choose independently their decision variables without taking

account of the global constraint (1.1). Thanks to the choice of p, the solutions v_i satisfy (1.1) anyway, and are a non cooperative equilibrium for the original problem.

However, in Management problems, (probably more than in Economics problems), the payoffs of the different decision makers are generally not independent. It is not difficult to describe concrete examples of such a situation in large decentralized firms. However, just to get an idea of how it could occur, even in apparently independent units, consider a large firm which is divided into m plants and which gives incentives to the different plants. Of course, the total amount of incentives which is available, is a function of the v_i 's. Denoting by $K(v_1, \dots , v_m)$ this redistributed revenue, and supposing that n°=1 gets a share equal to $\lambda_i K(v_1, \dots, v_m)$ (where $\lambda_i \geq 0$ and $\sum \lambda_i = 1$), then it is clear that the problem can be reformulated as follows

(1.2) find a non cooperative equilibrium for the m decision makers, each of them having a payoff equal to :
$$J_i(v_i) + \lambda_i K(v_1, \dots, v_m),$$
under the constraint of scarcity :
$$\sum_{i=1}^{m} B_i(v_i) \leq b.$$

The problem of price decentralization remains as before. Is it possible to set a price p for the scarce resource, such that leaving decision makers free to buy what they want and to take their decisions in a non cooperative fashion, then the final decisions satisfy the constraint of scarcity and also are a non cooperative equilibrium for problem (1.2).

The main advantage of the price decentralization approach remains the fact that it can be easily implemented in practical situations, even though the new payoffs

$$J_i(v_i) + \lambda_i K(v_1, \dots, v_m) + (p, B_i(v_i))$$

are still interrelated.

Setting of the problem

2.1 Notations :

Let $U_1, \dots, U_m$ be m vector spaces (finite dimensional) Let us set :

(2.1) $$\Psi = R^p$$

and let us consider m mappings $B_i : U_i \longrightarrow \Psi$, and also m subsets $U_1^{ad} \subset U_i$, $i = 1 \ldots m$.

Let us then consider m functionals $J_i : U_1 x \ldots x U_m \longrightarrow R$. The original problem can be formulated as follows :

$$(2.2) \quad \begin{cases} \text{Find a Nash point for the functionals } J_i(v_1, \ldots, v_m), \\ \ldots, J_m(v_1, \ldots, v_m) \\ \text{under the constraints :} \\ \qquad \sum_{j=1}^{m} B_j(v_j) \leqslant 0, \quad v_j \in U_j^{ad} . \end{cases}$$

Let us recall the definition of a Nash point, which is exactly the same thing as a non cooperative equilibrium. Let $u_1, \ldots, u_m$ be a Nash point. By definition, $u = (u_1, \ldots, u_m)$ must satisfy the property :

$$(2.3) \quad \begin{cases} \forall i = 1, \ldots, m, \quad J_i(u_1, \ldots, u_{i-1}, u_i, u_{i+1} \ldots, u_m) \leqslant \\ \qquad \leqslant J_i(u_1, \ldots, u_{i-1}, v_i, u_{i+1} \ldots, u_m) \\ \forall v_i \text{ such that } v_i \in U_i^{ad}, \quad \sum_{j \neq i} B_j u_j + B_i(v_i) \leqslant 0 \end{cases}$$

and

$$(2.4) \quad \forall i, \quad u_i \in U_i^{ad}, \quad \sum_{j=1}^{m} B_j u_j \leqslant 0.$$

2.2. <u>Decentralized non cooperative equilibrium</u> :

Let us now define what we call a <u>decentralized Nash point</u>. We shall say that $u = u_1 \ldots u_m$ is a decentralized Nash point (or a decentralized non cooperative equilibrium) if there exists a price vector p such that the following properties are satisfied :

$$(2.5) \quad \begin{cases} p \geqslant 0, \quad \sum_{j=1}^{m} B_j(u_j) \leqslant 0, \quad u_j \in U_j^{ad} \; \forall j, \quad (p, \sum_{j=1}^{m} B_j(u_j)) = 0, \\ J_i(u_1, \ldots, u_{i-1}, u_i, u_{i+1}, \ldots, u_m) + (p, B_i(u_i)) \leqslant \\ \leqslant J_i(u_1, \ldots, u_{i-1}, v_i, u_{i+1}, \ldots, u_m) + (p, B_i(v_i)) \\ \forall v_i \in U_i^{ad}, \quad \forall i = 1, \ldots, m. \end{cases}$$

<u>Remark 2.1</u> : As already emphasized in §1, the m decision makers buy the scarce resource at price p and take their deci-

sions in a non cooperative fashion. The price p is such that the global constraint :

$$\sum_{j=1}^{m} B_j(u_j) \leq 0$$

is automatically satisfied.

Remark 2.2 : We have taken $b = 0$, but it is not really an assumption, since we can write :

$$\sum_{j=1}^{m} B_j(u_j) \leq b, \text{ under the form :}$$

$$\sum_{j=1}^{m} (B_j(u_j) - b_j) \leq 0, \text{ where } b = \sum_{j=1}^{m} b_j,$$

Defining then :

$$\widetilde{B}_j(u_j) = B_j(u_j) - b_j,$$

one gets :

$$\sum_{j=1}^{m} \widetilde{B}_j(u_j) \leq 0.$$

Proposition 2.1 :

    <u>If $u = (u_1,\ldots,u_m)$ is a decentralized Nash point, then it is a Nash point for the original problem.</u>

Proof :

    Let u satisfying (2.5). Then of course, we get :

$$(2.6) \quad \sum_i J_i(u) \overset{(*)}{\leq} \sum_i J_i(u_1,\ldots u_{i-1}, v_i, v_{i+1}\ldots u_m) + \ldots + (p, B_i(v_i)).$$

Taking $v_j = u_j$ for $j \neq i$, it follows from (2.6) :

$$(2.7) \quad J_i(u) \leq J_i(u_1,\ldots,u_{i-1}, v_i, u_{i+1},\ldots u_m) + \ldots + (p, B_i(v_i)) + \sum_{j \neq i} B_j(u_j)).$$

Therefore, if v_i is such that $B_i(v_i) + \sum_{j \neq i} B_j(u_j) \leq 0$,

it follows from (2.7), since $p \geq 0$,

that $\quad J_i(u) \leq J_i(u_1,\ldots,u_{i-1}, v_i, u_{i+1}, \ldots u_m),$

which proves that u satisfies (2.3).

$$(*) \quad J_i(u) = J_i(u_1,\ldots,u_m).$$

The reverse property is not true in general, which means that it will not be able to make the price decentralization in all cases.

Existence of a decentralized Nash point.

3.1 Differential form of (2.5) :
We now make the following assumptions :

(3.1)   The functionals J_i are convex and differentiable in v_i, the other variables being fixed,

(3.2)   The applications B_i are convex and U^{ad} are convex sets.

One then gets :
Proposition 3.1 :
A necessary and sufficient condition for a pair (u,p) to satisfy (2.5) is that

$$(3.3) \quad p \geq 0, \quad u_i \in U_i^{ad}, \quad \sum_{i=1}^{m} B_i(u_i) \leq 0$$

$$\sum_{i=1}^{m} \left(\frac{\partial J_i}{\partial v_i}(u), v_i - u_i \right) + \left(p, \sum_{i=1}^{m} B_i(v_i) \right) \geq 0 \quad \forall v_i \in U_i^{ad}.$$

Proof :
Necessary conditions : Suppose that (u,p) satisfies (2.5). Then one gets :

$$(3.4) \quad \sum_{i=1}^{m} J_i(u) \leq \sum_{i=1}^{m} J_i(u_1, \ldots, u_{i-1}, v_i, u_{i+1}, \ldots, u_m) +$$

$$\ldots + \left(p, \sum_{i=1}^{m} B_i(v_i) \right).$$

Let us define $J_o(v)$, $v = v_1 x \ldots, x v_m$ by :

$$(3.5) \quad J_o(v) = \sum_{i=1}^{m} J_i(u_1, \ldots, u_{i-1}, v_i, u_{i+1}, \ldots, u_m)$$

and $K_o(v)$ by :

$$(3.6) \quad K_o(v) = \left(p, \sum_{i=1}^{m} B_i(v_i) \right).$$

Thus (3.4) can be rewritten as follows :

$$(3.7) \quad J_o(v) + K_o(v) \geq J_o(u) + K_o(u), \quad \forall v \in U_1^{ad} x \ldots x U_m^{ad}.$$

In the left member of (3.7), $J_o(v)$ is a convex differential functional and $K_o(v)$ is a convex (eventually non differentiable) functional. Since $U_1^{ad} \times \ldots \times U_m^{ad}$ is a convex subset of $U_1 \times \ldots \ldots \times U_m$, it follows from $(3.7)^m$ that :

$$(3.8) \quad (J_o'(u), v-u) + K_o(v) - K_o(u) \geq 0 \quad \forall v \in U_1^{ad} \times \ldots \times U_m^{ad} .$$

The property (3.8) is a classical property of convex functions, which are the sum of a differentiable and a non differentiable function (for the proof of (3.8), see for instance [2]). Calculating $J_o'(u)$ and taking account of the fact that $K_o(u) = 0$, one easily checks that (3.8) is nothing else than (3.3).

<u>Sufficient condition</u> : If (3.3) is verified, it follows from convexity properties of $J_o(v)$ that :

$$(3.9) \quad \sum_{i=1}^{m} J_i(u_1, \ldots, u_{i-1}, v_i, u_{i+1}, \ldots u_m) - \sum_{i=1}^{m} J_i(u) +$$
$$\ldots + (p, \sum_{i=1}^{m} B_i(v_i)) \geq 0 ,$$
$$\forall v_i \in U_i^{ad}$$

Taking then $v_i = u_i$ in (3.9), one gets :

$$(3.10) \quad (p, \sum_{i=1}^{m} B_i(u_i)) \geq 0$$

But from $p \leq 0$ it follows : and $\sum_{i=1}^{m} B_i(u_i) \leq 0,$

$$(3.11) \quad (p, \sum_{i=1}^{m} B_i(u_i)) \leq 0 .$$

Therefore, from (3.10) and (3.11), it immediately follows that :

$$(p, \sum_{i=1}^{m} B_i(u_i)) = 0.$$

<u>3.2</u> <u>Existence theorem of a decentralized Nash point</u> :

We shall give here an existence theorem in order that there exists a pair (u,p) satisfying (3.3). Setting :

$$(3.12) \quad U = U_1 \times \ldots \times U_m , \quad U^{ad} = U_1^{ad} \times \ldots \times U_m^{ad} ,$$

and defining the mapping $A : U \longrightarrow U$, by :

$$(3.13) \qquad A(v) = \left(\frac{\partial J_1}{\partial v_1}(v), \frac{\partial J_2}{\partial v_2}(v), \ldots, \frac{\partial J_m}{\partial v_m}(v) \right)$$

and $B : U \longrightarrow \Psi$ by :

$$(3.14) \qquad B(v) = \sum_{i=1}^{m} B_i(v_i) \ ,$$

it is clear that one can rewrite (3.3) as follows :

$$(3.3)' \qquad \begin{cases} p \in \Psi \ , \ p \geqslant 0 \ , \ B(u) \leqslant 0, \qquad u \in U^{ad}, \\ (A(u), v-u) + (p, B(v)) \geqslant 0 \qquad \forall v \in U^{ad}. \end{cases}$$

We shall now make the following assumptions :

$$(3.15) \qquad A \text{ is continuous,} \qquad B \text{ is continuous}$$

$$(3.16) \qquad (A(v) - A(w), v-w) \geqslant 0 \qquad v, w \ U^{ad}$$

$$(3.17) \qquad \exists v_0 \in U^{ad} \ , \text{ such that } B(v_0) \leqslant 0 \quad \text{and}$$

$$\frac{(A(v), v - v_0)}{\|v\|} \xrightarrow{\quad} +\infty \ , \text{ if } \|v\| \xrightarrow{\quad} \infty \qquad (*)$$

$$(3.18) \qquad U^{ad} \text{ is closed}$$

$$(3.19) \qquad \text{Slater's assumption : } \exists v_1 \in U^{ad} \text{ such that } B(v_1) < 0.$$

Theorem 3.1 : <u>Under assumptions</u> (3.1),(3.2),(3.15),(3.16),(3.17), (3.18) <u>and</u> (3.19), <u>there exists a pair</u> (u,p) <u>satisfying</u> (3.3)', i.e, there exists a decentralized Nash point.

Proof : Let us prove first that $\exists u \in U^{ad}$ such that

$$(3.20) \qquad (A(u), v-u) \geqslant 0, \ \forall v \in U^{ad}, \ B(v) \leqslant 0.$$

Setting $K \subset U$, defined by

$$(3.21) \qquad K = \left\{ v \in U \ \middle| \ v \in U^{ad}, \ B(v) \leqslant 0, \right\}$$

then K is a closed convex subset of U. The problem amounts to finding a $u \in K$ such that :

$$(3.22) \qquad (A(u), v-u) \geqslant 0, \qquad \forall v \in K.$$

---

* This assumption is not necessary when U^{ad} is bounded.

Thus u is solution of a <u>variational inequality</u>, the theory of which has been made by J.L. LIONS [2] ,[3] . From his results, it follows that, when (3.15), (3.16), (3.17) and (3.18) are satisfied, there exists u solution of (3.22).

Now from (3.20), it follows that u is also a solution of the convex mathematical program

$$(3.23) \qquad \underset{v \in U^{ad}}{\text{Min}} (A(u),v)$$
$$B(v) \leqslant 0.$$

Therefore p can be interpreted as a Lagrange multiplier for the convex program (3.23). But the existence of p is a well-known consequence of the Slater's assumption (3.19),(for instance [4]), which ends the proof of the theorem.

<u>Remark 3.1</u> : The possibility of decentralization for non cooperative equilibria occurs on relatively severe assumptions. However, they were expectable. Indeed, already in the case of independent payoffs, the known conditions of decentralization are severe. Moreover, since we are dealing with Nash points, condition (3.16) could be expected, since it already appears in Rosen's work on Nash equilibria 5 . Rosen calls Nash points satisfying property (3.16), <u>normalized Nash points</u>.
Few results are known in the case of non normalized Nash points.

<u>Decentralization algorithm.</u>

Assuming now that the decentralization is possible (for instance when the assumptions of Theorem 3.1 are satisfied), there remains a problem of realization since the price p is completely unknown. Anyhow, the right price p can be obtained through an iterative process, which can be looked at as a two level information exchange. (cf. figure).
The second level viewed as the managerial head of the firm, fixes the price of the scarce resource at p_n. Then the first level units react at this price, by trying to establish a non cooperative equilibrium. They inform the second level of their decisions, and the second level fixes a new price, $p_n + 1$, taking account of the first level decisions, and   so on.

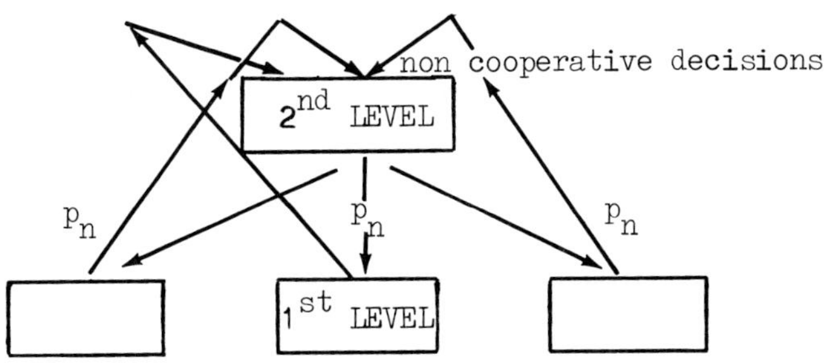

4.1 Description of the decentralization algorithm :

The decentralization algorithm works as follows :
Suppose a price p_n has been fixed by the second level, then :

Step 1 : Define a Nash point for the functionals

$$(4.1) \quad \begin{cases} J_1(v_1, \ldots, v_m) + (p_n, B_1(v_1)) \quad \text{with} \quad v_1 \in U_1^{ad} \\ \vdots \\ \vdots \quad\quad\quad\quad\quad\quad \Longrightarrow \quad (u_1^n, \ldots, u_m^n) \\ \vdots \\ J_m(v_1, \ldots, v_m) + (p_n, B_m(v_m)) \end{cases}$$

Step 2 : Change p_n into

$$(4.2) \quad p_{n+1} = Pr \left(p_n + \rho_n \sum_{i+1}^{m} B_i(u_i^n) \right), \quad \rho_n > 0.$$

where Pr means the projection on the positive orthant.
Go to step 1.

Remark 4.1 : In practice, a stopping test must be defined, depending for instance on the difference of

$$\sum_i \| u_i^{n+1} - u_i^n \|.$$

4.2 Convergence of the decentralization algorithm :

We shall obtain the following convergence theorem :

<u>Theorem 4.1</u> : <u>Under the assumptions of Theorem</u> 3.1, <u>with</u> (3.16) <u>changed into</u>

$$(3.16)' \qquad (A(v) - A(w), \ v-w) \geqslant \alpha \ \|v-w\|^2 , \ \forall v, \ w \in U^{ad}, \quad \alpha > 0,$$

<u>and</u>

$$(4.3) \qquad \|B(v) - B(w)\| \leqslant C\|v-w\| , \qquad \forall \ v, \ w \in U^{ad}$$

<u>then there exists a unique decentralized Nash point</u> u <u>and</u>

$$(4.4) \qquad u_n \longrightarrow u$$
$$n \longrightarrow \infty$$

(4.5) If p is a cluster point of the sequence p_n, then (u,p) is solution of (3.3).

<u>Proof</u> : Let us first prove the uniqueness of u . Suppose that there exist two decentralized Nash points, denoted by u^1 and u^2. According to (3.20) we can write

$$(4.6) \qquad \begin{aligned} (A(u^1), \ u^2 - u^1) &\geqslant 0 \\ (A(u^2), \ u^1 - u^2) &\geqslant 0 \end{aligned}$$

Then, substracting, one gets

$$(4.7) \qquad (A(u^1) - A(u^2), \ u^1 - u^2) \leqslant 0 \quad ,$$

which implies, using $(3.16)'$: $\|u^1 - u^2\| \leqslant 0$, i.e., $u^1 = u^2$.

Let us reinterpret the step 1, as in Proposition 3.1, using the convexity and differentiability properties of the functionals J_i. By definition of the Nash point, one will have

$$(4.8) \quad J_i(u_i^n, \ldots, u_{i-1}^n, \ v_i, \ u_{i+1}^n, \ldots, u_m^n) + (p_n, \ B_i(v_i)) \geqslant$$
$$\geqslant J_i(u^n) + (p_n, \ B_i(u_i^n)), \qquad \forall v_i \in U_i^{ad}, \ \forall i,$$

which is equivalent to

$$(4.9) \qquad (\frac{\partial J_i}{\partial v_i}(u^n), \ v_i - u_i^n) + (p_n, \ B_i(v_i) - B_i(u_i^n)) \geqslant 0$$
$$\forall v_i \in U_i^{ad} \quad , \ \forall i \quad .$$

But the m variational inequalities (4.9) can be summarized as follows :

$$(4.10) \qquad \sum_{i=1}^{m} \left(\frac{\partial J_i}{\partial v_i}(u^n), \, v_i - u_i^n \right) + \sum_{i=1}^{m} (p_n, \, B_i(v_i) - B_i(u_i^n)) \geqslant 0,$$

$$v_i \in U_i^{ad}$$

Using the definition of functions A and B, one easily checks that (4.10) is nothing else than

$$(4.11) \qquad (A(u^n), \, v-u^n) + (p_n, \, B(v) - B(u^n)) \geqslant 0 \, , \qquad \forall v \in U^{ad}.$$

Since Theorem 3.1 is true, there exists p such that (u,p) satisfies

$$(4.12) \qquad (A(u), \, v-u) + (p, \, B(v) - B(u)) \geqslant 0 \qquad \forall v \in U^{ad}$$

(note that $(p, \, B(u)) = 0$ and use $(3.3)'$).
Moreover p satisfies

$$(4.13) \qquad \begin{array}{c} p = \text{Pr}\,(p + \rho_n B(u)) \\ \text{since } \rho_n B(u) \leqslant 0, \quad p \geqslant 0 \end{array} \qquad \text{and } (p, B(u)) = 0.$$

Let us define $r_n = p_n - p$ and let us take $v = u^n$ in (4.12) and $v = u$ in (4.11). We get :

$$(4.14) \qquad \begin{array}{l} (A(u^n), \, u-u^n) + (p_n, \, B(u)-B(u^n)) \geqslant 0 \\ (A(u), \, u^n-u) + (p, B(u^n) - B(u)) \geqslant 0 \end{array}$$

from which it follows :

$$(4.15) \qquad (A(u^n) - A(u), \, u^n-u) + (p_n - p, \, B(u^n)- B(u)) \leqslant 0.$$

Thus, $(3.16)'$ implies

$$(4.16) \qquad - (p_n - p, \, B(u^n) - B(u)) \geqslant \alpha \|u^n - u\|^2 .$$

Now, $r_{n+1} = p_{n+1} - p = \text{Pr}(p_n + \rho_n B(u^n)) - \text{Pr}(p + \rho_n B(u)).$

Thus, using the fact that Pr is a contraction mapping, one gets
$$(4.17) \qquad \| r_{n+1} \| \leqslant \| r_n + \rho_n (B(u^n) - B(u)) \| .$$

Also $\| r_{n+1} \|^2 \leqslant \| r_n \|^2 + \rho_n^2 \| B(u^n) - B(u) \|^2 + 2 \rho_n (r_n, B(u^n)-B(u))$

From (4.16) and assumption (4.3), it then follows

$$\| r_{n+1} \|^2 \leqslant \| r_n \|^2 + \rho_n^2 \, c^2 \|u^n - u\|^2 - 2 \alpha \rho_n \|u^n-u\|^2$$

We now choose ρ_n such that

(4.18)
$$0 < \delta \leq 2\,\alpha\,\rho_n - \rho_n^2\, c^2 \quad, \quad \text{we obtain :}$$

(4.19)
$$\|r_{n+1}\|^2 + \delta\|u^n - u\|^2 \leq r_n^{\,2}.$$

Therefore $\|r_n\|$ is a decreasing and thus a convergent sequence.
The property (4.4) follows immediately from (4.19).
Let us now show that :

(4.20)
$$\rho_n\,(B(u^n),\ p_n) \longrightarrow 0.$$

Let γ be a cluster point of the sequence $\rho_n(B(u^n),\ p_n)$. Using
the fact that p_n remains in a compact subset of Ψ, and ρ_n
in a compact subset of R, one may extract a subsequence
such that

$$\rho'_n \longrightarrow \rho$$
$$p'_n \longrightarrow p + r \quad (\text{i.e } r'_n \longrightarrow r$$
$$\rho'_n\,(B(u^{n'}), p'_n) \longrightarrow \gamma.$$

Since $B(u^{n'}) \longrightarrow B(u)$, then one gets

(4.21)
$$\gamma = \rho\,(B(u),\ p+r) = \rho(B(u), r)$$

But $\rho > 0$ (since $\rho_{n'} > 0$) and $p+r \geq 0$ since $p_n \geq 0$. Thus,

(4.22)
$$\gamma \leq 0$$

Since $p_{n'}$ is the projection of $p_{n'-1} + \rho_{n'-1}\,B(u^{n'-1})$, one gets :

(4.23)
$$(p_{n'-1} + \rho_{n'-1}\,B(u^{n'-1}) - p_{n'}\ ,\ z - p_{n'}) \leq 0, \quad \forall z \geq 0.$$

Let us take $z = p$. Then (4.23) implies

(4.24)
$$-(r_{n'-1} - r_{n'} + \rho_{n'-1}\,B(u^{n'-1}),\ r_{n'}) \leq 0. \quad \text{Thus :}$$
$$\rho_{n'-1}(B(u^{n'-1}),\ r_{n'}) \geq (r_{n'},\ r_{n'} - r_{n'-1}) =$$
$$= \frac{1}{2}\,|r_{n'}|^2 - \frac{1}{2}\,|r_{n'-1}|^2 + \frac{1}{2}\,|r_{n'} - r_{n'-1}|^2$$
$$\geq \frac{1}{2}\,|r_{n'}|^2 - \frac{1}{2}\,|r_{n'-1}|^2.$$

Also :

(4.25)
$$(B(u^{n'-1}),\ r_{n'}) \geq \frac{1}{2\,\rho_{n'-1}}\,(|r_{n'}|^2 - |r_{n'-1}|^2).$$

Since the sequence $|r_n|^2$ is convergent, according to (4.19), and since $\rho_{n'-1}$ is bounded, the right member of (4.25) tends to 0, when $n' \longrightarrow \infty$. But $B(u^{n'-1}) \longrightarrow B(u)$, and $r_{n'} \to r$. Thus going to the limit in (4.25), one gets

$$(4.26) \qquad \gamma = (B(u),r) \geq 0.$$

Taking account of (4.22), we obtain $\gamma = 0$, which proves that 0 is the only cluster point of the sequence : $\rho_n(B(u^n), p_n)$ and thus proves (4.20).

Since the mapping A is continuous, one gets from (4.4), that

$$(4.27) \qquad A(u^n) \longrightarrow A(u).$$

Let now p be a **cluster** point of the sequence p_n. There exists a subsequence $p_{n'}$ converging towards $\bar{p}$. We can then go to the limit in (4.11) obtaining, thanks to (4.27) and (4.20)

$$(4.28) \qquad (A(u), v-u) + (\bar{p}, B(v)) \geq 0 \qquad \forall v \in U^{ad}.$$

Since $\bar{p} \geq 0$, it is clear that the pair $(u,\bar{p})$ is a solution of (3.3)'.

<u>Applications</u>

5.1 <u>The usual decentralization problem</u> :

It corresponds to the case

$$(5.1) \qquad J_i(v) = J_i(v_i).$$

Let us interpret the assumptions of Theorem 3.1 and 4.1 :
Assumption (3.1) becomes

$$(5.2) \qquad J_i(v_i) \text{ is convex and differentiable.}$$

Then
$$(5.3) \qquad A(v) = \frac{\partial J_1}{\partial v_1}(v_1), \frac{\partial J_2}{\partial v_2}(v_2), \ldots, \frac{\partial J_m}{\partial v_m}(v_m)).$$

Therefore (3.15) becomes

$$(5.4) \qquad v_i \longrightarrow \frac{\partial J_i}{\partial v_i}(v_i) \text{ is continuous on } U_i^{ad} ,$$

and (3.16) becomes :

$$(5.5) \qquad (-\frac{\partial J_i}{\partial v_i}(v_i) - \frac{\partial J_i}{\partial w_i}(w_i), \; v_i - w_i) \geq 0 \quad \forall v_i, \; w_i \in U_i^{ad} \; .$$

But (5.5) is automatically verified since J_i is convex.

Let us consider assumption (3.17) (which is relevant only when U_i^{ad} is unbounded). We can rewrite (3.17) as follows :

$$(5.6) \qquad \frac{\displaystyle\sum_{i=1}^{m} (-\frac{\partial J_i}{\partial v_i}, \; v_i - v_i^o)}{\left(\displaystyle\sum_{i=1}^{m} \| v_1 \|^2\right)^{\frac{1}{2}}} \longrightarrow + \infty \quad \text{when} \; \| v \| \longrightarrow \infty$$

Thanks to classical properties of convex functions, one gets

$$(5.7) \qquad (\frac{\partial J_i}{\partial v_i}, \; v_i - v^o) \geq J_1(v_1) - J_1(v^o) \; . \qquad \text{Therefore, if}$$

$$(5.8) \qquad \frac{J_i(v_i)}{\| v_i \|} \longrightarrow + \infty \quad \text{when} \; \| v_i \| \longrightarrow \infty \; \text{is satisfied}$$

for any i , then also

$$\frac{J_i(v_i) - J_i(v^o)}{\| v_i \|} \longrightarrow + \infty \quad \text{when} \; v_i \longrightarrow \infty$$

and

$$\frac{\displaystyle\sum_{i=1}^{m} J_1(v_i) - J_1(v^o)}{\| v \|} \longrightarrow + \infty \quad \text{when} \; v \longrightarrow \infty$$

which implies (5.6), using (5.7). Indeed, it can be proved $[2.]$ that (5.8) may be weakened into :

$$(5.9) \qquad J_i(v_i) \longrightarrow + \infty \qquad \text{when} \; \| v_i \| \longrightarrow + \infty \; .$$

5.2 A decentralization model with incentives :

Let us go back to the model of incentives, sketched in § 1, (for details $[6]$). We take :

$$(5.10) \qquad J_i(v_i) = (b_i, \; v_i) \qquad \text{(linear functional) and the}$$

redistributed revenue K (*) as a quadratic form :

$$(5.11) \qquad K(v) = (Kv, v) + 2 (f, v)$$

where K is a symmetric matrix, $K \in \mathcal{L}(U_1 x,..,xU_m; U_1 x,..,xU_m)$, $K = (K_{ij})$ $(K_{ij} \in \mathcal{L}(U_j ; U_i))$. Therefore, one gets :

$$(5.12) \qquad J_i(v) = (b_i, v_i) + \lambda_i (Kv,v) + 2\lambda_i (f,v).$$

Let us notice that if there was no extra revenue redistributed, there could have been no equilibrium, since the functionals (5.10) do not satisfy (5.9). Thus, the redistributed revenue can help to find out a decentralized equilibrium. Let us check the assumptions of Theorems 3.1 and 4.1. Assumption (3.15) is clearly satisfied. Noticing that

$$(5.13) \qquad \frac{\partial J_i}{\partial v_i} = b_i + 2\lambda_i\left(f_i + \sum_j K_{ij} v_j\right), \text{ thus,}$$

$$(5.14) \qquad A(v) = c + 2 \widetilde{K}v$$

where

$$c = \begin{pmatrix} b_1 + 2\lambda_1 b_1 \\ \vdots \\ b_m + 2\lambda_m f_m \end{pmatrix}$$

and

$$\widetilde{K} = \begin{pmatrix} \cdots \cdots \\ \lambda_i K_{ij} \\ \cdots \cdots \end{pmatrix}$$

Therefore (3.16) amounts to

$$(5.15) \qquad \widetilde{K} \text{ non negative definite and } (3.17)$$

amounts to

$$(5.16) \qquad \widetilde{K} \text{ positive definite (i.e., invertible)}$$

(*) Counted negatively, since we are minimizing. The real
     redistributed revenue can be something like $G - K(v)$.
It can be negative, and it is then a redistributed charge. In
any case, it is bounded above.

If (5.16) is true, then $(3.16)'$ is automatically satisfied.
Therefore the decentralization algorithm will work out, when
$\tilde{K}$ is positive definite.

Conclusions :

We have seen in this paper under what conditions price decentralization is possible in the case of interrelated payoffs, and
how it could be implemented, through an iterative process.

The incentives example has shown that it is possible to obtain
a decentralized equilibrium using a system of charges or redistributed revenue. This technique could help to surmount the
relatively severe assumptions, under which the decentralization
was possible.

References

1. R. Armand, Décentralisation par les prix, Metra.
2. J.L. Lions, Contrôle optimal, Dunod, Paris, (1968).
    English Translation, Springer Verlag, (1971).
3. R. Glowinski, J.L. Lions, R. Trémolières, Book on varia-
    tional inequalities, to appear.
4. Arrow, Hurwicz, Uzawa, Studies in Linear and non Linear
    programming, Stanford University Press.
5. Rosen, Existence and Uniqueness of Eq. pts for concave
    N-persons games, Econometrica, Vol. 33, 3, (1965).
6. R. Kleindorfer, Unpublished report on Incentives, M.I.T.
7. Bonini, Jaebicke, Wagner, Management Controls, New Direc-
    tions in Basic Research, Mac Graw-Hill, (1964).
8. J.J. Laffont, G. Laroque, Prise en compte des effets exter-
    nes dans la théorie de l'équilibre général, presented at
    Congrès Européen de la Société d'Econométrie, Barcelone,
    (Sept. 1971), to appear in "Les Cahiers du Sémainaire
    d'Econométrie", C.N.R.S., Paris, (1972).
9. J.J.Laffont, Allocation optimale en présence d'effets
    externes, Discussion Paper, University Dauphine, Paris,
    (1971).

A ONE SECTOR MODEL OF ECONOMIC GROWTH WITH
UNCERTAIN TECHNOLOGY: AN EXAMPLE OF
STEADY STATE ANALYSIS IN A STOCHASTIC
OPTIMAL CONTROL PROBLEM

William A. Brock and Leonard J. Mirman

University of Chicago
and
University of Massachusetts

1. Introduction

This work is an attempt to carry the well-known
results of Cass [1], Koopmans [3], et al. concerning
convergence of optimum levels of consumption and
investment to a steady state to the case where output or
technical progress is a random variable. We follow the
above writers in conducting our analysis in a one-sector
model. We depart from them by taking as our objective
the maximization of the expected value of the discounted
sum of utilities facing uncertain technology or technological progress and, furthermore, our analysis is
conducted in discrete time.

Using elementary mathematical technique we
establish a stochastic analogue of convergence to a
steady state--the modified golden rule.

This paper is essentially a blend of dynamic
programming and discrete time stochastic optimal
control theory. For those that are not interested in the
economic issues involved, we feel that this work should
be useful to illustrate some of the problems that arise in
the analysis of economic models, in particular, models
of economic growth.

Models of economic growth tend to possess a lot of nonlinear concavity in both the objective functional and the constraint functions. The focus of the analysis may take several directions. One direction is to examine steady state or limiting behavior of the optimal control and state variables. That is the direction we shall take here.

The paper is organized as follows: 1. Introduction, 2. A short review of literature we have found, 3. The model, 4. Lemmas and proofs, and 5. Remarks.

2. Review of the Literature

The structure of optimal growth problems is related to that of various inventory models. Although the inventory theory does not seem to be directly applicable, learning the technique used is payoff enough for the growth theorist to spend some time on inventory theory. For example of a stochastic stability theorem in inventory theory see Karlin [2]. However, Karlin's functional analytic techniques are beyond the reach of the average economist. Our technique is well within that bound and, furthermore, appeals to the intuition.

Mirman [6] and Mirrlees [7], [8] are the works we have discovered that seem to be most relevant for our purposes. Mirrlees formulates a continuous time, one sector neoclassical model where technical change is random, Harrod neutral, has exponentially growing mean, and its logarithm follows a Wiener process through time. His objective is maximization of the expectation of the discounted sum of utilities. In the main, he finds a differential equation to characterize optimal paths and proves existence of such. He does not investigate any kind of stochastic stability. Mirrlees operates in continuous time where the theory of stochastic processes is messy. We avoid the messier mathematics by working in discrete time.

Mirman [6] investigates a class of stochastic difference equations that represent capital accumulation. These stochastic processes do not arise from any optimal growth model--at least not explicitly. Hence, it is important, from an economic point of view, to see if some kind of convergence to a steady state can established for an <u>optimal</u> growth model.

3. The Model

The model is set forth as follows:

$$\text{Maximize} \quad E \sum_{t=0}^{\infty} \delta^t u(t)$$

s. t.

$$
\begin{aligned}
c_0 + x_0 &= S \\
c_1 + x_1 &= f(x_0, r) \\
c_2 + x_2 &= f(x_1, r)
\end{aligned}
\tag{1}
$$

. . .

where c_t, x_t, r, $f(x, r)$, S, δ, u, E are consumption at t, investiment at t, positive valued random variable, production function, initial stocks, discount factor, utility function, and expected value operator, respectively. We assume u, f, strictly concave, differentiable and increasing, $u'(0) = \infty$, $f'(o, r) = \infty$ and $f'(\infty, r) < 1$ for all values of r, and $0 < \delta < 1$. For simplicity we assume that $r: \Omega \to R$ where Ω is a finite set of elementary events and R is the reals. The assumption that Ω is finite avoids messy mathematics. A generalization appears in [10]. Examples follow.

For example, if investment is reversible,[1] then

$$C_{t+1} + K_{t+1} - K_t = F(K_t, L^t, r) \tag{2}$$

may be put in the form (1) by

$$\frac{C_{t+1}}{L^{t+1}} + \frac{K_{t+1}}{L^{t+1}} = \frac{1}{L}[F(\frac{K_t}{L^t}, 1, r) + \frac{K_t}{L^t}]$$

if F is linear homogeneous in the first two variables.

Define

$$c_t = \frac{C_t}{L^t}, \quad x_t = \frac{K_t}{L^t}, \quad f(x, r) = \frac{1}{L} F(x, 1, r) + 1$$

If u is homogeneous of some degree, and F satisfies the usual economic conditions, we are in the realm of (1). For another example, if technical change is random Harrod neutral, i.e.,

$$F(K, L, t) = F(K, AL)$$

where A is of the form $A = \lambda^t r$ where r is random. $r: \Omega \to [a, b]$. If F is linear homogeneous and u is homogeneous of some degree, then a change of units puts us in the reach of (1).

Some caution is needed in formulating a model of random technical progress. That is in a reasonable world one expects that the worst in the future should beat the best today, i.e., in the random Harrod neutral case

$$(3) \qquad \underset{\omega \in R}{\text{Min}} \ \lambda \, r(\omega) > \underset{\omega \in R}{\text{Max}} \ r(\omega)$$

Inequality (3) asserts that with probability one, there be technical progress - not regress and amounts to $\lambda > b/a$.

Although the above discussion was brief, we hope that we have convinced the reader that the model has some economic content.

Problem (1) is, in language that may be more familiar to some readers, is just a discrete time stochastic optimal control problem.

Solution of (1) produces, for each S, a sequence of optimal random variables $\bar{x}_0, \bar{x}_1, \bar{x}_2, \ldots$. Does this sequence converge in some sense? The most useful concept of convergence seems to be that of convergence in distribution. To explain, let $F_t(x) = \Pr\{\bar{x}_t < x\} \equiv$ probability that $\bar{x}_t < x$. F_t is called the distribution function of the random variable $\bar{x}_t$. The sequence of random variables $\bar{x}_0, \bar{x}_1, \bar{x}_2, \ldots$ converges in distribution if for all $x \in R$, $F_t(x) \to F(x)$, $t \to \infty$. Here, R denotes the real numbers. Our main result is: let F_t be the distribution function of $\bar{x}_t$ where $\{\bar{x}_t\}_{t=0}^{\infty}$ solves (1) then $F_t \to F$, $t \to \infty$ and F is independent of initial stock, S.

Much of the structure of our analysis can be displayed by first considering the simpler deterministic problem

$$\text{Maximize } \sum_{t=0}^{\infty} \delta^t \, u(c_t)$$

s.t.

(4)
$$c_0 + x_0 = S$$
$$c_1 + x_1 = f(x_0)$$
$$c_2 + x_2 = f(x_1)$$

By strict concavity we have, for each S, exactly one $\bar{x}_0$, $\bar{x}_1$, ... that solves (4). Hence $x_0 \equiv h(S)$. By straightforward dynamic programming considerations, e.g., see Levhari and Srinivasan [4] we have the existence of a policy function h such that $\bar{x}_t = h[f(\bar{x}_{t-1})]$. Furthermore, following [4] it is straightforward to prove that h is increasing, continuous, $h(0) = 0$. Proving the convergence theorem $\bar{x}_t \to \bar{x}$, $t \to \infty$ is equivalent to proving that h has only one positive fixed point $\bar{x}$ and that $\bar{x}$ is stable. To dispose of this examine the necessary condition of optimality

$$(5) \qquad a(x_{t-1}) \equiv u'[f(x_{t-1}) - x_t] = u'[f(x_{t-1}) - h(f(x_{t-1}))]$$

$$= \delta u'[f(x_t) - h(f(x_t))] \, f'(x_t) \equiv b(x_t)$$

Since optimal consumption $\bar{c}_{t+1} = f(\bar{x}_t) - h[f(\bar{x}_t)]$ increases in $\bar{x}_t$ (dynamic programming as in Levhari and Srinivasan [4], again) we see that a(x), b(x) decrease in x. Furthermore, $b(x)/a(x) = f'(x)$ decreases in x, $b(0)/a(0) = \infty$, $b(\infty)/a(\infty) = 0$. Hence there is only one positive $\bar{x}$ such that $a(\bar{x}) = b(\bar{x})$. If $a \equiv f^{-1}(S) < \bar{x}$ then $\bar{x}_t$ increases to $\bar{x}$, if $a = \bar{x}$ then $\bar{x}_t = \bar{x}$, if $a > \bar{x}$, then $\bar{x}_t$ decreases to $\bar{x}$.

The above analysis was terse and informal but its structural similarity should ease the reader's entry into the more difficult random case.

Analysis of the random case (1) proceeds as follows:

(a) We establish by dynamic programming the existence of a policy function h such that $\bar{x}_t = h[f(\bar{x}_{t-1}, r)] \equiv h_r(\bar{x}_{t-1})$, h is increasing and h(0) = 0. (b) Let $r: \Omega \to R$ have minimum value 1 and maximum value n where n is the number of points in Ω. Then we study the structure of fixed points for the pair of functions $h_1(x)$, $h_n(x)$, by using the necessary conditions of optimality. Using the stochastic process $\bar{x}_t = h_r(\bar{x}_{t-1})$ we derive a relationship between $F_t(x) \equiv Pr\{\bar{x}_t < x\}$ and $F_{t-1}(x)$ by

$$F_t(x) = \sum_{s=1}^{n} Pr[h_s(\bar{x}_{t-1}) < x] \, \alpha_s$$

$$(6) \qquad = \sum_{s=1}^{n} F_{t-1}[q_s(x)] \, \alpha_s$$

where q_s is the inverse function of h_s, and α_s = probability that r takes the value of s, s = 1, 2, ..., n. Hence any limit distribution function F must satisfy

$$(7) \qquad F(x) = \sum_{s=1}^{n} F[q_s(x)] \, \alpha_s$$

Under conditions imposed by optimality, and general assumptions on f(x, r), it will be shown that equation (7) has a unique solution and it is "stable". Unique solution of (7) amounts to convergence in distribution to the "stochastic modified golden rule". The details follow. First we put down all assumptions in one place.

<u>Assumptions</u>: $r: \Omega \to \{1, 2, ..., n\}$, minimum $r(\omega) = 1$, $\omega \in \Omega$

maximum $r(\omega) = n$; f(x, s) is (a) increasing in s or (b) $\omega \in \Omega$

decreasing in s; to fix matters assume (a). u', $\frac{\delta f}{\delta x}(x, s) =$ f'(x, s) both exist; u, f are strictly concave, u'(0) = ∞, f'(0, s) = ∞, f'(∞, s) $\leq$ 1 for s = 1, 2, ..., n.

4. Lemmas and Proofs

Lemma 1 Let x_0, x_1, ... solve (1). Then there is a policy function h such that $\bar{x}_t = h_r(\bar{x}_{t-1})$. Furthermore h is increasing, $h(0) = 0$, $g[f(x,r)] \equiv f(x,r) - h[f(x,r)]$ is continuous and increasing.

Proof: Follow [4]. Continuity is straightforward.

Lemma 2 There is $\epsilon > 0$ such that on $(0,\epsilon)$ we have $h_1(x) > x$.

Proof: Straightforward.

Lemma 3 If $x_s \equiv \{ \text{Max} \{ x | h_s(x) = x \}$ exists then for all implies $h_s(x) < x$. I.e., the "last" fixed point of each h_s is stable.

Proof: Straightforward.
This is another intuitively obvious lemma because if it were not true for some s_0 then we can obtain unbounded capital accumulations from finite initial stock.
$[a,b]$ is a stable interval if there are no fixed points of $h_1(x)$ greater than a, no positive fixed points of $h_n(x)$ less than $b, h_1(a) = a$ and $h_n(b) = b$. It is obvious from the definition that if a stable interval exists it is unique. Basically, a stable interval means that the Markov process $x_t = h(x_{t-1}, r_t)$ is ergodic and acylic. We give an heuristic definition of these terms. Ergodic means that the state space has only one interval $[a,b]$ possessing the property that once entered it is never left and no $[c,d] \overset{c}{=} [a,b]$ has that property. Acylic means that there are no cyclically moving subclasses. I.e., between any two non-degenerate subintervals I_1, I_2 of $[a,b]$ there is a positive probability path from I_1 to I_2. In other words, there is a time T such that the probability of moving from I_1 to I_2 in T steps is positive.
Lemma 4 is the most important lemma of this paper because we use information from the optimization process to demonstrate the existence of a stable interval.
Recall that $h_1(x) < h_n(x)$ for each x, because $h_s(x)$ increases in s for each x.

Lemma 4 There is a stable interval.
This lemma is the direct analogue of the uniqueness of the fixed point in the deterministic case.

Proof: Let $g[f(x, r)] = f(x, r) - h[f(x, r)]$. By optimality of $\bar{x}_0, \bar{x}_1, \ldots$ we have

$$(8) \quad u'[g(f(\bar{x}_{t-1}, r))] = \delta \sum_{s=1}^{n} u'(g[f(\bar{x}_t, s)] \, f'(\bar{x}_t, s) \, \alpha_s, \quad t = 1, 2, \ldots$$

To understand (8) recall that $\bar{x}_t : \Omega^t \to [0, \infty); r : \Omega \to R$ where Ω^t is the Cartesian product of ω with itself t times. Now (8) says, in full: For each sample path $(\omega_1, \omega_2, \ldots, \omega_t) \in \Omega^t$

$$u'[g(f(\bar{x}_{t-1}(\omega_1 \ldots, \omega_{t-1})), r(\omega_t))]$$

$$(9) \quad = \delta \sum_{s=1}^{n} u'[g(f(\bar{x}_t(\omega_1, \ldots, \omega_t), s)] \, f'[\bar{x}_t(\omega_1, \ldots, \omega_t), s] \, \alpha_s$$

$$t = 1, 2, \ldots$$

Needless to say, the compact form (8) is preferable to (9). Put $d[f(x, r)] = u'[g(f(x, r))]$. Consider the fixed points $x_i = h[f(s_i, i)]$ $i = 1, 2, \ldots, n$. Since the stochastic process described by the necessary condition of optimality (8) is the same as $\bar{x}_t = h[f(\bar{x}_{t-1}, r)]$ it is straightforward to prove for any fixed point $x_i = h[f(x_i, i)]$ that

$$(10) \quad d[f(x_i, i)] = \delta \sum_{s=1}^{n} d[f(x_i, s)] \, f'(x_i, s) \, \alpha_s$$

$$i = 1, 2, \ldots, n$$

Now d is decreasing, hence, $f(x_i, 1) < f(x_i, s) < f(x_i, n)$ implies $d[f(x_i, 1)] > d[f(x_i, s)] > d[f(x_i, n)] s = 2, 3, \ldots, n-1$. Therefore from (10), $d[f(x_1, 1)] < \delta \sum_{s=1}^{n} d[f(x_1, 1)] f'(x_1, s) \, \alpha_s$,

and $d[f(x_n, n)] > \delta \sum_{s=1}^{n} d[f(x_n, n)] f'(x_n, s) \, \alpha_s$

Thus[5]

$$(11) \quad 1 < \delta \sum_{s=1}^{n} f'(x_1, s) \, \alpha_s$$

and

$$1 > \delta \sum_{s=1}^{n} f'(x_n, s) \, \alpha_s$$

Suppose now that there exists another pair of fixed points x_1', x_n' such that $x_1 < x_n \le x_1' < x_n'$. Now x_1', x_n' must satisfy (11). Because $f'(x, s)$ decreases in x,

$$1 < \delta \sum_{s=1}^{n} f'(x_1', s) \, \alpha_s \le \delta \sum_{s=1}^{n} f'(x_n, s) \, \alpha_s < 1$$

a contradiction. The Lemma is established.

<u>Remark</u>: The proof of Lemma 4 for the case where $f(x, s)$ is decreasing in s is similar.

Lemma 4 assures us that x_1, x_n is a pair of fixed points such that there are no fixed points of h_1 to the right of x_n and no fixed points of h_n to the left of x_1.

Intuition plus a little reflection makes it plausible that the stochastic process $\bar{x}_t = h_r(\bar{x}_{t-1})$ will settle in the range $[x_1, x_n]$ as $t \to \infty$. Notice also that the range $[x_1, x_n]$ is independent of initial stock S.

<u>Theorem</u>: Let $F_t(x) = \Pr[\bar{x}_t < x]$. Then there is a distribution function F such that for all $x \in R$, $F_t(x) \to F(x)$, $t \to \infty$. Furthermore, F does not depend on S.

Proof: Let $\bar{x}_0$ be optimal initial investment. It is a straightforward job to prove that for any $\bar{x}_0 > 0$ the sequence

$$(12) \quad F_t(x) = \sum_{s=1}^{n} F_{t-1}[q_s(x)] \, \alpha_s, \quad t = 1, 2, \ldots$$

converges uniformly on the line R. Here $F_0(x)$ is the degenerate distribution corresponding to $\bar{x}_0$. I.e., $F_0(x) = 0$ for $x \le \bar{x}_0$ and $F_0(x) = 1$ for $x > \bar{x}_0$. For lack of space we must give an heuristic proof. To prove uniform convergence we show that the sequence $\{F_t\}_{t=0}^{\infty}$ defined by (12) is uniformly Cauchy, i.e., given $\epsilon > 0$ there is N_ϵ such that for all positive integers p, k, $p \ge N_\epsilon$ we have $\sup_{x \in R} \{ |F_{p+k}(x) - F_p(x)| \} < \epsilon$ where "sup" is supremum. To do this notice that by iterating (12) we get

$$(13) \quad F_{p+k}(x) - F_p(x) = \sum_{i_{p+k}, \ldots, i_{p+1}} \sum_{i_p, \ldots, i_1} \{F_0$$

$$[q_{i_{p+k}} q_{i_{p+k-1}} \ldots q_{i_p} \ldots q_{i_1}(x)] - F_0[q_{i_p} \ldots$$

$$q_{i_1}(x)]\} \, \alpha_{i_p} \ldots \alpha_{i_1} \alpha_{i_{p+k}} \ldots \alpha_{i_{p+1}}$$

where each i_j may take the values 1, 2, ... n. Hence there are n^{p+k} items summed in the inner brackets. Notice that entities such as $q_{i_p} \ldots q_{i_1}(x)$ are simply realizations of the stochastic process.

$$(14) \quad y_t = q(y_{t-1}, r) = q_r(y_{t-1}), \quad y_0 = x$$

which is the "inverse" process of $x_t = h[f(x_{t-1}, r)]$. The stochastic process (14) converges to 0 or ∞ in probability. To see this if we start with x then one period later $q_{i_1}(x)$

emerges with probability α_{i_1}, two periods later $q_{i_2} q_{i_1}(x)$

emerges with probability $\alpha_{i_2} \cdot \alpha_{i_1}$, etc. Notice that $[x_{1m}, x_{nM}]$ is transient, i.e., once it is left it is never returned to and it is left with positive probability. Hence, after a long period of time "most" realizations of (14) leave $[x_{1m}, x_{nM}]$. Now if the interval $[0, x_{1m}]$ is entered by a realization of (14), i.e., there is p, $i_1, \ldots, i_p$ such that $x_p = q_{i_p} \ldots q_{i_1}(x) \in [0, x_{1m}]$, then $x_{p+1} = \max_{1 \leq i \leq n} q_i(x_p) < x_p$ and the sequence $x_{p+j} = \max_{1 \leq i \leq n} q_i(x_{p+j-1})$ converges monotonely to 0 as $j \to \infty$. I.e., the process (14) is monotone decreasing to 0 on $[0, x_{1m}]$. Similarly the process (14) is monotone increasing to ∞ on (x_{nM}, ∞).

Let us now return to equation (13). If p is large "most" of the $q_{i_p} q_{i_{p-1}} \ldots q_{i_1}(x)$ are in $[0, x_{1m}]$ or (x_{nM}, ∞)

and if $q_{i_p} \ldots q_{i_1}(x)$ is in $[0, x_{1m})$ or (x_{nM}, ∞) then so is

$q_{i_{p+k}} \ldots q_{i_p} \ldots q_{i_1}(x)$ in $[0, x_{1m}]$ or (x_{nM}, ∞), respective-

ly. Thus if $\bar{x}_0 > 0$ and since $F_0(x) = 0$, $x \leqq \bar{x}_0$ and $F_0(x) = 1$, $x > \bar{x}_0$ then there will be a p_0 large enough so that for $p \geqq p_0$, $k = 0, 1, 2, \ldots$ "most" of the $F_0[q_{i_p} \ldots q_{i_1}(x)] = F_0$

$[q_{i_{p+k}} \ldots q_{i_1}(x)]$. Hence, the difference in (13) goes to

zero as $p \to \infty$. Furthermore, a little reflection will con-
vince one that this convergence is uniform.

Hence there is a limit, F, such that $F_t \to F$ uniformly. Thus F is nondecreasing. It is trivial to prove, by
exploitation of (12) that $F = 1$ on (x_{nM}, ∞) and $F = 0$ on
$[0, x_{1m}]$. Also F is continuous from the left as it is a
uniform limit of continuous from the left F_t. Hence F is
a distribution. F must satisfy

$$(15) \quad F(x) = \sum_{s=1}^{n} F\,[q_s(x)]\, \alpha_s$$

To see that the limit F is independent of initial stocks S we
need only show that there is only one distribution F that
satisfies (15).

The equation (15) has a unique solution. For if F, G
are two distributions that solve (15) then

$$(16) \quad H(x) = F(x) - G(x) = \sum_{i_1} H[q_{i_1}(x)]\alpha_{i_1} = \sum_{i_1, i_2} H[q_{i_2} q_{i_1}(x)]\alpha_{i_2}\alpha_{i_1} = \sum_{i_p, \ldots, i_1} H[q_{i_p} \ldots q_{i_1}(x)]\alpha_{i_p} \ldots \alpha_{i_1}$$

Now, if F, G solve (15) it is trivial to prove that $F = G = 0$
on $[0, x_{1m}]$ and $F = G = 1$ on $[x_{nM}, \infty)$. Hence, an argument
exactly like the one given for the convergence of the F_t
given by (12) gives us $H(x) = 0$ for all x. For that argument shows that the right hand side of (16) goes to zero
uniformly in x as $p \to \infty$.

5. Remarks:

Mirman [9] has extended his work in [6] to include the case of processes with a stable interval. Thus, his class of stochastic processes in [9] include processes arising from optional growth models of the type studied in this paper. A generalization of the results of this paper to general random variables is in [10]. The case $\delta = 1$ is treated in [11] for a one good model like the one in this paper. Patrick Jeanjean [12] has extended some of these results to multisector models.

In this paper we chose to take the self contained route in order to delineate the ideas and to reach a wider audience. However, one may apply tools from probability theory to prove ergodicity and acyclicity. See [9] for this and a list of references to the relevant Markov theoretic articles.

Footnotes

1. Let us expand upon this point. If capital cannot be eaten into after it is installed, then the constraint $K_{t+1} \geq K_t$, $t = 0, 1, \ldots$ must be added to the optimization problem. In this case it does not follow that the problem may be put into the form (1) and give the same optimal solution as (1) does. By assuming that investment is reversible we allow the possibility that $K_{t+1} < K_t$ for some time t. We conjecture that our results can be proved in essentially the same form for the case where the constraint $K_{t+1} \geq K_t$ is present. We are working on this now.

2. Notice that (11) is the stochastic analogue to the condition $1 = \delta f'(\bar{x})$ which is the equation determining the modified golden rule in the deterministic case.

References

[1]   Cass, D., "Optimal Growth in an Aggregative Model of Capital Accumulation," Review of Economic Studies, 1965.

[2]   Karlin, S., "Steady State Solutions," Studies in the Mathematical Theory of Inventory and Production, pp. 233-69, 1958.

[3]   Koopmans, T., "On the Concept of Optimal Economic Growth," Pontificae Academiae Scientiarum Scripta Varia, pp. 225-300, 1965.

[4]   Levhari D., and T. Srinivasan, "Optimal Savings Under Uncertainty," Review of Economic Studies, 1969.

[5]   Loeve, M., Probability Theory, Van Nostrand, Princeton, 1963.

[6]   Mirman, L., "Two Essays on Uncertainty and Economics," Unpublished Ph.D. thesis, University of Rochester, 1970.

[7]   Mirrlees, J., "Optimum Economic Policies Under Uncertainty," Unpublished, 1965.

[8]   Mirrlees, J., "Optimum Accumulation Under Uncertainty," 1965, Unpublished.

[9]   Mirman, L., "The Steady State Behavior of a Class of One Sector Growth Models with Uncertain Technology," University of Massachusetts, 1971.

[10] Brock, W., and Mirman, L., "Optimal Economic Growth and Uncertainty: The Discounted Case," Journal of Economic Theory (Forthcoming) 1971.

[11] Brock, W., and Mirman, L., "Optimal Economic Growth and Uncertainty: The Ramsey-Weiszacker Case," MSSB Working Paper 7, Dept. of Economics, University of California at Berkeley, Summer 1971.

[12] Jeanjean, P., Ph.D. Thesis, University of California at Berkeley, Dept. of Economics, 1971.

LINEAR AND NONLINEAR PROGRAMMING

DUAL PROBLEMS OF OPTIMAL CONTROL

R. Tyrrell Rockafellar*

The close connection between duality and convexity is
well known. It is no surprise, therefore, that the strongest properties of duality in optimal control are displayed
by problems which are especially "convex" in nature. Such
problems arise commonly in various situations, for example
economic applications, where duality may be interpreted
in terms of price behavior. They can also arise theoretically as local or global convexifications of more general
problems. The possible use of duality in the construction
of algorithms is another motivation for studying them.

In what follows, we indicate briefly some of the main
results that have been obtained for convex problems of
Bolza [1,2,3,4,5]. To simplify the discussion and to make
clearer the relationship with control problems as they are
usually formulated, we limit ourselves here to the "autonomous" case and choose a model in which the control variables
appear explicitly. Nevertheless we use the device of incorporating constraints into the cost functions by means of
infinite penalties, since this is not only very convenient
in theory, but essential if the basic ideas are not to be
obscured.

The model problem consists of minimizing

$$(1) \qquad \int_0^1 f(x(t),u(t))dt + l(x(0),x(1))$$

over all the absolutely continuous arcs $x:[0,1] \to R^n$ and
measurable functions $u:[0,1] \to R^n$ satisfying

$$(2) \qquad \dot{x}(t) = Ax(t) + u(t) \qquad a.e.,$$

---

*This work was supported in part by the Air Force Office of
Scientific Research under grant AFOSR - 71-1994.

where it is assumed that f and l are lower semicontinuous, convex functions from $R^n \times R^n$ to $(-\infty, +\infty]$, not identically $+\infty$. Note that $f(x,u)$ is to be convex jointly in x and u, rather than just convex in u, as would be a more common assumption. The problem can be described in the context of functional analysis as that of minimizing a certain Bolza functional

$$(3) \qquad \Phi(x) = \int_0^1 f(x(t), \dot{x}(t) - Ax(t))dt + l(x(0),x(1))$$

over a Banach space $\mathcal{A}$ consisting of all the absolutely continuous functions $x:[0,1] \to R^n$.

Since in a problem of minimization the points where the cost function has the value $+\infty$ do not compete for the optimum, our model contains implicit constraints on the controls, states and endpoints. For almost every t, the control vector $u(t)$ should belong to the set $U(x(t))$, where

$$(4) \qquad U(x) = \{u\varepsilon R^n \mid f(x,u) < +\infty\}$$

(implicit control region), and thus the state vector $x(t)$ should belong to the set

$$(5) \qquad X = \{x\varepsilon R^n \mid \exists u, f(x,u) < +\infty\}.$$

The endpoint pair $(x(0),x(1))$ should belong to the set

$$(6) \qquad C = \{(a_0,a_1)\varepsilon\ R^n \times R^n \mid l(a_0,a_1) < +\infty\}.$$

As an illustration, consider the problem of minimizing (for some $\lambda \geq 0$)

$$\lambda \int_0^1 |x(t)|dt + \int_0^1 |u(t)|^2 dt$$

subject to $\dot{x} = Ax + u$, $|u| \leq 1$, $x(0) = a$ and $x(1)\varepsilon R^n_+$ (nonnegative orthant). This corresponds to

$$f(x,u) = \lambda|x| + |u|^2 \quad \text{if} \quad |u| \leq 1,$$
$$= +\infty \quad \text{if} \quad |u| > 1,$$

$$l(a_0,a_1) = 0 \quad \text{if} \quad a_0 = a \quad \text{and} \quad a_1\varepsilon R^n_+,$$
$$= +\infty \quad \text{if} \quad a_0 \neq a \quad \text{or} \quad a_1\notin R^n_+.$$

The example serves to emphasize that no differentiability is assumed in the cost functions, even with respect to x. It shows also that, although we speak formally of problems of Bolza, other classes of problems, such as those of Lagrange, are aptly covered by the same notation.

In working with $+\infty$, it is necessary to take a somewhat different approach than usual to a number of technical questions concerning measurability, integrability, and so forth. This is particularly true in the case, not discussed here, where f and A depend on t. Fortunately, the theory of measurable multifunctions, as developed extensively by Castaing and others, comes to our aid. At the same time, convexity leads to many simplifications. Thus it can be shown under our assumptions that the Bolza functional $\Phi: \mathcal{a} \to (-\infty, +\infty]$ is not only well-defined and convex, but lower semicontinuous in the weak (and strong) topologies [1]. In fact, if the generalized Hamiltonian function

$$(7) \qquad H(x,p) = \sup_{u} \{p \cdot (Ax+u) - f(x,u)\}$$

nowhere has the value $+\infty$, then Φ has the remarkable property that its level sets

$$\{x \varepsilon \, \mathcal{a} \mid \Phi(x) \leq \alpha\}, \qquad \alpha \quad \text{real},$$

are (closed and) locally compact in the weak topology [3]. This can be used to deduce the existence of optimal arcs in the control problem [4]. However, the existence also follows from duality theorems stated below. Results on necessary conditions for optimality also follow from the duality theorems, so that, for convex problems of Bolza, the latter really play the central role.

Duality is obtained by passing to the convex functions f* and l* conjugate to f and l , or rather to slightly modified forms of these functions. We define the dual cost functions g and m by

$$(8) \qquad g(p,w) = f^*(w,p)$$
$$= \sup_{.x,u} \{w \cdot x + p \cdot u - f(x,u)\},$$

$$(9) \qquad m(b_0,b_1) = l^*(b_0,-b_1)$$
$$= \sup_{a_0,a_1} \{a_0 \cdot b_0 - a_1 \cdot b_1 - l(a_0,a_1)\}.$$

Then g and m are again lower semicontinuous, convex functions from $R^n \times R^n$ to $(-\infty,+\infty]$, not identically $+\infty$. Furthermore, f and l may be recovered in turn as the duals of g and m:

$$f(x,u) = \sup_{p,w}\{u \cdot p + x \cdot w - g(p,w)\},$$
$$l(a_0,a_1) = \sup_{b_0,b_1} \{a_0 \cdot b_0 - a_1 \cdot b_1 - m(b_0,b_1)\}.$$

The dual control problem is taken to be that of minimizing

$$(10) \qquad \int_0^1 f(p(t),\, w(t))dt + m(p(0),\, p(1))$$

over all the absolutely continuous arcs $p:[0,1] \to R^n$ and measurable functions $w:[0,1] \to R^n$ satisfying

$$(11) \qquad \dot{p}(t) = -A^*p(t) + w(t) \qquad a.e.,$$

where A^* is the transpose of the matrix A. In view of the symmetric relationship between f and g, and between l and m, the problem which is dual to the dual problem is the primal (i.e. original) problem. The dual, like the primal, contains implicit constraints on the controls $w(t)$, states $p(t)$ and endpoint pair $(p(0),p(1))$. It can be regarded as the problem of minimizing the lower semicontinuous, convex Bolza functional

$$(12) \qquad \Psi(p) = \int_0^1 g(p(t),\, \dot{p}(t) + A^*p(t))dt + m(p(0),p(1))$$

over the Banach space $\mathcal{A}$.

In the example given earlier, one calculates easily from (8) and (9) that the dual problem consists of minimizing

$$\int_0^1 \theta(|p(t)|)dt + a \cdot p(0)$$

subject to $\dot{p} = -A^*p + w$, $|w| \le \lambda$ and $p(1) \in R^n_+$, where

$$\theta(s) = s^2/4 \qquad \text{if } s \leq 2,$$

$$= s - 1 \qquad \text{if } s \geq 2.$$

The definitions of the Bolza functionals imply that the inequality

$$\Phi(x) + \Psi(p) \geq 0$$

is valid, and consequently that

(13)   [inf in primal] $\geq$ - [inf in dual].

A fundamental question in duality theory is whether, or rather under what conditions, equality holds in (13). It usually happens that, when equality can be established, one obtains by the same argument the existence of a minimizing arc for Φ or Ψ, so that "inf" can be replaced by "min" in one of the two problems. In general, the study of (13) involves the convex functionals on $\mathcal{Q}*$ conjugate to Φ and Ψ, and these can be described in terms of the behavior of the control problems with respect to certain perturbations of the data [1]. To obtain sharper results, which proceed from "readily verifiable" assumptions on the functions f and l and the matrix A, an argument based on the separation of convex sets has been devised [3]. This argument is made complicated by the fact that the convex sets belong to the space $\mathcal{Q}* \times R^2$. One must show, despite the underlying nonreflexivity, that the separating hyperplane corresponds to an element of $\mathcal{Q} \times R^2$, rather than $\mathcal{Q}** \times R^2$.

The main duality theorems depend on finiteness assumptions on the Hamiltonian function (7) and attainability assumptions on the control systems. An endpoint pair $(x(0),x(1))$ is said to be <u>attainable</u> for the primal problem if it arises from an arc satisfying (2) and

$$u(t) \in U(x(t)) \qquad \text{a.e.,}$$

where $U(x(t))$ is the implicit control region in (4). The set of all such pairs is convex in $R^n \times R^n$. The <u>attainability condition</u> is said to be satisfied if the relative interior of this set meets the relative interior of the convex set of all feasible endpoint pairs, that is, the set C in (6).

DUALITY THEOREM 1[3]. If the primal problem satisfies
the attainability contion and $H(x,p) > -\infty$ everywhere (in
other words, there are no real state constraints on x),
one has

(14)   [inf in primal] = -[min in dual].

DUALITY THEOREM 2[3]. If the dual problem satisfies
the attainability condition and $H(x,p) < +\infty$ everywhere
(a growth condition of Nagumo-Tonelli type on the function
$u \to f(x,u)$), one has

(15)   [min in primal] = -[inf in dual].

The dual attainability condition on endpoint pairs
$(p(0),p(1))$ can be expressed equivalently as a growth
condition on the Bolza functional Φ in the primal problem
[3].

The fact that the attainment of the infimum in the dual
problem seems to require the absence of state constraints
in the primal problem can be explained from the role that
dual optimal arcs have in the statement of necessary
conditions for the primal, as seen below. It is known
that, when state constraints are present, the necessary
conditions ought to involve jumps in the costate vector
$p(t)$. However, an optimal arc for the dual problem is
by definition absolutely continuous. This suggests that,
in order to obtain a better duality theory in the case of
state constraints, one should pass to a generalized dual
problem where the arcs p are allowed to be discontinuous.
The idea has been worked out in [4] for a large class of
problems in which, roughly speaking, the effects of the
state constraints on x can be kept separate from the
effects of the other constraints. The dual problem then
consists of minimizing an extended Bolza functional, not
over $\mathcal{A}$, but over a Banach space $\mathcal{B}$ consisting of all the
functions $p:[0,1] \to R^n$ of bounded variation.

Nevertheless, the theory is still incomplete, since it does not cover many problems, encountered for example in economics, where the effects of the state constraints cannot be kept separate. Also, it would seem desirable ultimately, at least for the sake of symmetry, to formulate the primal, as well as the dual, in terms of functions of bounded variation, deriving continuity properties of the optimizing arcs in a given case from the necessary conditions for optimality.

We conclude by describing the necessary conditions that correspond to the situation in Theorem 1. These depend on the fact that the Hamiltonian $H(x,p)$ is not only convex in p, as follows immediately from the definition (7), but also concave in x. The latter property is a consequence (indeed, virtually an equivalent form) of our assumption that $f(x,u)$ is convex jointly in x and u. Recall that in dealing with convex functions one can replace the usual notion of differentiability, that of a tangent hyperplane to the graph of the function, by the notion of a supporting hyperplane to the epigraph of a function. Specifically, if φ is an extended real-valued, convex function on R^n, we define a <u>subgradient</u> of φ at z to be a vector y such that the inequality

$$\varphi(z') \geq \varphi(z) + y \cdot (z' - z)$$

holds for all $z' \varepsilon R^n$. The set of such subgradients is denoted by $\partial\varphi(z)$. Subgradients of concave fundtions are defined analogously, with the opposite inequality. Applying this idea to $H(x,p)$ as a function of x and p separately, we obtain a generalized Hamiltonian system of differential equations:

$$(16) \qquad \dot{x} \varepsilon\, \partial_p H(x,p) \quad \text{and} \; - \dot{p} \varepsilon\, \partial_x H(x,p) \qquad \text{a.e.}$$

Subgradients can also be used to formulate a generalized transversality condition:

$$(17) \qquad (-p(0),p(1)\, \varepsilon\, \partial l(x(0),x(1)).$$

We refer to (16) and (17) as the fundamental <u>optimality conditions</u> for a convex problem of Bolza. Of course, in particular cases the many theorems available for the calculation of subgradients can be used to express these conditions in a less abstract manner [1]. The main result is the following:

THEOREM 3. Let the assumptions of Theorem 1 be satisfied. Then, in order that an arc $x \in \mathcal{A}$ be optimal for the primal problem, it is necessary and sufficient that there exist an arc $p \in \mathcal{A}$ for which conditions (16) and (17) are satisfied. Such an arc p is optimal for the dual problem.

The generalized Hamiltonian system is quite amenable to study, despite its "multivaluedness". For example, under the assumption that H is finite, it has been shown that local solutions exist, and along such solutions the value of H is constant [2]. The behavior of the system near a saddle-point of H (in the minimax sense) has also been investigated and shown to be relevant to certain optimal control problems where $[0,1]$ is replaced by an infinite time interval [5].

Theorem 3 has been extended in [4] to a class of problems with state constraints through an appropriate definition of what is meant by the Hamiltonian condition (16) in the case where p is not absolutely continuous, but merely of bounded variation.

REFERENCES

1. R.T. Rockafellar, "Conjugate convex functions in optimal control and the calculus of variations," J. Math. Anal. Appl. 32(1970), 174-222.

2. ________, "Generalized Hamiltonian equations in convex problems of Lagrange," Pacific J. Math. 33(1970), 411-427.

3. ________, "Existence and duality theorems for convex problems of Bolza," Trans. A.M.S. 159(1971), 1-40.

4. ________, "State constraints in convex control problems of Bolza," S.I.A.M. Journal Control, to appear.

5. _________, "Saddle points of Hamiltonian systems in convex problems of Lagrange," J. Opt. Theory Appl., to appear.

Induced constraints for stochastic optimization problems

Roger Wets

1.   Introduction

Stochastic optimization problems are essentially <u>sequential</u> decision problems. Given that at a particular stage a decision must be made one should take into account the fact that given the decision selected and given the possible disturbances which are due to random elements, the problem must remain a feasible problem. In other words the here and now decision must take into account the fact that we must satisfy the constraints to be encountered at the further stages of the problem whatever be the outcome of the random variables. Constraints generated by such future considerations are called <u>induced constraints.</u> In general such constraints are hard to compute since deriving them involves a backing-up process. Moreover for constraints involving random elements it is usually sufficient to satisfy them with probability 1 and not necessarily for all points of the distribution.

The purpose of this paper is to formulate a rather general framework for studying induced constraints and obtain some very weak conditions under which it is possible to pass to a somewhat more constructive descriptions of these induced constraints. We also give some indications how the general theory applies in the fields of stochastic programming and stochastic optimal control.

2. Theory

We consider the following problem: Let $(\Xi, \mathcal{F}, \mu)$ be a probability space where $\mathcal{F}$ is the σ–algebra and μ is a probability measure. We shall assume that a topology compatible with $\mathcal{F}$ has been defined on Ξ. The random variable is denoted by $\underset{\sim}{\xi}$ and points of Ξ by ξ. Let $\mathcal{X} \times \mathcal{Y}$ be a topological space such that every subspace of $\mathcal{X}$ is separable in the induced topology, we shall say that $\mathcal{X}$ is <u>hereditarily separable.</u> Let f be a mapping from $X \times Y \times \Xi$ into R^m where X and Y are subsets of $\mathcal{X}$ and $\mathcal{Y}$ respectively. The question is to characterize the set

$$D = \{x \mid o \in f(x, Y, \underset{\sim}{\xi}) \text{ a.s.}\}$$

where a.s. stands for almost surely. The set D is obviously well defined in this manner, the goal is to find a somewhat more constructive description which would enable us to actually compute the set of acceptable x's.

In terms of our stochastic optimization problem we view the space $\mathcal{X}$ as representing the space from which a decision must be selected <u>here and now.</u> The space $\underset{\sim}{\mathcal{Y}}$ corresponds to the space of future decisions whereas $\underset{\sim}{\xi}$ stands for all future random events not yet observed. Thus finding the set D is to determine the set of acceptable here and now decisions which with probability 1 will enable us to find a feasible solution to the subsequent decision problems that we shall have to face.

We shall not consider the most general possible class of mapping f. Those for which we have been able to derive some results satisfy the following assumptions:

(A.1) $f(x, y, \xi)$ is continuous in x (for fixed y and ξ).

(A.2) $f(x, y, \xi)$ is continuous in ξ (for fixed x and y).

(A.3) $f(x, Y, \xi)$ is a upper semicontinuous with respect to ξ for each fixed x in X, i.e. the graph of $f(x,Y,\xi)$ is closed in the product space $\Xi \times R^m$ for each x in X.

We define a _relation_ between the points of X and Ξ as follows: we say that x is related to ξ, which we write as

$$x \sim \xi, \quad \text{if} \quad o \in f(x,Y,\xi).$$

The set of all points in X related to a given ξ in Ξ is denoted by

$$\omega(\xi) = \{x \in X \mid x \sim \xi\}$$

and thus

$$\omega(\Xi) = \{x \in X \mid x \sim \xi \text{ for all } \xi \text{ in } \Xi\} = \bigcap_{\xi \in \Xi} \{x \in X \mid x \sim \xi\}$$

or more generally, for Σ an arbitrary subset of Ξ

$$\omega(\Sigma) = \bigcap_{\xi \in \Sigma} \{x \in X \mid x \sim \xi\}.$$

Similarly, we introduce the inverse map

$$\omega^{-1}(x) = \{\xi \in \Xi \mid x \sim \xi\}$$

and

$$\omega^{-1}(S) = \bigcap_{x \in S} \{\xi \in \Xi \mid x \sim \xi\}$$

for S an arbitrary subset of X. In order to avoid some pathological situations which are of little interest, we assume that

(A.4)   For fixed x is $f(x,Y,\xi)$ is a measurable subset of $\Xi \times R^m$ where the measurable sets are obtained as the σ-product of $\mathcal{F}$ and the Borel algebra for R^m. From this it follows immediately that

Proposition 1. For all $x \in X$, the set $\omega(x)$ is measurable, i.e. $\omega^{-1}(x) \in \mathcal{F}$.

The mappings ω and ω^{-1} satisfy natural duality relations such as exhibited by the following proposition

Proposition 2: $\omega\,\omega^{-1}(S) \supset S$ and $\omega^{-1}\,\omega(\Sigma) \supset \Sigma$ for S a subset of X and Σ a subset of Ξ.

Proof: Suppose $x \in S$, then $\omega^{-1}(x) = \{x \mid x \sim \xi\}$ and thus $x \in \omega[\omega^{-1}(x)]$ since the $\sim$ relation is reflexive. Note that this holds even if $\omega^{-1}(x)$ is empty since $\omega(\phi) = X$. The other inclusion is obtained in the same manner.

From the properties of f we also get the following properties of ω and ω^{-1}.

Proposition 3: For every ξ in Ξ, $\omega(\xi)$ is a closed subset of X relative to the induced topology.

Proof: Fix ξ, then $x \in \omega(\xi)$ if and only if $o \in f(x,Y,\xi)$. The proposition follows from (A.1) the continuity of f with respect to x.

Proposition 4: Suppose Σ is a subset of Σ such that the restriction of $f(x,Y,\xi)$ is upper semicontinuous on Σ, then $\omega^{-1}(x) \cap \Sigma$ is closed in Σ with respect to the induced topology.

Proof: This follows immediately from (A.3) the closure of the graph of $f(x,Y,\Sigma)$ in $\Sigma \times R^m$ and (A.2) the continuity of $f(x,y,\xi)$ with respect to ξ.

We can now reformulate the basic problem in terms of the maps ω and ω^{-1} and ignore how they were generated.

Problem. Given a probability space $(\Xi, \mathcal{F}, \mu)$ with a topology on Ξ compatible with $\mathcal{F}$, a map ω defined on a subset X of an hereditarily separable space with range into subsets of Ξ and ω^{-1} defined from Ξ into subsets of X satisfying:

(A.i) $\Sigma \subset \Sigma'$ implies $\omega(\Sigma) \supset \omega(\Sigma')$ and $S \subset S'$ implies $\omega^{-1}(S) \supset \omega^{-1}(S)$ in particular $\omega\,\omega^{-1}(S) \supset S$, $\omega^{-1}\omega(\Sigma) \supset \Sigma$ for $S \subset X$ and $\Sigma \subset \Xi$ (Prop. 2),

(A.ii) For all ξ in Ξ, $\omega(\xi)$ is closed in X (Prop. 3),

(A.iii) $\omega^{-1}(x) \cap \Sigma$ is a measurable closed subset of Σ for all x in X (Prop. 1 and 4),

characterize the set
$$D = \{x \in X \mid \mu[\omega^{-1}(x)] = 1\}.$$

The following lemma is useful to establish the key theorem.

Lemma. Suppose Σ is a subset of Ξ of measure 1 then $D \supset \omega(\Sigma)$.

Proof: $\{x\} \subset \omega(\Sigma)$ implies $\omega^{-1}(x) \supset \omega^{-1}\omega(\Sigma) \supset \Sigma$ by (A.i) and thus $x \in D$ since $\mu[\omega^{-1}(x)] \geqslant \mu(\Sigma) = 1$.

Theorem 1. There always exists a subset Σ of measure 1 contained in Ξ such that $\omega(\Sigma) = D$. Moreover Σ can be selected to be $\omega^{-1}(D)$.

Proof: First assume that D is empty then $\omega^{-1}(D) = \Xi$ and $\omega(\Xi) = \omega\,\omega^{-1}(D) \supset D = \phi$, i.e. Ξ is the appropriate set. Now consider the case when D is nonempty. Take S a countable dense subset of D. The existence of such a set follows from the hereditary separability of X. Set $\Sigma = \omega^{-1}(S)$. If $x \in S \subset D$ it follows from the definition of D that $\mu[\omega^{-1}(x)] = 1$ thus $\mu(\Sigma) = \mu[\omega^{-1}(S)] = \mu[\bigcap_{x \in S} \omega^{-1}(x)] = 1$ since Σ can be written as the countable intersection of set of measure 1. We now use the lemma from which it follows that since $\Sigma \subset \Xi$ and $\mu(\Sigma) = 1$ we have that $\omega(\Sigma) \subset D$. To obtain the other direction first observe that for all ξ in Σ, $\omega(\xi) \supset S$ but by (A.ii) $\omega(\xi)$ is closed and consequently $\omega(\xi) \supset D$ for all ξ in Σ. Thus $\omega(\Sigma) \supset D$.

Corollary 1. The set D is closed.

Proof: It suffices to observe that $D = \omega(\Sigma)$ for some set Σ and that for all ξ in Σ, $\omega(\xi)$ is closed by (A.ii).

Corollary 2. If for all ξ in Σ, $\omega(\xi)$ is convex then D is a closed convex subset of X.

Proof: Similar to corollary 1.

Corollary 3. If with its topology Ξ is a Lindelöf space then the set D can be expressed as a countable intersection of sets of the type $\omega(\xi)$. Moreover if Ξ is a discrete set then it can be written as a finite intersection.

Proof: These characterizations of Ξ follow immediately from the definition of Lindelöf in one case and the fact that Σ is a subset of Ξ.

It is worthwhile to observe that so far we have not made use of (A.iii), which in terms of the original problem followed from (A.3) and (A.4). But although we have been able to establish that there exist some set Σ such that $D = \omega(\Sigma)$ we have no way to compute Σ except by knowing already D itself. The following theorem breaks this vicious circle. By $\widetilde{\Xi}$ we shall denote the <u>support of μ</u>, i.e. the smallest closed subset of Ξ of measure 1. This set is usually available at the outset or can easily be obtained from the original data.

Theorem 2. $D = \omega(\widetilde{\Xi})$.

Proof: Suppose $x \in D$ then $\mu[\omega^{-1}(x)] = 1$. By (A.iii) $\omega^{-1}(x) \cap \widetilde{\Xi}$ is a closed set of measure 1. Since $\omega^{-1}(x) \cap \Xi$ is closed in Ξ (A.iii) and $\widetilde{\Xi}$ is the smallest closed set of measure 1 it follows that $\omega^{-1}(x) \supset \widetilde{\Xi}$ for all x in D. By (A.i) we have that $\omega\,\omega^{-1}(x) \subset \omega(\widetilde{\Xi})$ for all x in D and also that $D \subset \omega\,\omega^{-1}(D) \subset \omega(\widetilde{\Xi})$. To obtain the inclusion in the other direction it suffice to invoke once more the lemma, thus $D = \omega(\widetilde{\Xi})$.

Before we turn to some application we rephrase the last theorem in terms

of the mapping f.

Theorem 2′: $D = \{x \mid o \in f(x,Y,\underset{\sim}{\xi}) \text{ a.s.}\} = \{x \mid o \in f(x,Y,\xi)\ \forall\ \xi \in \widetilde{\Xi}\}$

provided f satisfies assumptions (A.1),. . .,(A.4).

3.　Applications

a. <u>Linear Stochastic Programs</u>.　We consider here the so-called stochastic

program with recourse model.　In view of the mathematical equivalence

between stochastic programs with recourse and stochastic programs with

chance constraints established in [7] it is rather straightforward to derive

similar results for the latter class of problems.　The following constraints

appear in such problems

$$\underset{\sim}{T}x + \underset{\sim}{W}y - \underset{\sim}{p} = o \qquad y \in K$$

where $x \in R^n$ is the here and now decision, the matrices T, W and p are

random matrices whose elements will be known only after x is selected

but before the subsequent decision $y \in R^{\bar{n}}$ has to be chosen.　The set K

represents the amalgamation of the constraints imposed on y by future

decisions and those which were explicitly present in the formulation of

the problem.

We can string out the elements of $\underset{\sim}{T}, \underset{\sim}{W}$ and $\underset{\sim}{p}$ and write them as one

random variable

$$\underset{\sim}{\xi} = (t_{11},\ldots,t_{m,n},\ w_{11},\ldots,w_{m,n},\ p_1,\ldots,p_m).$$

We have thus the function

$$f(x,y,\underset{\sim}{\xi}) = \underset{\sim}{T}x + \underset{\sim}{W}y - \underset{\sim}{p}$$

with $X = \mathbf{R}^n$ and $Y = K$. It is obvious that this function satisfies (A.1) (A.2) and (A.4) provided K be any reasonable type of set. However (A.3) is not satisfied in general. What we need to verify is that as $\xi = (T,W,p)$ varies on Ξ and x is fixed, the multivalued map

$$T_\xi x + W_\xi(K) - p_\xi$$

is upper semicontinuous as a function of ξ where

$$W_\xi(K) = \{t \mid t = W_\xi y, \; y \in K\}.$$

The only real task is to verify if $W_\xi(K)$ is upper semicontinuous since for the other terms there is no problem. If $W_\xi = W$ is fixed and $W(K)$ is closed then obviously (A.4) is satisfied and thus by theorem 2′ we can obtain the constraints on x as an intersection of the form

$$D = \bigcap_{\xi \in \widetilde{\Xi}} \{x \mid T_\xi x - p_\xi \in W(K)\}.$$

If K and $\widetilde{\Xi}$ are polyhedral so is D and one can derive the linear constraints induced on K. This is also the case when only p_ξ is random and has arbitrary distribution and K is polyhedral, see [9].

In general when $W_\xi(K)$ depends on ξ, one has to study the properties of $W_\xi(K)$. The only work in this direction that we know of has been done when K belongs to certain classes of convex sets see e.g. [2] [4] [11] for general results and [5] [8] for results which lead us to a more constructive approach when K possesses some specific properties.

b. <u>Linear Discrete Stochastic Control Problems.</u> A discrete linear control problem can always be formulated as a multistage stochastic program with recourse. A general framework for the formulation of control problems with discrete observations is given in [6] and pratical formulations can be found in [1] and [3] for example. We can thus apply the theory for multistage stochastic programs as derived in [10] where we see that to compute the induced constraints from one stage to the preceding one, the problem is identical to the one encountered for linear stochastic program.

c. <u>Nonlinear stochastic programs</u>. When we use the term nonlinear we refer to the constraints since if the constraints are linear and only the objective is nonlinear we can use the results for linear stochastic programs. We only consider the following problem

$$T(x) + W(y) - \xi \leq o \qquad y \in K$$

where $x \in R^n$ is the here and now decision; T and W are convex maps from R^n and $R^{\bar{n}}$ respectively into R^m, i.e. $T_i(x) + W_i(y) - \xi_j$ are convex functions in x and y for $i=1,\ldots,m$; y is the decision to be selected after observing ξ. Again assumptions (A.1) (A.2) and (A.4) present no difficulties. To verify (A.4) it suffices to check if $W(K)$ is a closed set. Obviously no general theory is available here and to derive specific conditions we must rely on the specific properties of W and K. If in fact $W(K)$ is closed we can here too derive the set D as an intersection of the same type than the one obtained for linear stochastic program, one should expect however that the task of obtaining explicit expressions for the induced constraint set D might be much more complicated.

d. <u>Stochastic Optimal Control Problems</u>. To extend this study to the stochastic control problem where observations occur continuously is would be necessary to extend the theory to the situation where the range of the map f is infinite dimensional. The extension seems to present little or no difficulties but to obtain usable results, it seems that one should work with a different set of assumptions since these seem very difficult to verify.

References.

[1]   R. S. Ellis and R. W. Rishel: An application of stochastic optimal control theory to the optimal rescheduling of airplanes. JACC Conf., St. Louis (1971).

[2]   J-L. Jolly: Une famille de topologie et de convergences sur l'ensemble des fonctionelles convexes. Thesis (1970). Faculté des Sciences de Grenoble.

[3]   J. Midler: Investment in network expansion under uncertainty Rand Corporation report RM—5920—PR, (1969).

[4]   B. Van Cutsem: Elements aléatoires á valeurs convexes compactes. Thesis (1971). Université Scientifique et Médicale de Grenoble.

[5]   B. Van Cutsem: Random variables with convex compact values, applications to stochastic programming (1971), this Volume.

[6]   R. Van Slyke and R. Wets: Stochastic Programs in abstract spaces, in Stochastic Optimization and Control, ed. F. Karreman, Wiley (1968), 25—45.

[7]   D. Walkup and R. Wets: Stochastic programs with recourse: Special forms Proc. Princeton Symposium on Mathematical Programming, ed. H. Kuhn. Princeton Press (1970), 139—161.

[8]   D. Walkup and R. Wets: Continuity of some convex—cone valued mappings, Proc. Amer. Math. Soc. 18(1967), 229—235.

[9]   R. Wets: Programming under uncertainty: the solution set, SIAM J. Appl. Math., 14(1966), 1143—1151.

[10] R. Wets: Stochastic programs with recourse: A basic theorem for
multistage problems, Z. Wahrscheinlichkeitstheorie verw.
Gebiete (1962),

[11] R. Wijsman: Convergence of sequence of sets, cones and functions I,
Bull. Amer. Math. Soc. 70(1964), 186—188. II. Trans. Amer.
Math. Soc. 123(1966), 32—45.

Department of Mathematics
University of Kentucky
Lexington, Kentucky 40506

PROBLEMS OF CONVERGENCE IN STOCHASTIC LINEAR PROGRAMMING

Bernard VAN CUTSEM

Maître de Conférences à l'Université Scientifique et
Médicale de GRENOBLE (France)

SUMMARY

The purpose of this lecture is to give an example of the use
of set valued functions to study the convergence of the solution of a sequence of optimization problems to the solution set of the limit problem.

INTRODUCTION

The $\mathbb{R}^m$ spaces which we shall use, will be supposed referred
to their natural basis and will suppose the usual order
relation

$$\forall x \in \mathbb{R}^m, \; \forall y \in \mathbb{R}^m, \; x \leq y \iff \forall i \in \{1,2,\ldots,m\} , \; x_i \leq y_i$$

where x_i and y_i are the ith component of x and y in the
chosen basis.
Let us consider the following problem of linear programming

$$K = \{x \in \mathbb{R}^m \mid \mathcal{A} x \leq b\}$$

$$\max\{<x',x> \mid x \in K\}$$

where $\mathcal{A}$ is a matrix with p rows and m columns, $b \in \mathbb{R}^p$ and x'
is a linear form which takes $\mathbb{R}^m$ into $\mathbb{R}$. The resolution of
such a problem involves the determination of

$$\alpha = \max\{<x',x> \mid x \in K\}$$
and of
$$K'(x') = \{x \in K \mid <x',x> = \alpha\}$$

In many practical problems, the terms of $\mathcal{A}$,b,x' are estimated from observations or from measurements with random errors. Without going into the details of statistical considerations, we shall stand by considering that $\mathcal{A}$,b x' are values
of random elements A,B,X' defined on a probability space
$(\Omega,\mathcal{A},P)$. Therefore, to each $\omega \in \Omega$ is associated the linear
program defined by

$$\Gamma(\omega) = \{x \in \mathbb{R}^m \mid A(\omega)x \leq B(\omega)\}$$

$$\max \{<X'(\omega),x> \mid x \in \Gamma(\omega)\}$$

and its solutions

$$\alpha = \max\{<X'(\omega),x> \mid x \in \Gamma(\omega)\}$$

$$\Gamma'(X'(\omega),\omega) = \{x \in \Gamma(\omega) \mid <X'(\omega),x> = \alpha(\omega)\}$$

(for the sake of simplicity, we shall note $\Gamma(X',\omega)$ instead of $\Gamma(X'(\omega),\omega)$).
Rockafellar [1] proved that the measurability of the transformations :

$$A : \Omega \to \mathbb{R}^{m+p} \quad , \quad B : \Omega \to \mathbb{R}^p \quad , \quad X' : \Omega \to \mathbb{R}^m$$

for the Borel σ-algebras on the spaces $\mathbb{R}^{m+p},\mathbb{R}^p,\mathbb{R}^m$ and the σ-algebra $\mathcal{Q}$ on Ω, involves the measurability of $\alpha(.)$ and of the set valued functions $\Gamma(.)$ and $\Gamma'(X',.)$.

In fact the statistical methods often lead to the use of sequences of estimators of $A(.),B(.)$ and $X'(.)$, which converge either in probability, or in mean or almost surely. Then we are concerned, in this case, with sequences of stochastic linear programs :

$$\Gamma_n(\omega) = \{x \in \mathbb{R}^m \mid A_n(\omega)x \leq B_n(\omega)\}$$

$$\max\{<X'_n(\omega),x> \mid x \in \Gamma_n(\omega)\}$$

defined by the sequences of random elements $A_n(.),B_n(.)$ and $X'_n(.)$. As before, we associate to them the sequences of random elements

$$\alpha_n(\omega) = \max\{<X'_n(\omega),x> \mid x \in \Gamma_n(\omega)\}$$

$$\Gamma'_n(X'_n,\omega) = \{x \in \Gamma_n(\omega) \mid <X'_n(\omega),x> = \alpha_n(\omega)\}$$

So we intend to define the convergence of sequences of measurable set valued functions and to study conditions of convergence for the sequences $\alpha_n(.)$ and $\Gamma'_n(X'_n,.)$. Before we study the convergence of sequences of convex compact non empty sets in $\mathbb{R}^m$.

I – CONVERGENCE OF SEQUENCES OF CONVEX COMPACT SETS

We shall denote by $CK(\mathbb{R}^m)$, the set of convex compact non empty subsets of $\mathbb{R}^m$. Three notions of distance are used on $CK(\mathbb{R}^m)$: we briefly give their definitions. Let $\{K_n\}_{n \in \mathbb{N}}$ be a sequence of elements of $CK(\mathbb{R}^m)$ and K an element of

$CK(\mathbb{R}^m)$.

I.1 - <u>Définition</u> : We say that $\{K_n\}_{n\in\mathbb{N}}$ <u>converges</u> to K <u>in</u> <u>the Hausdorff sense</u>, if

$$\lim_{n\to\infty} \Delta(K_n,K) = 0$$

where $\Delta(.,.)$ is the Hausdorff distance between two subsets of $\mathbb{R}^m$:

$$\Delta(A,B) = \max\{e(A,B), e(B,A)\}$$

$$e(A,B) = \sup\{d(x,A) \mid x \in B\}$$

and where $d(.,.)$ is the euclidian distance on $\mathbb{R}^m$.

I.2 - <u>Définition</u> : We say that $\{K_n\}_{n\in\mathbb{N}}$ <u>converges</u> to K <u>in</u> <u>the Kuratowski sense</u> if the two following sets are identical to K :

$$\underline{\lim} \; K_n = \{x\in \mathbb{R}^m \mid \exists\{x_n\}_{n\in\mathbb{N}}, \forall n, \; x_n\in K_n, \; \lim_{n\to\infty} d(x_n,x)=0\}$$

$$\overline{\lim} \; K_n = \{x\in \mathbb{R}^m \mid \exists\{x_n\}_{n\in\mathbb{N}}, \; \lim_{n\to\infty} d(x_n,x) = 0$$

$$\exists\Psi:\mathbb{N} \to \mathbb{N} \text{ strictly increasing, } \forall n, \; x_n\in K_{\Psi(n)}\}$$

Then we shall let $K = \lim K_n$.

I.3 - <u>Définition</u> : We say that $\{K_n\}_{n\in\mathbb{N}}$ <u>scalarly converges</u> <u>to</u> K, if

$$\forall x' \in \mathbb{R}^m, \; \lim_n \varphi(x',K_n) = \varphi(x',K)$$

where φ is the support function of K, i.e :
$$\varphi : x' \to \varphi(x',K) = \max\{<x',x> \mid x \in K\}$$

I.4 - <u>Definition</u> : <u>On</u> $CK(\mathbb{R}^m)$, <u>the three previous notions</u> <u>of convergence are equivalent.</u>
The proof of this proposition follows for instance from
Mosco [1] for the equivalence of the convergences in the
sense of Haudorff and Kuratowski, and from Valadier [1] (lem-
me 1.24 and 1.25) and the formulas

$$\Delta(K_1,K_2) = \sup\{|\varphi(x',K_1)-\varphi(x',K_2)| \mid \|x'\| \le 1\}$$

(Hörmander [1])for the equivalence of the scalar convergence and the Hausdorff one.

Then we shall say that K_n converges to K without indicating which of the tree previous notions of convergence is used.

I.5 - <u>Proprosition</u> : <u>Let</u> $\{K_n\}_{n\in\mathbb{N}}$ *be a sequence of elements of* $CK(\mathbb{R}^m)$ *converging to* $K\in CK(\mathbb{R}^m)$. *Let* $\{x'\}_{n\in\mathbb{N}}$ *be a sequence of non null elements of* $\mathbb{R}^m$ *converging to* $x'\in\mathbb{R}^m\setminus\{0\}$. *Then*

$$\overline{\lim_n} \; K'_n(x'_n) \subset K'(x').$$

Proof : Let $x\in \overline{\lim_n} \; K'_n(x'_n)$. So there exists a sequence $\{x_n\}_{n\in\mathbb{N}}$ converging to x and a strictly increasing transformation Ψ from $\mathbb{N}$ to $\mathbb{N}$ such that $\forall n$, $x_n \in K'_{\Psi(n)}(x_{\Psi(n)})$. It is then easy to see that
$\langle x',x\rangle = \varphi(x',K)$ and that $x \in K$, i.e. $x \in K'(x')$.

So we can see that if $K'_n(x'_n)$ converges, then

$$\overline{\lim} \; K'_n(x'_n) \subset K'(x').$$

Generally, we have $\overline{\lim} \; K'_n(x'_n) \neq K'(x')$, as the following example shows

$$K_n = \{ \begin{bmatrix} x \\ y \end{bmatrix} \in \mathbb{R}^2 \mid 0 \leq x \leq 1, \; \lambda_n x \leq y \leq 1\}$$

$$K = \{ \begin{bmatrix} x \\ y \end{bmatrix} \in \mathbb{R}^2 \mid 0 \leq x \leq 1, \; 0 \leq y \leq 1\}$$

If $\lim\limits_{n\to\infty} \lambda_n = 0$, it is obvious that K_n converges to K.
Nevertheless, if $u' = \begin{bmatrix} 0 \\ -1 \end{bmatrix}$, then

$$K'_n(u') = \{\begin{bmatrix} 0 \\ 0 \end{bmatrix}\} \text{ and } K'(u') = \{\begin{bmatrix} x \\ 0 \end{bmatrix}\mid 0 \leq x \leq 1\}.$$

I.6 <u>Proposition</u> : *Let* $K\in CK(\mathbb{R}^m)$ *and* $x' \in \mathbb{R}^m\setminus\{0\}$. *Then there is an increasing, continuous function* $h: \mathbb{R}_+ \to \mathbb{R}_+$ *for which* $h(0)=0$ *and* $h(\rho)\geq\rho$ *and also :*

$$\forall K_1 \in CK(\mathbb{R}^m), \; \forall \rho \geq 0,$$

$$\Delta(K,K_1) \leq \rho \Rightarrow e(K'(x'),K'_1(x')) \leq h(\rho).$$

<u>proof</u> :The proposition is obvious if the thickness of K is null in the direction x'. Then we suppose this thickness to be different from zero.
Let Π denote the supporting plane of K in the direction x', and define

$$B_\rho = \{x \in \mathbb{R}^m \mid d(x,0) \leq \rho\}$$

$$K+B_\rho = \{x \in \mathbb{R}^m \mid d(x,K) \leq \rho\}$$
$$\Pi+B_\rho = \{x \in \mathbb{R}^m \mid d(x,\Pi) \leq \rho\}$$
$$F_\rho = (K+B_\rho) \cap (\Pi+B_\rho)$$

It is easy to see that

$$\Delta(K,K_1) \leq \rho \implies K_1'(x') \subset F_\rho$$

Then we define

$$h(\rho) = \sup\{d(x,K'(x')) \mid x \in F_\rho\} = e(K'(x'),F_\rho)$$

The derivation of the above properties of h is obvious. We shall say that h is related to K and x'.

Example :—Let K be a convex compact non empty polyhedron in $\mathbb{R}^m$ and $x' \in \mathbb{R}^m \setminus \{0\}$. It is obvious that the function h related to K and x' is a piece-wise linear function and that its expression in a neighbourhood of the origin is $h(\rho) = \alpha\rho$; the value of α is related to the minimum angle between the supporting plane $\Pi(x')$ and the edges of the polyhedron which pass through the extremal points of $K'(x')$.

I.7 —Proposition : Let $\{K_n\}_{n\in\mathbb{N}}$ be a sequence of elements of $CK(\mathbb{R}^m)$ which converges to $K \in CK(\mathbb{R}^m)$. Let $x' \in \mathbb{R}^m \setminus \{\emptyset\}$. Let $h(.)$ be the function $\mathbb{R}_+ \to \mathbb{R}_+$ defined in the proposition I.6 by the direction x' and the set K_n. Then the two following assertions are equivalent :

1) $\sup_n h_n(.)$ is continuous at the origin

2) $\lim_{n\to\infty} (K_n'(x'),K'(x')) = 0.$

The prove that 1 => 2, we first remark that

$$\forall \varepsilon>0, \ \exists\rho_1>0, \ \forall\rho<\rho_1, \ \Delta(K_n,K)<\rho \implies K_n'(x') \subset K'(x')+B_\varepsilon$$

and moreover that the continuity at the origin of $\sup_n h_n(.)$ involves

$$\forall \varepsilon>0, \ \exists\rho_2<0, \ \forall\rho<\rho_2, \ \exists h\in\mathbb{N}, \ \forall n\geq k : \Delta(K_n,K)\leq\rho \text{ and}$$

$K'(x') \subset K_n'(x')+B_\varepsilon$ and the assertion 2 follows immediately.

Inversely, to prove 2=>1, we prove that if $\lim_{\rho\to 0} \sup_n h_n(\rho)=a >0$ then $\overline{\lim} \ K_n'(x')$ is strictly included in $K'(x')$. The derivation of this result is too long to be given here. It is described in Van Cutsem [1].

I.8 – <u>Remark</u> – The preceding results of convergence have been established only for convex compact non empty sets in $\mathbb{R}^m$. In fact, it can be prouved that they are still true for sequences of convex closed non empty polyhedra in $\mathbb{R}^m$ (with a finite number of faces).

I.9 – <u>Remark</u> – If we apply the preceding results to sequences of convex closed non empty polyhedras

$$K_n = \{x \in \mathbb{R}^m \mid \mathcal{A}_n x \leq b_n\} \quad b_n \in \mathbb{R}^p$$

such that $\lim_{n\to\infty} \mathcal{A}_n = \mathcal{A}$ and $\lim_{n\to\infty} b_n = b$.
We get the results of Dantzig, Folkman, Shapiro [1] on the convergence of the sequence K_n to the polyhedron

$$K = \{x \in \mathbb{R}^m \mid Ax \leq b\}.$$

II – C ONVERGENCE OF SEQUENCES OF SET VALUED FUNCTIONS

Let G be the set of measurable set valued functions defined on $(\Omega, \mathcal{Q}, P)$ with values in $CK(\mathbb{R}^m)$.

II.1 <u>Lemma</u> – *Let $\Gamma \in G$ and $x' \in \mathbb{R}^m \setminus \{0\}$. Let $h(.,\omega)$ be the function related to the set $\Gamma(\omega)$ and to the direction x'. Then for each $\rho > 0$, $h(\rho,.)$ is a measurable function. If moreover Γ is an integrable set valued function, then for each $\rho > 0$, $h(\rho,.)$ is integrable.*

The measurability of $h(\rho,.)$ follows immediately from theorems of Rockafellar [1]. The integrability of $h(\rho,.)$ is also an easy consequence of the integrability of Γ. (for the definition and properties of integration of set valued functions see, for instance, Aumann [1] or Valadier [1]).

II.2 – <u>Definition</u> : Let $\{\Gamma_n\}_{n \in \mathbb{N}}$ be a sequence of elements of G and $\Gamma \in G$. We shall say that

a) Γ_n *converges in probability to Γ if*

$$\forall \varepsilon > 0, \quad \lim_{n\to\infty} P\{\omega \mid \Delta(\Gamma_n(\omega), \Gamma(\omega)) > \varepsilon\} = 0$$

b) Γ_n *converges in mean to Γ if*

$$\lim_{n\to\infty} \int_\Omega \Delta(\Gamma_n, \Gamma) dP = 0$$

c) - Γ_n *converges almost surely to* Γ *is* $\Delta(\Gamma_n(.),\Gamma(.))$ *converges almost surely to zero.*

d) - Γ_n *scalarly converges in probability to* Γ *if*

$$\forall \varepsilon > 0, \ \forall x' \in \mathbb{R}^m, \ \lim_{n \to \infty} P\{|\varphi(x',\Gamma_n(.))-\varphi(x',\Gamma(.))|>\varepsilon\} = 0$$

e) - Γ_n *scalarly converges in mean to* Γ *if*

$$\forall x' \in \mathbb{R}^m, \ \lim_{n \to \infty} \int_\Omega |\varphi(x',\Gamma_n(.))-\varphi(x',\Gamma(.))| dP = 0$$

f) - Γ_n *scalarly converges almost surely to* Γ *if*

$$\forall x' \in \mathbb{R}^m, \ \exists N \in \mathfrak{a}, P(N)=0, \ \forall \omega \in \Omega \smallsetminus N, \ \lim_{n \to \infty} \varphi(x',\Gamma_n(\omega))=\varphi(x',\Gamma(\omega))$$

All these convergences can be compared. So if Γ_n converges in one of the preceding senses, then Γ_n scalarly converges in the same sense. Moreover, it is easily possible to prove that scalarly almost sure convergence and almost sure convergence are equivalent. Therefore, the classical theorems in the theory of probability on the comparison between almost sure convergence, convergence in mean and convergence in probability and on the extraction of sub-sequences with stronger properties are still true for set valued functions.

The main results of this paper are the following.

II.3 - <u>Proposition</u> : *Let* $\{\Gamma_n\}_{n \in \mathbb{N}}$ *be a sequence of elements of* G *and* $\Gamma \in G$. *Let* $x' \in \mathbb{R}^m \smallsetminus \{0\}$.

a) *if* Γ_n *converges in probability to* Γ *then*

$$\forall \delta > 0, \ \lim_{n \to \infty} P\{\omega|\Gamma_n'(x',\omega) \subset \Gamma'(x',\omega)+B_\delta\} = 0$$

b) *if* Γ_n *converges in mean to* Γ, *if there exists* $Q \in CK(\mathbb{R}^m)$ *such that*

$$\forall n \in \mathbb{N}, \ \forall \omega \in \Omega, \ \Gamma_n(\omega) \subset Q, \ \Gamma(\omega) \subset Q$$

then

$$\lim_{n \to \infty} \int_\Omega e(\Gamma'(x',\omega),\Gamma_n'(x',\omega)) dP = 0$$

c) *If* Γ_n *converges almost surely to* Γ, *then*

$$\exists N \in \mathfrak{a}, \ P(N) = 0, \ \forall \omega \in \Omega \smallsetminus N, \ \overline{\lim}\ \Gamma_n'(x',\omega) \subset \Gamma'(x',\omega).$$

This proposition results of the properties of the function $h(.,\omega)$ related to $\Gamma(\omega)$ and x'.

II.4 - <u>Proposition</u>: *Let* $\{\Gamma_n\}_{n \in \mathbb{N}}$ *be a sequence of elements of* G *and* $\Gamma \in G$. *Let* $x' \in \mathbb{R}^m \smallsetminus \{0\}$. *Let* $h(.,\omega)$ *be the functions related* $\Gamma_n(\omega)$ *and to* x', *in proposition I.6.*

a) *if Γ_n converges in probability to Γ, and if also uniformly in $n \in N$,*

$$\forall \varepsilon > 0, \; \lim_{p \to \infty} P\{h_n(\rho,.) > \varepsilon\} \; = 0,$$

then $\Gamma'_n(x',.)$ converges in probability to $\Gamma'(x',.)$

b) *If there exist $Q \in CK(\mathbb{R}^m)$ and $\lambda \in L^2(\Omega, \mathcal{Q}, P)$ such that*

$$\forall n \in \mathbb{N}, \; \forall \omega \in \Omega, \; \Gamma_n(\omega) \subset \lambda(\omega)Q, \; \Gamma(\omega) \subset \lambda(\omega)Q$$

and if Γ_n converges in mean to Γ, and if

$$\forall \varepsilon > 0, \; \exists \alpha < 0, \; \forall a < \alpha, \; \sup_n \int_\Omega h_n(.,a)dP < \varepsilon,$$

then $\Gamma'_n(x',.)$ converges in mean to $\Gamma'(x',.)$.

c) *If Γ_n converges almost surely to Γ, and if $\sup_n h_n(\rho,.)$ converges in probability to zero when $\rho \to 0$, then $\Gamma'_n(x',.)$ converges almost surely to $\Gamma'(x',.)$.*

Proof : a) Setting, for every $\varepsilon > 0$,

$$A^n_\varepsilon = \{\omega \in \Omega \mid \Gamma'_n(x',\omega) \subset \Gamma'(x',\omega)+B_\varepsilon\}$$
$$B^n_{\rho,\varepsilon} = \{\omega \in \Omega \mid h_n(\rho,\omega) \le \varepsilon \}$$
$$C^n_\rho = \{\omega \in \Omega \mid \Delta(\Gamma_n(\omega),\Gamma(\omega)) \le \rho\}$$

we can show that

$$P(\Delta(\Gamma'_n(x',.),\Gamma'(x',.)) > \varepsilon) \le P(\overline{A^n_\varepsilon}) + P(\overline{B^n_{\rho,\varepsilon}}) + P(\overline{C^n_\rho})$$

where $\overline{A^n_\varepsilon}, \overline{B^n_{\rho,\varepsilon}}, \overline{C^n_\rho}$ are respectively the complementary sets of $A^n_\varepsilon, B^n_{\rho,\varepsilon}, C^n_\rho$. It is then sufficient to prove that, for a fixed $\varepsilon > 0$,

$$\forall \varepsilon_1 > 0, \; \exists k \in \mathbb{N}, \; \exists \rho_o > 0, \; \forall \rho < \rho_o, \; \forall n \ge k$$

$$P(\overline{A^n_\varepsilon}) + P(\overline{B^n_{\rho,\varepsilon}}) + P(\overline{C^n_\rho}) \le \varepsilon_1$$

and this follows easily from the hypotheses.

b) We show here, with the above notations, that

$$\forall \rho > 0, \; \int_\Omega e(\Gamma'(x',\omega),\Gamma'_n(x',\omega)dP \le \int_{C^n_\rho} e(\Gamma'(x',\omega),\Gamma'_n(x',\omega))dP$$
$$+ \int_{\overline{C^n_\rho}} e(\Gamma'(x',\omega),\Gamma'_n(x',\omega))dP$$

from which we obtain

$$\int_\Omega e(\Gamma'(x',\omega)\Gamma'_n(x',\omega))dP \leq \int_\Omega e(\Gamma'(x',\omega),\Gamma'_n(x',\omega))dP +$$
$$+ \int_{C^n_R} e(\Gamma'(x',\omega),\Gamma'_n(x',\omega))dP$$

where R is the diameter of Q. It is sufficient to prove that both terms of the right hand side of the preceding inequality tend to zero when n approaches infinity.

c) On a set of probability one we get

$$\Gamma_n(\omega) \to \Gamma(\omega)$$

and then, on the same set : $\Gamma'_n(x',\omega) \to \Gamma(x',\omega)$

II.5 - <u>Remark</u> - Let, for each $n \in \mathbb{N}$, $\Gamma_n(\omega)=\{x \in \mathbb{R}^m | A_n(\omega)x \leq \mathbf{B}_n(\omega)\}$ and $\Gamma(\omega) = \{x \in \mathbb{R}^m | A(\omega)x \leq B(\omega)\}$ be convex closed non empty polyhedra of $\mathbb{R}^m$, such that Γ_n converges to Γ in one of the three preceding senses : in probability, in mean or almost surely. Let $x' \in \mathbb{R}^m \setminus \{0\}$. The above proposition allows then to give sufficient conditions under which the sequences

$$\alpha_n(\omega) = \max\{<x',x> \mid x \in \Gamma_n(\omega)\}$$
$$\Gamma'_n(x',\omega) = \{x \in \Gamma_n(\omega) \mid <x',x> = \alpha_n(\omega)\}$$

converge in the same sense to

$$\alpha(\omega) = \max\{<x',x> \mid x \in \Gamma(\omega)\}$$
$$\Gamma'(x',\omega) = \{x \in \Gamma(\omega) \mid <x',x> = \alpha(\omega)\}$$

We can thus generalise to <u>stochastic</u> linear programming same results analoguous to those which are known in the deterministic case (see remark I.9).

REFERENCES

R.J. AUMANN - Integrals of set valued functions
Journal of Math. Anal. and Appl. Vol. 12
(1965) p. 1-12.

G.B. DANTZIG - J. FOLKMAN - N. SHAPIRO
On the continuity of the minimum set of a continuous function.
Journal of Math. and Appl. Vol. 17 (1967).
p. 519-546.

L. HÖRMANDER - Sur la fonction d'appui des ensembles convexes dans un espace localement convexe.
Arkiv för Mathematik. Vol. 3 (1964).
p. 181-186.

V. MOSCO - Convergence of convex sets and of solutions of
variational inequalities.
Adv. in Math. Vol. 3 (1969) p. 510-585.

T. ROCKAFELLAR - Measurable dependance of convex sets and
functions on parameters.
Journal of Math. Anal. and Appl. Vol. 28
(1969) p. 4-25.

M. VALADIER - Contribution à l'analyse convexe.
Thèse Paris 1970.

B. VAN CUTSEM - Eléments aléatoires à valeurs convexes
compactes.
Thèse Grenoble 1971.

B. VAN CUTSEM
Mathématiques Appliquées
CEDEX 53
38 - GRENOBLE-Gare
FRANCE

COORDINATION OF TWO-LEVEL ORGANIZATIONS
WITH MULTIPLE OBJECTIVES*

A. M. Geoffrion and W. W. Hogan

University of California
Los Angeles, California

Abstract

We consider a class of organizations composed of a
coordinating agency and several semi-autonomous operating
divisions, each pursuing one or more objectives. It is desired to coordinate the divisions (as by allocating scarce
resources to them) in a way that optimally achieves the objectives of the organization as a whole. Utility functions
are assumed to exist both at the divisional and coordinating levels, but need not be explicitly known. A robustly
convergent iterative procedure is derived that interactive-
ly leads to a sequence of coordination decisions which improves as rapidly as possible in a certain sense. In terms
of the kind of information that must be exchanged between
the coordinating agency and the operating divisions, the
procedure seems well suited to a number of possible applications.

1. Introduction

This paper addresses the problem of coordinating an
organization composed of several semi-autonomous operating
divisions or departments, each of which may be pursuing
different goals within the charter of the parent organization. We are particularly interested in non-market organizations such as are found in government and education.
Since not all of the goals pursued in such organizations
are commensurate with one another, the allocation of scarce

---

*Sections 1 and 2 of this paper are based largely on
[3] and [4], while Section 3 draws heavily from [7]. This
work was supported partially by the National Science Fndn.
and by the Ford Fndn.

resources and other coordinating decisions must strive for the most preferred balance or tradeoff possible between the performances of the various divisions. This task is complicated by the fact that the coordinating agency must reckon with the vagaries of how each division exercises its allowed decision prerogatives.

The coordination problem is formally stated as follows:

$$\text{Maximize}_{y \in Y} \; P[v^1(y),\ldots,v^N(y)],$$

where $v^i(y)$ is the optimal performance of Division i given y, namely

$$v^i(y) \equiv \underset{x^i}{\text{Maximum}} \; f^i(x^i,y) \text{ subject to } g^i(x^i,y) \leqq 0, \qquad (D^i y)$$

and

$y \equiv$ vector of decision (e.g., resource allocation) variables under the control of the Coordinator

$Y \equiv$ set of allowable choices for y as constrained by the resources and policies of the Coordinator

$P \equiv$ Coordinator's preference function, defined on divisional performance outcomes

$x^i \equiv$ vector of decision variables under the control of Division i

$g^i(x^i,y) \equiv$ m^i-vector of constraints to be observed by Division i, as determined by y and its policies and technological constraints

$f^i \equiv$ objective function of Division i

The arbitrary terms "Coordinator" and "Division" are adopted to personify the two levels of the hierarchy.

It is <u>assumed throughout</u> that P is a differentiable concave increasing function known only implicitly by the Coordinator, that Y is a nonempty compact convex set, and that each component function g_k^i of g^i, and each function $-f^i$, are all jointly convex and differentiable in (x^i,y). Economic interpretations of these assumptions, which are commonly employed in the theory of convex mathematical programming, can be found, for instance, in [10].

Allowing P to be implicit, however, is far from common and distinguishes this study from others of its general type (see [9] for a recent survey). We feel that this is necessary in order to avoid a serious pitfall facing other optimization-based coordination methods for multi-level organizations, namely the need to conjure up an explicit objective function for the Coordinator.

The only notational convention requiring special mention is that an overbar is used to denote a function evaluated at a particular point, e.g., $\nabla_x \bar{f}^i$ stands for the gradient of $f^i(x^i,y)$ taken with respect to the x^i-variables evaluated at $(\bar{x}^i,\bar{y})$.

2. The Idealized Coordination Procedure

In the interest of pedagogical clarity, this section presents the proposed coordination procedure in idealized form. Technical matters pertaining to convergence, and several practical features which enable the procedure to accomodate error in the implementation of several of the steps, are all postponed to Sec. 3.

The procedure should be viewed as a structured dialogue between the Coordinator and the Divisions. Loosely speaking, a typical round of the dialogue consists of the Coordinator asking the Divisions what would happen if y were set at $\bar{y}$, to which the Divisions respond by giving some local information concerning the optimal solution of the corresponding divisional operating problems. The Coordinator then uses this information in a prescribed manner to determine a revised trial setting for y.

2.1 Outline of the Idealized Procedure

Step 1 (Coordinator). The Coordinator selects an initial trial $\bar{y} \in Y$ and communicates it to the Divisions.

Step 2 (the Divisions). Each Division determines an optimal solution $\bar{x}^i$ of its corresponding operating problem $(D^i\bar{y})$. It then communicates to the Coordinator the values and first derivatives of f^i and g^i, all evaluated at $(\bar{x}^i,\bar{y})$.

Step 3 (Coordinator). The Coordinator assesses the direction of the gradient of the preference function P at the point $\bar{v} \equiv [v^1(\bar{y}),\ldots,v^N(\bar{y})]$ via N-1 tradeoff assessments.

Step 4 (Coordinator's Technical Staff). Compute the feasible direction $\overline{\Delta y}$ in which $P[v^1(y),\ldots,v^N(y)]$ initially increases the most rapidly as y departs from $\bar{y}$.

Step 5 (Coordinator). Select a step-size $\bar{\theta}$ which maximizes

$P[v^1(\bar{y}+\theta\overline{\Delta y}),\ldots,v^N(\bar{y}+\theta\overline{\Delta y})]$ subject to the requirement that $\hat{y} \equiv \bar{y}+\theta\overline{\Delta y} \in Y$. Communicate $\hat{y}$ to the Divisions and return to Step 2 (with $\hat{y}$ in place of $\bar{y}$).

2.2 Further Details

Step 1

For an existing organization, the force of precedent usually dictates that the initial $\bar{y}$ be taken not too far from the allocation vector actually implemented in the previous decision period.

Step 2

Note that the divisions never give the Coordinator _full_ information about divisional objectives or technology, but only _local_ information. Each Division may solve its operating problem by any convenient means. It is shown in Sec. 3.1 that it is actually only necessary for $\bar{x}^i$ to be a suitably near-optimal solution of $(D^i\bar{y})$.

Step 3

It bears emphasis that only the _direction_ of the gradient of P at $\bar{v}$ is required, and not its _magnitude_. (The direction is, of course, invariant under any strictly increasing transformation of P; in this sense P is employed as an ordinal rather than cardinal preference function.) This direction can be obtained by selecting one Division, say the first, to play a distinguished role as a "reference" Division; and then assessing the marginal rates of substitution at $\bar{v}$:

$$\bar{p}^j \equiv \left(\frac{\partial P}{\partial v^j}\right) \Big/ \left(\frac{\partial P}{\partial v^1}\right)\bigg|_{\bar{v}}, \quad j = 1,\ldots,N.$$

The vector $(1,\bar{p}^2,\ldots,\bar{p}^N)$ obviously has the same direction as $\nabla P[\bar{v}]$.

The assessment of these marginal rates of substitution is not difficult in principle, since they have a natural "indifference tradeoff" interpretation. The quantity $\bar{p}^j$ is, in a limiting sense, the reciprocal of the amount of change in the level of v^1 required to "exactly compensate" the Coordinator for a unit decrease in the level of v^j, with all other divisional performance levels remaining unchanged. See [5] for details and practical illustrations of the assessment of marginal rates of substitution which are beyond the scope of this paper.

Step 4

The principal difficulty encountered here is a technical one, namely that the $v^i(y)$ functions are not differentiable. Thus the Coordinator's objective function is not differentiable. Fortunately, however, $v^i(y)$ is <u>directionally</u> differentiable, and this is enough to enable Step 4 to be carried out. The following key theorem gives a constructive characterization of $\dot{v}^i(\bar{y};\Delta y)$, the directional derivative of v^i at $\bar{y}$ in the direction Δy.

Theorem 1. Assume that $\bar{y}$ is in the interior of the set $\{y\,|\,g^i(x^i,y) < 0 \text{ for some } x^i\}$, and that $(D^i\bar{y})$ has an optimal solution $\bar{x}^i$. Then for all directions Δy we have

$$\dot{v}^i(\bar{y};\Delta y) = \nabla_y \bar{f}^i \Delta y$$
$$+ \underset{w^i}{\text{Max }} \nabla_x \bar{f}^i w^i \text{ subject to}$$
$$\nabla_x \bar{g}^i_k w^i \leq -\nabla_y \bar{g}^i_k \Delta y, \text{ all } k: \bar{g}^i_k = 0,$$

where w^i is a column vector of the same dimension as x^i.

This result leads to the following optimal direction-finding problem for implementing Step 4 in the spirit of a natural generalization of the Frank-Wolfe algorithm for convex programming [2] (generalizations of other feasible-directions algorithms [11] are also possible).

$$\underset{z,w^1,\ldots,w^N}{\text{Maximize}} \sum_{i=1}^{N} (\bar{p}^i \nabla_y \bar{f}^i (z-\bar{y}) + \bar{p}^i \nabla_x \bar{f}^i w^i)$$
$$\text{subject to } \nabla_x \bar{g}^i_k w^i \leq -\nabla_y \bar{g}^i_k (z-\bar{y}), \text{ all } k: \bar{g}^i_k = 0$$
$$\text{and } z \in Y.$$

Note that the optimal direction-finding problem becomes a linear program when Y is given by linear constraints. In any case, if $(\bar{z},\bar{w}^1,\ldots,\bar{w}^N)$ is an optimal solution, then $\Delta y \equiv \bar{z}-\bar{y}$ is the desired optimal feasible direction.

Notice that the optimal solution is not altered if the $\bar{p}$ vector is scaled by an arbitrary positive constant. That is why it was sufficient at Step 3 to assess only the direction of $\nabla \bar{P}$.

Step 5

Ideally the Coordinator seeks a step-size $\bar{\theta}$ which maximizes $P[v^1(\bar{y}+\theta\overline{\Delta y}),\ldots,v^N(\bar{y}+\theta\overline{\Delta y})]$ over $0 \leq \theta \leq 1$. In practice, however, this would usually be very difficult to do

even with the benefit of considerable extra communications
with the Divisions (as a function of θ, only the point val-
ue and first derivative at $\theta = 0$ are known for $v^i(\overline{y}+\theta\overline{\Delta y})$).
Fortunately, as we shall see in Sec. 3.2, there is wide
latitude for the choice of step-size within the requirements for global convergence of the overall coordination
procedure. An actual implementation would therefore probably let the Coordinator rely upon experience and judgment,
rather than upon extra communication with the Divisions, in
selecting the step-size.

2.3 Multiple Criteria at the Divisional Level

An interesting extension of the organizational model
treated above would be to permit multiple criteria and an
implicit preference function for some or all of the Divisions as well as for the Coordinator. The divisional operating problem given y would then become

$$\underset{x^i}{\text{Maximize}} \ U^i[F^i(x^i,y)] \text{ subject to } g^i(x^i,y) \leqq 0,$$

where F^i is an r^i-dimensional vector-valued function of
utility-relevant operating characteristics and U^i is a
utility function defined on the range of F^i. The optimal
performance function for Division i, $v^i(y)$, would be the
optimal value of this operating problem. It will be assumed that each component function of F^i is jointly concave
and differentiable in (x^i,y), and that U^i is concave, increasing, and differentiable.

If U^i is explicitly known, then of course the con-
flicts between criteria are explicitly resolved and the
coordination procedure sketched above is directly applicable (the composition of U^i with F^i is the explicit concave
divisional objective function). If U^i is not explicitly
known, however, then the coordination procedure must be
generalized.

A chief difficulty caused by letting U^i be implicit is
that the gradient of the composition $U^i(F^i)$ with respect to
(x^i,y) is not immediately at hand for the divisions to communicate to the Coordinator at Step 2. Unfortunately, the
coordination procedure requires knowledge of the magnitude
as well as the direction of this gradient; this is in contrast to the fact that the procedure requires knowledge
only of the direction of the implicit preference function
P. This requirement necessitates an additional assumption
such as the following: $U^i(F^i)$ can be expressed as

$$F_1^i + \varphi^i(F_2^i, \ldots, F_{r^i}^i),$$

where φ^i is a concave, increasing, differentiable function
known only implicitly to Division i. Thus the utility
function U^i, which if interpreted ordinally allows an arbitrary strictly increasing transformation, has been assumed
linear and separable in one of the divisional operating
characteristics (taken without loss of generality to be the
first). This operating characteristic can be thought of as
an "independent" one, a Division's desire for which is not
influenced by the attained levels of the other characteristics, and conversely. Often F_1^i will be defined to coincide
with the primary mission of Division i. See [1] for further discussion of this class of utility functions.

With this assumption on U^i, the first component of the
gradient of U^i is always unity. Thus knowledge of the direction of ∇U^i is equivalent to knowledge of ∇U^i itself
(just rescale the direction vector so that the first component is unity). A Division may therefore numerically assess the gradient of its utility function by assessing its
marginal rates of substitution

$$\left(\frac{\partial U}{\partial F_j^i}\right) \Big/ \left(\frac{\partial U}{\partial F_1^i}\right)\Bigg|_{\bar{x}^i, \bar{y}} \, , \ j = 2, \ldots, r^i$$

in exactly the same manner as the Coordinator assesses its
marginal rates of substitution at Step 3.

We mention in passing that it is possible to perform
a consistency check on the assessed gradient of U^i. This
check may be performed by Division i or, if the first derivatives of F^i at $(\bar{x}^i, \bar{y})$ are communicated to the Coordinator, by the Coordinator. It is not difficult to show,
from the Kuhn-Tucker optimality conditions for the divisional operating problem, that the assessed gradient should
be in the set S^i of all nonnegative r^i-dimensional row vectors δ such that $\delta_1 = 1$ and the following conditions have a
nonnegative solution in the m^i-vector u:

$$\delta \nabla_x \bar{F}^i - u \nabla_x \bar{g}^i = 0,$$

$$u_k = 0, \text{ all } k : \bar{g}_k^i < 0.$$

The set S^i is the smallest one containing $\nabla \bar{U}^i$ that can be
deduced with no direct knowledge of φ^i. It is a simple
matter computationally to check whether the assessed vector
is in S^i as it is supposed to be.

The above discussion shows how the gradient of the
divisional objective function can be determined at Step 2,
but there is still the difficulty that the optimal value of
the divisional objective function is unknown. The inabil-
ity to communicate this value to the Coordinator induces a
difficulty in the tradeoff assessment at Step 3, for the
numerical value of $\bar{v}$ may well influence the Coordinator's
marginal rates of substitution. This difficulty disappears
in the important case where the Coordinator's preference
function P is linear over the relevant range, for then the
marginal rates of substitution do not depend on $\bar{v}$ at all.
Of course, this amounts to assuming that P is essentially
known as well as linear, since one execution of Step 3
yields a positive scalar multiple of P. If P is not linear,
then it appears necessary to assume that the Coordinator
can somehow deduce or intuit the ratios $(\bar{v}^1/\bar{v}^i)$ for i =
2,...,N. It follows that the required marginal rates of
substitution equal those for percentagewise deviations from
the current $\bar{v}$ multiplied by the corresponding ratio defined
above. The percentagewise marginal rates of substitution
can be assessed directly.

This completes the discussion of conditions under
which an implicit preference function and multiple criteria
can be accommodated at the divisional level. These conditions are fairly restrictive, but not so restrictive as to
preclude possible applications--especially if relatively
few of the Divisions are subject to this added complexity.

3. Convergence and Robustness of the Coordination Procedure

Attention now turns to the technical details of convergence and robustness of convergence of the idealized coordination procedure described in the previous section.
Sec. 3.1 explains a technical modification to the direction-
finding problem of Step 4 which appears necessary for the
procedure to be generally convergent. Sec. 3.2 weakens
Steps 2, 3 and 5 so as to accommodate some error in the
strict requirements of the idealized procedure. Suboptimal
solutions are allowed at Steps 2 and 5, and error is permitted in the assessment of marginal rates of substitution
at Step 3 or at Step 2 (when there are multiple criteria at
the divisional level). A convergence theorem is given
which simultaneously accommodates all of these changes and
improvements in the idealized procedure.

It will be convenient in this section to distinguish any divisional constraints that happen to be linear equalities. The divisional constraints will henceforth be written as $g^i(x^i,y) \leqq 0$ and $h^i(x^i,y) = 0$, where h^i is a vector-valued (but possibly null) linear function. When we refer to Sec. 2, it will be with the understanding that the obvious and necessary changes are made to accommodate this new convention.

3.1 Modification of the Direction-Finding Problem of Step 4

The coordination procedure as stated in Sec. 2 can be viewed as a natural generalization of the Frank-Wolfe algorithm. It is subject, however, to a convergence pathology [6] which makes it necessary to modify the previously stated direction-finding problem. This modification is based on substituting a relaxed approximation to the true directional derivative $\dot{v}^i(\overline{y};\Delta y)$. The following parameterized function $\dot{v}^i_r$ is used in place of the characterization given in Th. 1:

$$\dot{v}^i_r(\overline{y};\Delta y) \equiv \nabla_y \overline{f}^i \Delta y$$

$$+ \operatorname*{Sup}_{w^i} \nabla_x \overline{f}^i w^i \text{ subject to}$$

$$r\overline{g}^i + \nabla_x \overline{g}^i w^i \leqq -\nabla_y \overline{g}^i \Delta y$$

$$\nabla_x \overline{h}^i w^i = -\nabla_y \overline{h}^i \Delta y$$

$$\|w^i\|_\infty \leq r\sigma,$$

where σ is a "large" positive constant and the scalar r $(1 \leq r < \infty)$ is the relaxation parameter.

The close relationship between $\dot{v}^i_r(\overline{y};\Delta y)$ and $\dot{v}^i(\overline{y};\Delta y)$ is stated in Lemma 1 below, which follows from Lemma 3 in [7]. The following new notation is required:

$$V_0 \equiv \{y \mid g^i(x^i,y) < 0 \text{ and } h^i(x^i,y) = 0 \text{ for some } x^i, \text{ all } i\}$$

$$M^i(y) \equiv \{x^i \mid g^i(x^i,y) \leqq 0, \ h^i(x^i,y) = 0, \text{ and } f^i(x^i,y) \geq v^i(y)\}.$$

The set V_0 is the collection of the Coordinator's decisions which imply the existence of a feasible solution to each Division's operating problem that is interior except for the linear equality constraints. The point-to-set map $M^i(y)$ is just the optimal solution set for the ith Division's operating problem at a given y.

Lemma 1. Suppose that Y is in the relative interior of V_0, $\overline{x}^i \in M^i(\overline{y})$, $M^i(\overline{y})$ is bounded, $r \geq 1$, and σ is greater than the diameter of $M^i(\overline{y})$. Then the following conditions hold:

(a) $\dot{v}_r^i(\bar{y};y-\bar{y}) \leq \dot{v}^i(\bar{y};y-\bar{y})$ for all $y \in \mathrm{aff}(V_0)$;

(b) $\dot{v}_r^i(\bar{y};\Delta y)$ is nondecreasing in r, and for each $y \in \mathrm{aff}(V_0)$ there is a number $r(y) < \infty$ such that $\dot{v}_r^i(\bar{y};y-\bar{y}) = \dot{v}^i(\bar{y};y-\bar{y})$ for all $r \geq r(y)$;

(c) The following statements are equivalent:

    (c.1) $\sup_{y\in Y} \dot{v}_r^i(\bar{y};y-\bar{y}) = 0$ for all $r \geq 1$;

    (c.2) $\sup_{y\in Y} \dot{v}^i(\bar{y};y-\bar{y}) = 0$.

The following variant of Step 4 results when $\dot{v}_r^i$ is used in place of $\dot{v}^i$ (it is understood that some value for r greater than 1 was selected at Step 1).

Step 4. Solve the following direction-finding problem for an optimal solution $(\bar{z},\bar{w}^1,\ldots,\bar{w}^N)$:

$$\underset{z,w^1,\ldots,w^N}{\text{Maximize}} \sum_{i=1}^{N} \left(\bar{p}^i \nabla_y \bar{f}^i(z-\bar{y}) + \bar{p}^i \nabla_x \bar{f}^i w^i\right)$$

$$\text{subject to } r\bar{g}^i + \nabla_x \bar{g}^i w^i \leq -\nabla_y \bar{g}^i(z-\bar{y})$$

$$\nabla_x \bar{h}^i w^i = -\nabla_y \bar{h}^i(z-\bar{y})$$

$$\|w^i\| \leq r\sigma, \quad i = 1,\ldots,N$$

$$z \in Y.$$

Let $\overline{\Delta y} = \bar{z}-\bar{y}$.

For large values of r this variant of Step 4 will perform nearly identically to the previously given version. The directions produced may even prove to be superior in practice since consideration is now given to those divisional constraints which are not exactly binding at $(\bar{x}^i,\bar{y})$. In any event, this variant leads to a generally convergent procedure as will be made precise in Theorem 2 below.

3.2 More Practical Versions of Steps 2, 3 and 5

In practice it is only reasonable to require the Divisions to determine ϵ-optimal solutions to their operating problems at Step 2. Let $M^i(y,\epsilon)$ denote the set of all ϵ-optimal solutions to $(D^i\bar{y})$,

$$M^i(y,\epsilon) \equiv \{x^i \mid g^i(x^i,y) \leq 0, \ h^i(x^i,y) = 0, \text{ and } f^i(x^i,y) \geq v^i(y)-\epsilon\}.$$

Then one would prefer the following more practical version of Step 2.

Step 2. Each Division determines a point $\bar{x}^i \in M^i(\bar{y},\bar{\epsilon})$, and communicates to the Coordinator the values and first derivatives of f^i, g^i and h^i evaluated at $(\bar{x}^i,\bar{y})$. Replace $\bar{\epsilon}$ by $\bar{\epsilon}/2$.

The object of replacing $\bar{\epsilon}$ by $\bar{\epsilon}/2$ is to ensure that the sequence of $\bar{\epsilon}$'s employed at successive executions of Step 2 goes to 0. The particular rule suggested here is arbitrary, and plays no special role in the proof of our main convergence theorem. The initial (nonnegative) value of $\bar{\epsilon}$ could be selected at Step 1 or at the first execution of Step 2.

A similar practical consideration arises in connection with the optimization required by Step 5. It turns out, fortunately, that it is enough that the Coordinator be able to select a step-size which achieves some fixed percentage of the available gain. Specifically, the following weakened version of Step 5 may be employed. It is assumed that a constant $0 < \beta \le 1$ is chosen at Step 1.

<u>Step 5</u>. Select a step-size $0 \le \bar{\theta} \le 1$ such that

$$P[v(\overline{y}+\overline{\theta}\overline{\Delta y})] \ge \beta\left\{\max_{0\le\theta\le1} P[v(\overline{y}+\theta\overline{\Delta y})]-P[v(\overline{y})]\right\}+P[v(\overline{y})].$$

Communicate $\hat{y} = \overline{y}+\overline{\theta}\overline{\Delta y}$ to the Divisions and return to Step 2 (with $\hat{y}$ in place of $\overline{y}$).

Step 1 becomes as follows when the initializations required by the new versions of Steps 2, 4 and 5 are incorporated.

<u>Step 1</u>. The Coordinator selects an initial trial $\overline{y} \in Y$ and $\bar{\epsilon} \ge 0$, and communicates them to the Divisions. Choose scalars $r \ge 1$ and $0 < \beta \le 1$.

We are now ready to state the main convergence theorem.

<u>Theorem 2</u>. Suppose that Y is in the relative interior of V_0, that the union over Y of the feasible solutions for the divisional problems is compact, and that σ is greater than the diameter of this set. Then every accumulation point of the sequence of $\overline{y}$'s produced by the coordination procedure with the revised versions of Steps 1, 2, 4 and 5 is an optimal solution of the Coordinator's problem.

This theorem follows directly from results in [7].

One final practical consideration remains, namely the robustness of convergence when there is error in the assessment of the direction of the gradient of P at Step 3-- or in the assessment of the gradient of U^1 at Step 2 when there are multiple criteria at the divisional level as discussed in Sec. 2.3. Fortunately, the methods of [6] can be readily blended into the proof of Th. 2 to show that such errors do not upset convergence so long as they approach zero in the limit with successive iterations through the procedure.

References

1. Feinberg, A., "Partially Separable Utility Functions," discussion paper, Graduate School of Management, UCLA, August 1970.

2. Frank, M. and P. Wolfe, "An Algorithm for Quadratic Programming," NRLQ, 3, 1956.

3. Geoffrion, A. M., "Vector Maximal Decomposition Programming," Working Paper No. 164, Western Management Science Institute, UCLA, September 1970.

4. ________________, "Resource Allocation in Decentralized Non-Market Organizations with Multiple Objectives," multilith, presented at the Second World Congress of the Econometric Society, Cambridge, England, September 1970.

5. Geoffrion, A. M., J. S. Dyer, and A. Feinberg, "An Interactive Approach for Multi-Criterion Optimization, with an Application to the Operation of an Academic Department," Working Paper No. 176, Western Management Science Institute, UCLA, July 1971.

6. Hogan, W. W., "Convergence Results for Some Extensions of the Frank-Wolfe Method," Working Paper No. 169, Western Management Science Institute, UCLA, January 1971.

7. ________________, "Directional Derivatives for Extremal Value Functions with Applications to the Completely Convex Case," Working Paper No. 177, Western Management Science Institute, UCLA, July 1971.

8. ________________, "Optimization and Convergence for Extremal Value Functions Arising from Structured Nonlinear Programs," Working Paper No. 180, Western Management Science Institute, UCLA, September 1971.

9. Jennergren, L. P., "Studies in the Mathematical Theory of Decentralized Resource-Allocation," Ph.D. Dissertation, Graduate School of Business, Stanford University, June 1971.

10. Nikaido, H., Convex Structures and Economic Theory, Academic Press, New York, 1968.

11. Zoutendijk, G., Methods of Feasible Directions, Elsevier, Amsterdam, 1960.

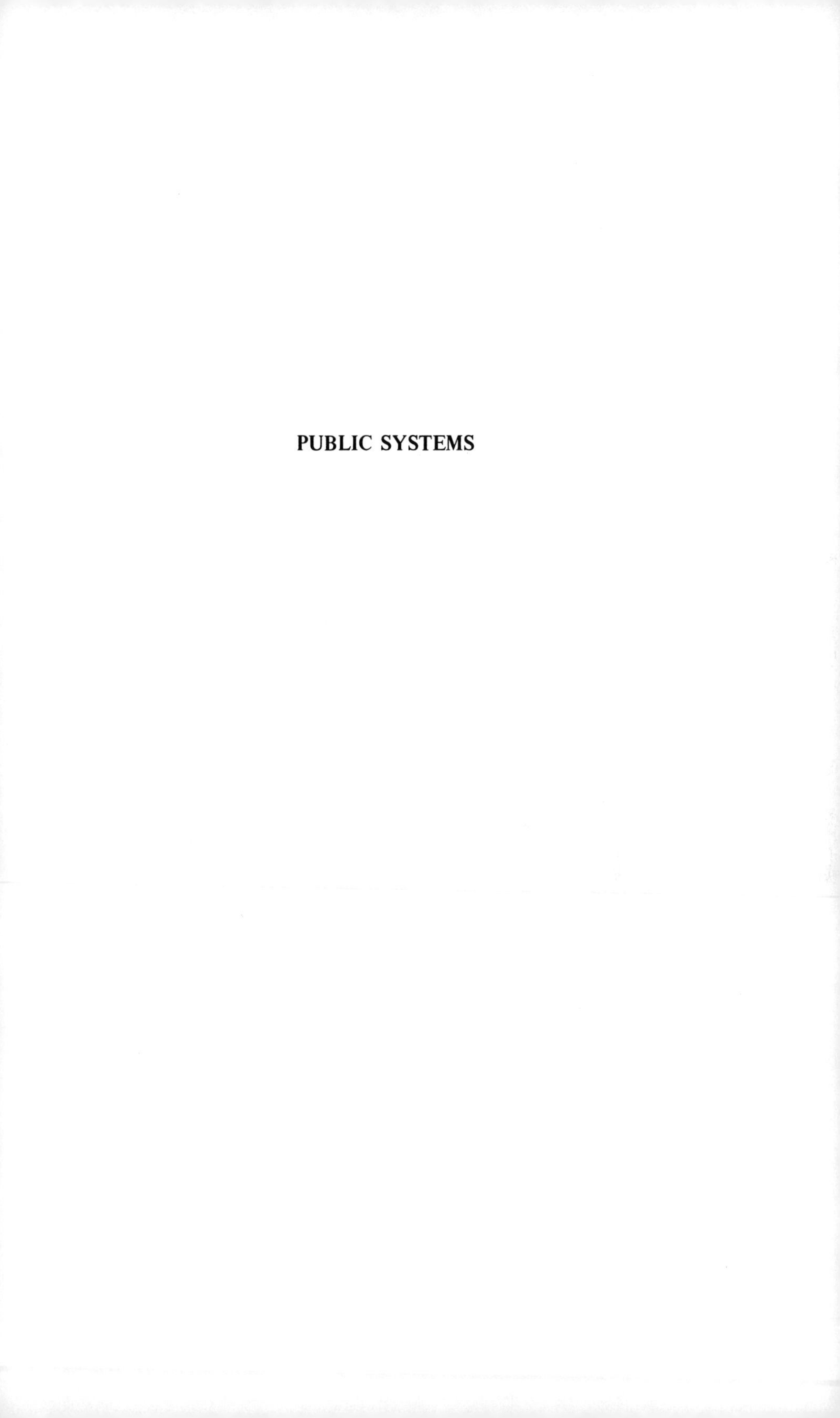

PUBLIC SYSTEMS

THE OPTIMAL NUMBER OF SERVERS
IN A MANY SERVER QUEUEING SYSTEM

Jos H.A. de SMIT

Center for Operations Research and Econometrics
Catholic University of Louvain
Heverlee, Belgium

1. Introduction.

Classical queueing theory may be described as a branch
of applied probability theory dealing with a model denoted
by $GI|G|s$ and with special cases and variants of this model.
In the model $GI|G|s$ we assume that customers arriving
according to a renewal process want to be served by one of
s identical servers, requiring service times which are
independent and identically distributed. If customers
cannot be served immediately they join a queue to wait for
service. Several queue disciplines (i.e. the order in which
customers are served) may be considered. Moreover one can
think of many different configurations (networks) of servers,
in parallel or in series or both, which are of practical
interest.

For such models one usually studies the distributions
of quantities like the waiting time, the queue length
and the length of a busy period. In this approach the
question of optimality does not come up immediately. In
practice, however, queueing models are applied for making
good or preferably optimal decisions. These decisions may
concern the queueing system itself, e.g. one may vary one
or more parameters or change the queue discipline or the
configuration of servers; they may also influence the
arrival or queueing behaviour of the customers.

In the present paper we concentrate on one system parameter, viz. the number of servers. We assume that this is
the only parameter we can vary in order to satisfy some
optimality criterion. All servers are identical and each
of them has the same fixed service capacity. We shall
consider the dimensioning problem, i.e. we assume that the
size of the system can be fixed only once and cannot be

changed easily. Under certain conditions the results for
the dimensioning problem may also be useful for the control
of a queueing system in which the number of servers can
be adapted continually to the traffic conditions. A general
study of the latter class of problems, however, does not
immediately follow from our approach.

The motivation of this paper is the availability of very
recent results in the theory of the many server queue which
have not yet been made operational for making optimal
decisions. Therefore we shall first give a review of some
important developments in queueing theory leading to solutions for a very wide class of many server queues. Next we
show how these results can be used to find a cost function
depending on the number of servers s. Since s is a positive
integer the optimization problem becomes trivial once we
have found that cost function.

It is often stated that complicated queueing systems can
be approximated by much simpler models and that the study
of more general models only leads to refinements which are
not of practical interest. Such an approach, however, may
lead to conclusions which are entirely wrong, e.g. it can
be shown numerically that in most cases exponential models
are not acceptable as approximations for non-exponential
models. Therefore it is essential to apply the most general
results that are available.

2. Many server queues : recent developments.

The first important results for many server queues were
obtained by Erlang in the beginning of this century. He
found solutions for the stationary behaviour of systems
with Poisson arrivals and exponential or deterministic
service times. During the past fifty years relatively little
attention was given to many server queues as compared to
the huge literature on the single server queue. Nevertheless
the many server queue is of great practical interest. The
analysis of the many server queue, however, is very complicated and troublesome and not very attractive from a mathematical point of view, whereas the single server queue
admits an elegant mathematical treatment.

The most powerful method leading to explicit solutions
for a wide class of many server queues is the analytic
method developed by Pollaczek (1961). We give a very short
outline of a generalized version of this theory (see de Smit,
1971 a).

To study the general model with s servers we consider the s-dimensional vector process $\{\theta_n = (\theta_{n,1}, \theta_{n,2}, \ldots, \theta_{n,s})$, $n=1,2,3,\ldots\}$, which is defined as follows :
$\theta_{n,j}$ is the amount of work the j-th server still has to do just before the arrival of the n-th customer. Hence, if t_n is the moment of arrival of the n-th customer, then $\theta_{n,j}$ is a random variable such that the j-th server would become idle at time $t_n + \theta_{n,j}$ if the n-th and subsequent customers would not arrive. The random variable $\theta_{n,j}$ may also be negative, meaning that just before t_n the j-th server has been idle during a period of length $-\theta_{n,j}$.
The customers are served in the order of their arrivals so that the n-th customer will be served by the server whose $\theta_{n,j}$ is minimal and for the waiting time w_n of that customer we obviously have, using the notation $x^+ = \max(0,x)$,

$$w_n = \left[\min_{1 \le j \le s} \theta_{n,j} \right]^+ . \tag{2.1}$$

Let σ_{n+1} be the interarrival time between the n-th and the (n+1)-th customer, i.e. $\sigma_{n+1} = t_{n+1} - t_n$; and let τ_n be the service time of the n-th customer. If the k-th server is the one who just before t_n has the smallest amount of work still to do, or

$$\theta_{n,k} = \min_{1 \le j \le s} \theta_{n,j} ,$$

then the following recurrence relations hold :

$$\theta_{n+1,j} = \theta_{n,j} - \sigma_{n+1}, \qquad \text{for } j \ne k,$$

$$\tag{2.2}$$

$$\theta_{n+1,k} = \theta_{n,k}^+ + \tau_n - \sigma_{n+1}.$$

Since it is assumed that the random variables σ_n, $n=2,3,\ldots$; and τ_n, $n=1,2,\ldots$; are independent, it is immediately seen that the s-dimensional vector process $\{\theta_n = (\theta_{n,1}, \theta_{n,2}, \ldots, \theta_{n,s})$, $n=1,2,\ldots\}$ is a vector Markov process. Using this fact the recurrence relations (2.2) for the random variables lead to recurrence relations for their distributions or for the

Laplace-Stieltjes transforms of these distributions.

Let i_1, i_2,...,i_s be a permutation of the numbers 1,2, ...,s such that

$$\theta_{n,i_1} \leq \theta_{n,i_2} \leq \cdots \leq \theta_{n,i_s}.$$

The partial sums of the sequences σ_n, n=2,3,...; and τ_n, n=1,2,...; are denoted by

$$a_n = \sum_{i=1}^{n} \sigma_{i+1} \quad \text{and} \quad b_n = \sum_{i=1}^{n} \tau_i.$$

We define the multivariate Laplace-Stieltjes transforms

$$Z_k(r,\phi,\psi,\eta,\rho) =$$

$$= \sum_{n=1}^{\infty} r^n \, E\{\exp(-\phi\theta^+_{n,i_1} - \psi\theta^+_{n,i_k} - \eta a_{n-1} - \rho b_{n-1})\},$$

for k = 2,3,...,s; Re $\phi \geq 0$, Re $\psi \geq 0$, Re $\eta \geq 0$, Re $\rho \geq 0$ and $|r| < 1$.

It appears that we can find the function $Z_k(\ldots)$ whenever the service time distribution has a rational Laplace-Stieltjes transform. In the simplest case, i.e. if the service time distribution is exponential we find an explicit result for $Z_k(\ldots)$. For more complicated service time distributions we can give an algorithm which yields numerical solutions for many quantities of interest. The class of distributions with rational Laplace-Stieltjes transforms includes the Erlang distributions and the hyper-exponential distributions and these may be fitted to almost any service time distribution occurring in practical situations.

Once we know the function $Z_k(\ldots)$ we can find expressions for the distributions of many interesting quantities. We mention the following examples :

(i) For the waiting time of the n-th customer we obviously have

$$w_n = \theta^+_{n,i_1}$$

which implies

$$\sum_{n=1}^{\infty} r^n E\{\exp(-\phi w_n)\} = Z_j(r,\phi,0,0,0).$$

In the cases in which we can find the functions $Z_k(\ldots)$ (service time distribution with rational Laplace - Stieltjes transform) it appears that the functions $Z_k(\ldots)$ are rational functions of ϕ and consequently the Laplace-Stieltjes transform is easily inverted and an expression is obtained for the generating function

$$\sum_{n=1}^{\infty} r^n P\{w_n < x\}.$$

Using Abel's theorem this expression gives us immediately the limiting distribution of the actual waiting times. Moreover we can study the asymptotic behaviour of $P\{w_n < x\}$ for $n \to \infty$.

(ii) If x_n is the number of waiting customers just before the arrival of the n-th customer, the following events are identical

$$\{x_{n+j+1} \leq j\} = \{t_n + w_n < t_{n+j}\}.$$

As we know the distribution of w_n expressed by $Z_k(\ldots)$ we find from this identity the distribution of x_n and for $|r| < 1$, $|q| \leq 1$, we have

$$\sum_{n=1}^{\infty} r^n E\{q^{x_n}\} = \frac{r}{1-qr} + \frac{(1-q)r}{2\pi i} \int_{-i\infty+0}^{i\infty+0} \frac{Z_k(r,\xi,0,0)A(-\xi)}{1-qrA(-\xi)} \frac{d\xi}{\xi} \ ,$$

where $A(\phi)$ is the Laplace-Stieltjes transform of the interarrival time distribution and the integral is a principal value integral resulting from the application of the inversion formula for Laplace-Stieltjes transforms. The expression is valid and meaningful if for every n we have $P\{t_n+w_n=t_{n+j+1}\} = 0$ and if there exists some real $\delta > 0$ such that $A(-\xi)$ is analytic for $\mathrm{Re}\ \xi < \delta$.

(iii) The busy cycle c is defined as the time that elapses between two successive moments at which an arriving customer finds the system empty. The number of customers arriving during such a busy cycle is denoted by m. So if the first customer arrives at $t_1 = 0$ we have

$$m = \min_{n \geq 1} \{n : \theta_{n+1,i_s} \leq 0\}$$

and

$$c = t_{m+1}$$

From renewal theory (cf. Cohen, 1969) it follows that

$$E\{r^m \exp(-\eta c)\} = 1 - \frac{r}{\sum_{n=1}^{\infty} r^n E\{(\theta_{n,i_s} \leq 0) \exp(-\eta t_n)\}}$$

$$= 1 - \frac{r}{\lim_{\psi \to \infty} Z_s(r,0,\psi,\eta,0)}$$

(iv) The virtual waiting time at time t is denoted by v_t and the number of customers arriving during $[0,t)$ by n_t; then we have

$$v_t = [\min(\theta_{n_t,i_1} + \tau_{n_t} - t + t_{n_t}, \ \theta_{n_t,i_2} - t + t_{n_t})]^+,$$

so that

$$v_t = \sum_{n=1}^{\infty} (n_t=n)\lceil\min(\theta^+_{n,i_1} + \tau_n - t + t_n, \ \theta^+_{n,i_2} - t + t_n)\rceil^+.$$

The joint distribution of θ^+_{n,i_1} and θ^+_{n,i_2} is given by the function $Z_2(\ldots)$ and it is possible to express the distribution of v_t in terms of the function $Z_2(\ldots)$ and the Laplace-Stieltjes transforms of inter-arrival time and service time distributions.

3. The dimensioning problem.

Assume that we have to design a service system given the (stochastic) arrival process and the distribution of the service times required by arriving customers. We have to determine the optimal size of that system because afterwards it will be difficult and expensive to change the size. Typical examples of such systems are telephone exchanges and computer systems. In any such system the optimal design will be the one that minimizes a cost function depending on the number of servers and the waiting process. The dependence on the number of servers may be or may not be linear (large systems may be relatively cheaper). The expected cost resulting from waiting may be a function of the average waiting time but also of the variance or the probability that the waiting time will exceed some critical value; similar considerations apply to the cost caused by the presence of a queue.

In the design of telephone exchanges one often uses very simple criteria to solve the dimensioning problem. One of them is Moe's principle which was already applied by Erlang in 1924 (see Syski, 1960). It postulates that the cost of an additional server is the same irrespective of the size of the system. If $F_s(\alpha,\beta)$ is the improvement of traffic conditions obtained by increasing the number of servers from s-1 to s in a queueing system with mean inter-arrival time α and mean service time β and the additional cost of one server is δ; then one should choose s such that

$$F_{s-1}(\alpha,\beta) \geq \delta > F_s(\alpha,\beta).$$

Erlang applied this principle under the assumption of exponential interarrival times and service times and that the improvement of traffic conditions consisted of a term proportional to the decrease of the mean waiting time and a term proportional to the decrease of the probability of delay. Even in this simple case the functions $F_s(\alpha,\beta)$ do not take a simple analytic form. Therefore telephone engineers generally use tables and graphs giving these functions numerically for a wide range of parameter values.

Next we will show how the above mentioned method can be improved by considering models which can be fitted much better to practical situations. We first introduce several restrictions which are not essential but which are imposed to keep the exposition simple without being trivial. Afterwards we discuss the consequences of dropping these restrictions.

Let us consider the queueing system $GI|M|s$, i.e. the system with a general arrival process, exponential service times and s servers. We mention the following results (cf. de Smit, 1971 b). By w_n we denote the actual waiting time of the n-th customer and we assume that $w_1 = 0$. We then have for $|r| < 1$, $\mathrm{Re}\,\phi \geq 0$,

$$\sum_{n=1}^{\infty} r^n E\{\exp(-\phi w_n)\} = \frac{r}{1-r}\left\{1 - \frac{f(r)}{\mu(r)}\frac{\phi}{\phi+\mu(r)/\beta}\right\},$$

where $\mu(r)$ is the unique zero in the right semi-plane of

$$z-s + srA(z/\beta),$$

with $A(\phi)$ the Laplace-Stieltjes transform of the inter-arrival time distribution and β the mean service time, while

$$f(r) = \left[\sum_{\lambda=0}^{s-1} \binom{s-1}{\lambda} \frac{1}{(\lambda-\mu(r))(\mu(r)-\lambda-1)} \prod_{i=1}^{\lambda} \left\{ \frac{1-rA(i/\beta)}{rA(i/\beta)} \right\} \right]^{-1} .$$

If x_n is the queue length just before the arrival of the n-th customer we have for $|r| < 1$, $|q| \leq 1$,

$$\sum_{n=1}^{\infty} r^n E\{q^{x_n}\} = \frac{r}{1-r} \left\{ 1 - \frac{f(r)}{\mu(r)} \frac{(1-q)(s-\mu(r))}{s(1-q)+q\mu(r)} \right\} .$$

Applying Abel's theorem for generating functions we find for the limiting distributions, assuming that they exist,

$$\lim_{n\to\infty} E\{\exp(-\phi w_n)\} = 1 - \frac{f(1)}{\mu(1)} \frac{\phi}{\phi+\mu(1)/\beta} ,$$

$$\lim_{n\to\infty} E\{q^{x_n}\} = 1 - \frac{f(1)}{\mu(1)} \frac{(1-q)(s-\mu(1))}{s(1-q)+q\mu(1)} .$$

It can be proved that the limiting distributions exist if $a = \dfrac{\beta}{\alpha s} < 1$.

Let the queueing system be in a stationary state then the waiting time and queue length distributions are given by the above expressions. Evidently the transforms can easily be inverted. Here we only use them to calculate moments. It follows that

$$\lim_{n\to\infty} E\{w_n\} = \frac{\beta}{\mu(1)^2} f(1),$$

$$\lim_{n\to\infty} E\{x_n\} = \frac{s-\mu(1)}{\mu(1)^2} f(1) .$$

The $\mu(1)$ and $f(1)$ are very easily calculated numerically.

Let $C(s,\alpha,\beta)$ be the cost function associated with this system, defined by

$$C(s,\alpha,\beta) = c_0 s + c_1 \lim_{n\to\infty} E\{w_n\} + c_2 \lim_{n\to\infty} E\{x_n\},$$

then after inserting the expressions for the means we have

$$C(s,\alpha,\beta) = c_0 s + \frac{f(1)}{\mu(1)^2} [c_1 \beta + c_2(s-\mu(1))].$$

From the definition of $f(r)$ we see that $C(s,\alpha,\beta)$ is not a simple function of s. For the model $M|M|s$, i.e. Poisson arrivals it somewhat simplifies because then we have $\mu(1) = s(1-a)$ and

$$f(1) = \left[\frac{(s-1)!}{(sa)^s} \sum_{\lambda=0}^{s} \frac{(sa)^\lambda}{\lambda!} + \frac{a}{s(1-a)}\right]^{-1}.$$

The numerical calculation of $C(s,\alpha,\beta)$ is easy and thus we can find the value of s for which it is minimal.

Until now we have considered models with exponential service times. Next we assume that the service time distribution is a linear combination of exponential distributions. The coefficients may be complex as long as the resulting distribution is real-valued. It has been shown (cf. de Smit, 1971 b) that the waiting time distribution will then be of the form

$$\sum_{n=1}^{\infty} r^n E\{\exp(-\phi w_n)\} = \frac{r}{1-r} \left\{1 - \sum_{i=1}^{N} a_i(r) \frac{\phi}{\phi+\mu_i(r)/\beta}\right\},$$

where $N = \binom{m+s-1}{s}$ if m is the number of terms of the service time distribution; and the $\mu_i(r)$ are the zeros in the right semi-plane of some determinant. We can give an algorithm for calculating the $\mu_i(r)$ and $a_i(r)$ numerically. Clearly the above expression generalizes the result for queues with exponential service times. Similar generalizations can be given for the queue length and other quantities. As a consequence, the new cost function $C(s,\alpha,\beta)$ will be a linear combination of functions of the form described before.

If we would define a quite different cost function $C(s,\alpha,\beta)$ the analysis remains essentially the same.

The assumption that the queueing system is in a stationary state is not always realistic. There are some quantities describing the non-stationary behaviour which we can also calculate, such as the busy period, the busy cycle, the relaxation times for the waiting time and the queue length.

4. <u>Concluding remarks</u>.

It seems that the present paper is the first attempt to apply recent results from queueing theory to the problem of optimizing the number of servers. In future publications we intend to publish the algorithms mentioned in the previous section together with numerical results obtained by them. The most relevant form of the cost function should follow from practice; we have tried to show that we can deal with a wide class of such functions.

Our study is restricted to the dimensioning problem. The important problem of the optimal control of many server queues requires quite different models. We have some hope, however, that we might be able to attack such models with the same powerful techniques, developed for the classical model of the many server queue.

In many applications we have to deal with networks consisting of a number of many server queues. The cost function will then depend on several variables.

<u>References</u>.

Cohen, J.W. (1969), The Single Server Queue, North-Holland, Amsterdam.

de Smit, J.H.A. (1971 a), Some general results for many server queues, CORE Discussion Paper 7132.

de Smit, J.H.A. (1971 b), On the many server queue with exponential service times, CORE Discussion Paper 7134.

Pollaczek, F. (1961), Théorie analytique des problèmes stochastiques relatifs à un groupe de lignes téléphoniques avec dispositif d'attente, Gauthier-Villars, Paris.

Syski, R. (1960), Introduction to Congestion Theory in Telephone Systems, Oliver and Boyd, London.

ALLOCATION OF EMERGENCY UNITS: RESPONSE AREAS

Jan M. Chaiken

The New York City-Rand Institute
545 Madison Avenue, New York, N. Y. 10022

Abstract

The average travel time for emergency units such as
fire engines, ambulances, and police patrol cars, which
respond to spatially distributed incidents, is not necessarily minimized by always dispatching the closest available
unit(s) to each incident. Methods are described for
changing response areas so as to reduce average travel time
and also reduce the imbalance of workload among units.

In recent years optimization techniques have begun to
be applied to problems of urban emergency service systems
such as fire departments, police patrol systems, and
ambulance services. This paper describes an example of
such an application which has led to new principles for
designing the response areas of emergency units.
To set the stage, I will give a brief description of
the dispatch operations of a typical emergency service. A
member of the public ordinarily reports an emergency by
activating an alarm box located on the street or in a
building, or by calling an emergency telephone number. Each
alarm box has a code number which indicates its location,
and when the box is activated the code number registers at
the appropriate dispatch center. (Some of these boxes
permit voice communication between the caller and the dispatcher, so that the dispatcher can obtain additional
information about the nature of the emergency.) Once the
box number is known, the dispatcher consults a file giving,
in the case of mobile units such as police patrol cars, the
identification of the unit in whose patrol area the box is
located, or, in the case of units dispatched from fixed
facilities such as fire houses, the identity of the closest

facility or a sequenced list of facilities in order of their distance from the box.

If the call is reported by telephone rather than by alarm box, the dispatcher determines the address or approximate location of the incident and then consults a file which identifies the patrol area or alarm box closest to the address. Once the dispatcher knows the alarm box number, he can find the list of facilities as if he had received a box alarm.

The selection of units which will be dispatched to the incident is essentially determined by the information obtained from the file by the dispatcher. In the case of mobile units, if the patrol area found in the file contains an available unit, that unit will be dispatched to the incident. (There is, of course, no guarantee that this unit is actually closer to the incident than any other unit.) If the emergency is a fire, ordinarily several units of different types will be dispatched. When n_i units of type i are supposed to be dispatched, the first n_i units on the sequenced list will be dispatched, if they are available for service. If one or more of the first n_i are unavailable, the dispatcher may in some cases dispatch those units which are available, while in other cases he may search further down the list for additional available units to dispatch.

In any event, the units dispatched to the incident will always be the ones which the dispatcher believes are closer to the incident than any other available units of the same type. The objective underlying this dispatching protocol is to minimize each component of the vector which gives the travel times of units responding to incidents. (It is useful to note that the dispatching method described here allows us to define a response area for each unit; it consists of all locations to which the unit will be dispatched if all units are available.)

The possibility of changing the standard dispatching strategy arose as part of our work for the New York City Fire Department. This analysis was undertaken by Grace Carter, Edward Ignall, and myself. The fire department faced the problem that some parts of the city had so many fire alarms that the units in those areas might fight as many as 20 fires in a busy night, leading to a condition in which the firefighters felt overworked. However, nearby areas, sometimes not more than a mile or two away, might have a much lower level of activity.

A natural idea for helping to relieve the excessive workloads of some units is to contract the response areas for the busiest units and expand the response areas for the least busy ones. This would tend to distribute the workload more evenly among the units. But we felt we should find out what would happen to travel times if such a change were made. Presumably travel times would increase, and the problem was to estimate the magnitude of the increase. Thus, we wished to calculate expected travel times and workloads of units as functions of the response areas.

Because the travel times and workloads depend on which units are dispatched to each alarm, and the choice of units to dispatch depends on the availability of units, in general one finds that the quantitites of interest are functions of the arrival rates for alarms, the service times of units, and the dispatching policy. The arrival process can be modeled by assuming that there are several types of alarms, with type m alarms arriving in any region A according to a Poisson process with rate $\lambda_m(A)$. Then each λ_m is a measure on the region under consideration. (Poissonicity is an excellent approximation, but in practice the arrival rates are not actually constant over time. The alarm "types" may distinguish building fires from brush fires, or telephone alarms from box alarms, or other appropriate characteristics of the incidents.)

The service times may be modeled by assuming that if a group G of units is dispatched to an alarm at location $\underline{x}$, then the service times are given by a matrix $F_m(\underline{x},G)$ whose ij-component is the service-time distribution of the jth unit of type i. The dispatching policy can then be thought of as a function $H_n(\underline{x},\lambda,F,\underline{a})$ which specifies the group of companies to dispatch to a type n alarm at $\underline{x}$ if the λ_m's are the arrival rates, the F_m's are the service times, and the current availabilities of units are given by the vector $\underline{a}$ (whose components are either zero or an unexpired service time, depending on whether the corresponding unit is available or not). Typical dispatching strategies actually in use, as described above, are independent of λ and F and consider all units as being either available **or** unavailable, ignoring unexpired service times.

If F and H are sufficiently simple, it is possible to calculate the workloads of the units and the expected travel time for the kth-arriving unit at an incident. To take a simple example, let us suppose that there are only two units in a region B, that they are located at fixed facilities, and that exactly one of them will be dispatched to each alarm unless both are unavailable (in which case the alarm is served by some other unit which does not concern us for the moment). Then the "group" G to be dispatched is either {unit 1} or {unit 2}. We assume further that there is only one type of alarm, with arrival rate measure λ, and that the service-time distribution $F(\underline{x},G)$ does not depend on $\underline{x}$ or G and has finite mean $1/\mu$.

The dispatching policy in this example is as follows: We select a response area A for unit 1. If an alarm arrives in A when both units are available, unit 1 is dispatched. If the alarm arrives in the complement B-A when both are available, unit 2 is dispatched. If an alarm arrives when one unit is available, that unit is dispatched.

Given these assumptions, one can calculate the steady-state probabilities P_{ij} of the states

$$00 = \text{both units available}$$
$$10 = \text{unit 1 busy, unit 2 available}$$
$$01 = \text{unit 2 busy, unit 1 available}$$
$$11 = \text{both units busy.}$$

These depend on the service-time distribution only through its mean $1/\mu$, as is proved in Reference 2.

The workload W_j of unit j can be defined as the steady-state probability that unit j is busy, and thus $W_1 = P_{10} + P_{11}$, $W_2 = P_{01} + P_{11}$. The workload difference $|W_1 - W_2|$ can be calculated to be $P_{00}|\lambda(A) - \lambda(B-A)|/(\lambda(B) + \mu)$, which is minimized when A is selected so that half of the alarms arrive in A, and half in B-A.

The expected travel time can be calculated as follows: Of those alarms which arrive when the state is 00, a fraction $\lambda(A)/\lambda(B)$ will be served by unit 1 and have average response time $T_1(A)$ = average time to travel from the location of unit 1 into A. Similarly, a fraction $\lambda(B-A)/\lambda(B)$ will be served by unit 2 and have average response time $T_2(B-A)$. Those alarms which arrive when the state is P_{01} (resp. P_{10}) will be served from location 1 (resp. 2) and have average response time $T_1(B)$ (resp. $T_2(B)$). Thus, the expected travel time, given A is the response area for unit 1, is

$$\overline{T}(A) = P_{00}(T_1(A)\lambda(A) + T_2(B-A)\lambda(B-A))/\lambda(B)$$

$$+ P_{01}T_1(B) + P_{10}T_2(B) + P_{11}\tau,$$

where τ is the average travel time for calls served by units outside B.

After performing a calculation which is given in detail in Reference 1, one finds that

$$\overline{T}(A) = \frac{P_{00}}{\lambda(B)} \int_A (t_1 - t_2 - s_0)\,d\lambda + \alpha$$

where $t_j(\underline{x})$ is the travel time from location j to point $\underline{x}$, $s_0 = \lambda(B)(T_1(B) - T_2(B))/(\lambda(B) + \mu)$, and α is a constant (i.e., independent of A).

It is a property of the integral on the right that it attains its minimum when A is the set

$$X = \{\underline{x}: t_1(\underline{x}) - t_2(\underline{x}) < s_0\}$$

Unless $s_0 = 0$, this means that average travel time is not minimized by always dispatching the closest available unit to each alarm. In fact, if the policy of dispatching the closest available unit, i.e., of selecting A to be the set $\{\underline{x}: t_1(\underline{x}) < t_2(\underline{x})\}$, produces a situation where unit 2 has a greater workload than unit 1, it will commonly happen that $s_0 > 0$. Thus, the response area for unit 1 can be expanded in such a way that not only is the workload of unit 2 reduced but also the expected travel time decreases. The intuitive notion that one must pay a penalty in travel time in order to improve the balance in workloads is not correct.

It should be noted that the dispatch policy which minimizes expected travel time will dispatch unit 1 to $\underline{x}$ if and only if

    (a) both units are available and $\underline{x}$ is in the set X defined above, or

    (b) only unit 1 is available.

This policy depends on λ and F as well as the availabilities, since s_0 is a function of λ and μ.

The insights derived from this example suggest that in realistic models having more than 2 units it may be desirable to find the response areas which minimize expected travel time, since these may also reduce workload imbalance, compared to the usual response areas. If several units are to be dispatched to each alarm, one may have to substitute

some single objective function for the vector of travel
times for the kth-arriving unit at a type m alarm, since it
is not necessarily possible or desirable to find a dispatching policy which minimizes the expectation of each component
of the vector. This objective function would be a function
of the travel-time vector, for example a weighted average
of the components of the vector, and therefore we call it a
generalized travel time. We denote by $t_G(\underline{x},m)$ the generalized travel time if group G is dispatched to a type m alarm
at $\underline{x}$, which includes the weight attached to a type m alarm.

Methods have been developed for minimizing the expected
generalized travel time in special cases. The example which
follows is a case in which linear programming has been
applied. This model was developed by Edward Ignall. We
assume that there are a total of 2N units, of which N will
be dispatched to each alarm. Under these circumstances,
the dispatching policy is determined by specifying which
group responds to an alarm at $\underline{x}$ if all units are available,
since an alarm which arrives when only N units are available will be served by all of them. We denote by $y(\underline{x},m,G)$
a function which is 1 if group G is to be dispatched to a
type m alarm at $\underline{x}$ when all units are available, and is 0
otherwise. We continue to assume that the service times
$F_m(\underline{x},G)$ are independent of $m,\underline{x}$, and G and have finite mean
$1/\mu$; all units in a group complete service at the same time.
In addition, we assume that the alarm rate measures λ_m are
concentrated at a finite set of points; if $\underline{x}$ is one of
these points, we denote $\lambda_m(\{\underline{x}\})$ by $\lambda(\underline{x},m)$. Then λ denotes
the total alarm rate in the region.

In analogy with the average response times $T_j(B)$ in
the 2-unit sample given above, we introduce

$$T_G = \sum_{\underline{x},m} t_G(\underline{x},m)\lambda(\underline{x},m)/\lambda ,$$

which is the expected generalized response time if all
alarms are served by group G. One can then show (Reference
3) that with dispatching policy y, the expected generalized
travel time is

$$\overline{T} = \sum_{\underline{x},m,G} \frac{\lambda(\underline{x},m)y(\underline{x},m,G)}{\lambda(\lambda+\mu)/\mu}\left[t_G(\underline{x},m) + T_{G'}\beta + \frac{(T_G+T_{G'})\beta^2}{2(1-\beta)}\right] + \alpha$$

where G' denotes the group of all units not in G,

$\beta = \lambda/(\lambda + \mu)$, and α is a constant independent of y. A direct method of minimizing $\overline{T}$ is to choose, for each $\underline{x}$ and m, $y(\underline{x},m,G) = 1$ for the group G which minimizes the quantity in brackets on the right. However, one may wish to minimize $\overline{T}$ subject to constraints on the workloads of units.

The steady-state probability that unit j is working can be shown to be

$$W_j = P_{00} \sum_{\underline{x},m,G} \frac{\lambda(\underline{x},m)}{\lambda+\mu} \, y(\underline{x},m,G)\chi(j,G) + C$$

where $\chi(j,G) = 1$ if unit j is in group G, zero otherwise, and C is a constant, independent of y and j. One need only note that the y's enter into the expressions for $\overline{T}$ and W_j linearly to see that a linear programming formulation can be developed to minimize $\overline{T}$ subject to a constraint on the maximum workload difference among units. The optimal values of the policy variables y may turn out to be between 0 and 1, in which case one interprets $y(\underline{x},m,G)$ as specifying the probability of dispatching group G to a type m alarm at $\underline{x}$. Analysis of the results from using such a program is still in progress at this time.

References

1. Carter, G., J. Chaiken, and E. Ignall, "Response Areas for Two Emergency Units," Operations Research, to appear, 1972.

2. Chaiken, J., and E. Ignall, "An Extension of Erlang's Formulas which Distinguishes Individual Servers," J. Appl. Prob., to appear, 1972.

3. Ignall, E., "Response Areas For Groups of Fire-Fighting Units, I," unpublished mimeo, The New York City-Rand Institute, 1971.

OPTIMIZATION OF POWER SYSTEMS

Hermann W. Dommel

Bonneville Power Administration
Portland, Oregon

Abstract

Techniques for the economic dispatch of power plants in electric utility systems are reviewed. The power flows in the network for a given load and generation schedule have traditionally been analyzed separately. It is shown that power flow analysis and optimization can be solved simultaneously. Exploitation of matrix sparsity has made it possible to solve optimization problems of very high dimensionality.

Introduction

Optimization of power plant operation is not new [1-3], but its implementation has become more feasible with the advent of digital computers. Most optimization techniques in use today are based on certain simplifying assumptions which lead to practical rules for the so-called economic dispatch. However, they do not give any information about the flows in the network. The power flows are normally analyzed beforehand for expansion planning studies as well as for expected modes of operation. Future control centers will utilize the on-line computer not only for economic dispatch, but also for predicting power flows. It is shown that a fast power flow solution program can easily be extended to perform optimization functions as well. Such an optimal power flow program can use the best available estimate of the state of the power system instead of relying upon assumptions of "typical" modes of operation as in present schemes.

Review of Economic Dispatch Techniques

The optimal allocation of power to individual units within a power plant, or to closely coupled power plants within a system having negligible transmission losses, can be formulated as the minimization of

$$f = \sum_{i=1}^{n} K_i(P_i), \tag{1}$$

subject to the equality constraint

$$P_{LOAD} - \sum_{i=1}^{n} P_i = 0 \tag{2}$$

where $\quad P_i \qquad$ = power output of unit i,

$\quad K_i(P_i)$ = costs for producing P_i,

$\quad P_{LOAD}$ = total power load to be met.

Setting the derivatives of the Lagrange function

$$\mathcal{L} = f + \lambda (P_{LOAD} - \sum_{i=1}^{n} P_i) \tag{3}$$

with respect to P_i equal to 0 gives the necessary conditions

$$\frac{dK_i}{dP_i} = \lambda, \quad i = 1, \ldots .n. \tag{4}$$

Eq. (4) is the well-known rule that all units must operate at equal incremental costs dK_i/dP_i at the optimum. The solution of Eq. (4) and Eq. (2), which is usually obtained by adjusting λ iteratively until Eq. (2) is fulfilled, will give a minimum if all incremental cost curves are continuous and monotonically increasing and if P_i is not constrained. These sufficient conditions can be relaxed [4].

A simple example, optimal allocation to two units, illustrates the limitations of Eq. (4). Fig. 1 shows the cost curves and its derivatives for two units A and B with constrained output

$$P_{i\,min} \leq P_i \leq P_{i\,max} ,$$

and Fig. 2 the total costs as a function of P_A with P_{LOAD} as parameter. Eq. (4) leads to the minimum only if it lies in the interior of the permissible region, in this case for $P_{LOAD} > 60$ MW, but not if it lies at the boundary, in this case at $P_B = 5$ MW for $P_{LOAD} \leq 60$ MW. Notice that Eq. (4) could also lead into local maxima.

To determine the cost curves from measurements or design data is not at all easy [7]. Frequently used cost curves are shown in Fig. 3a and their derivatives in Fig. 3b. Curve A is typical for modern steam turbines with valve regulation; the discontinuities in the derivative are caused by throttle losses in the opening valve. Cost curves are often approximated by segments of quadratic functions (curve B), with the derivatives becoming straight-line segments. Straight-line segments are also used to approximate cost curves (curve C), with the derivatives becoming "staircase" functions. In the latter case, power is allocated in order of increasing incremental costs, with each unit loaded to its upper limit before power at the next unit is increased from its lower limit ("merit order scheduling").

The economic dispatch of power plants in large interconnected systems must take transmission losses P_{LOSS} into consideration. In this case, the equality constraint is

$$P_{LOAD} + P_{LOSS} - \sum_{i=1}^{n} P_i = 0, \tag{5}$$

which leads to the necessary conditions

$$\frac{dK_i}{dP_i} \cdot \frac{1}{1 - \dfrac{\partial P_{LOSS}}{\partial P_i}} = \lambda; \quad i = 1,\dots n. \tag{6}$$

Eq. (6) differs from Eq. (4) by a loss factor. Most optimization techniques in use express the losses P_{LOSS} as a quadratic function

$$P_{LOSS} = \sum_{i=1}^{n} \sum_{k=1}^{n} P_i B_{ik} P_k. \tag{7}$$

This quadratic form is an approximation. The coefficients B_{ik} are determined from "typical" power flow studies [5,6]; they are not truly constants but depend on the network configuration as well as on the load distribution and load characteristics (power factor). It is difficult to assess the accuracy of the loss formula (7) and to decide how often the coefficients should be recomputed. These difficulties are avoided if optimal power flow solution techniques are used.

The economic dispatch must be reevaluated periodically because the load changes as a function of time. Also, Eqs. (4) and (6) assume that the n units considered are already operating ("committed" to the system). The problem of unit commitment and spinning reserve as well as some present dispatch practices in interconnected power pools are discussed elsewhere [7].

The operation of power systems with a mixture of thermal and hydro plants can no longer be optimized with static optimization techniques. Instead, the optimization must be carried out for a period of time, either short-range (day, week) or long-range (month, year). Optimization of hydro-thermal systems is quite complicated because of the stochastic nature of load predictions and stream flows, because of navigational and environmental constraints on stream flows and water level fluctuations in reservoirs, because of hydraulic coupling between hydro plants, etc.

The principle of hydro-thermal system optimization is easy to explain with the following simplifications: The load variation is given as a deterministic function of time, the water discharge W_i of hydro unit i is only a function of output P_i (water head variations negligible) and the total amount of water which can be used over the time period T is specified as

$$\int_0^T W_i(P_i)dt = c_i, \quad i = n + 1,\dots m. \tag{8}$$

The objective is to minimize

$$I = \int_0^T \sum_{i=1}^{n} K_i(P_i)\, dt, \tag{9}$$

subject to the equality constraints of Eqs. (5) and (8). With the calculus of variations, using Lagrange multipliers for the equality constraints, the optimum is found as the extremum of the integral

$$J = \int_o^T \left\{ \sum_{i=1}^n K_i(P_i) + \sum_{i=n+1}^m \gamma_i W_i(P_i) + \lambda\left(P_{LOAD} + P_{LOSS} - \sum_{i=1}^m P_i\right) \right\} dt. \quad (10)$$

The Euler-Lagrange equations are simply

$$\frac{\partial f}{\partial P_i} = 0$$

in this case with f being the functional within braces of Eq. (10), giving the following necessary conditions:

$$\frac{dK_i}{dP_i(t)} \cdot \frac{1}{1 - \dfrac{\partial P_{LOSS}(t)}{\partial P_i(t)}} = \lambda(t), \qquad i = 1, \dots n \qquad (11a)$$

and

$$\gamma_i \frac{dW_i}{dP_i(t)} \cdot \frac{1}{1 - \dfrac{\partial P_{LOSS}(t)}{\partial P_i(t)}} = \lambda(t), \qquad i = n+1, \dots m. \qquad (11b)$$

The optimization rule Eq. (11) for any moment of time is basically the same as Eq. (6), except that the incremental discharge rate dW_i/dP_i is converted to an incremental cost rate with the conversion factor γ_i. The assignment of hypothetical costs to hydro plants is not surprising, because water is basically worth as much as the thermal generation which would have to be scheduled if the water were not available. The Lagrange factor γ_i has, therefore, physical meaning. The Lagrange multiplier $\lambda(t)$ must be chosen (in practice: iteratively adjusted) to fulfill Eq. (5) at all times and γ_i to fulfill Eq. (8). The solution can be simplified by treating P_{LOAD} as a "staircase" function instead of a continuous function of time, e.g., by assuming changes only at the end of every hour.

Power Flow Solutions

It is important to know the distribution of power flows in the network, at least for some "typical" generation and load schedules, to assess the reliability and security of power transmission. Much effort is devoted to such power flow studies, for normal operation as well as for disturbances (stability studies). They have been made on network analyzers (specialized analog computers) since about 1930, and nowadays primarily on large digital computers. Power flow studies are far from trivial due to the large size of interconnected high voltage transmission systems. Studies involving networks of up to 2,000 nodes and 3,000 branches are no longer uncommon.[*]

The steady-state behavior of a network having N nodes is described by a system of linear, algebraic equations,

---

[*] Nodes represent power plants and switching stations; branches represent overhead transmission lines, cables and transformers.

$$\sum_{m=1}^{N} \overline{Y}_{km}\overline{V}_m = \overline{I}_k \ , \quad k = 1, \ldots N \tag{12}$$

with $\quad \overline{Y}_{km}$ = complex element of the nodal admittance matrix,
$\quad\quad \overline{V}_m$ = complex node voltage,
$\quad\quad \overline{I}_k$ = complex current injected into node k.

All diagonal elements $\overline{Y}_{kk}$ are nonzero; however, off-diagonal elements $\overline{Y}_{km}$ are only nonzero if a branch exists which connects nodes k and m. Consequently, the nodal admittance matrix is very sparse.

In power flow studies, the real power P_k and the reactive power Q_k--or alternately the voltage magnitude $|\overline{V}_k|$--are specified on N-1 nodes, which are related to current by

$$P_k - jQ_k = \overline{I}_k\,\overline{V}_k^* \tag{13}$$

(asterisk indicates conjugate complex.) On the remaining node ("slack node"), the complex voltage is given. Eq. (13) makes the problem nonlinear, since

$$\mathrm{Re}\left\{ \overline{V}_k^* \Sigma\, \overline{Y}_{km}\overline{V}_m \right\} - P_k = 0 \tag{14a}$$

must be fulfilled wherever P_k is specified, and

$$\mathrm{Im}\left\{ \overline{V}_k^* \Sigma\, \overline{Y}_{km}\,\overline{V}_m \right\} + Q_k = 0 \tag{14b}$$

wherever Q_k is specified.

Experience has shown that the best solution technique is Newton's method if the algorithm exploits the sparsity of the associated matrix [8]. By using Eq. (14a) for all nodes with P_k specified and Eq. (14b) for all nodes with Q_k specified, a system of equations[*]

$$[g([x])] = 0 \tag{15}$$

is formed, with the vector [x] containing the unknown variables (θ_k for all nodes with P_k specified and V_k for all nodes with Q_k specified, with $V_k = V_k e^{j\theta_k}$). In practice, additional equations must be fulfilled to satisfy area interchange constraints, transformer control constraints, etc., which simply increases the dimension of [g] and [x][9]. Other versions of Newton's method are possible (see discussion by H. W. Dommel in [8], which also gives a review of other solution techniques). It is not possible to use Newton's method with complex variables in this case because the complex power equations are not analytic. Therefore, the Cauchy-Riemann differential equations are not fulfilled and no complex derivative exists (hint given to the author by F. D. Byrnes, Bonneville Power Administration).

---

[*] Brackets are used for vectors and matrices.

Newton's method: An improved approximation to the solution is found in iteration step h by solving the system of linear equations,

$$\left[\frac{\partial g}{\partial x}^{(h-1)}\right]\left[\Delta x\right] = -\left[g^{(h-1)}\right] \qquad (16)$$

for $[\Delta x]$ by Gaussian elimination (triangular factorization). Then

$$[x^{(h)}] = [x^{(h-1)}] + [\Delta x]. \qquad (17)$$

Sufficient accuracy is normally obtained after 3 to 4 iteration steps. The Jacobian matrix $[\partial g/\partial x]$ is very sparse. This sparsity is preserved as much as possible in the elimination algorithm by using optimal ordering schemes and by processing and storing the nonzero terms only [10, 11]. The idea of sparsity exploitation was pioneered in the power industry [12] and is gaining attention in many other fields now. It has made it possible to solve networks of 1,000 nodes, which is approximately equivalent to 2,000 nonlinear equations, with less than 14,000 words required for storing the values and indices of the upper triangular matrix (compared to approx. 2,000,000 for the values of a full upper triangular matrix!) in approx. 30 seconds on a CDC 6400 (time for optimal ordering plus 3 Newton steps).

Optimal Power Flow Solutions

Fast power flow solution techniques make it possible to solve the static optimization problem more accurately by observing the power flow equations (15) as equality constraints. The advantage over traditional optimization techniques lies not so much in higher accuracy. More important is the ability to check the power flows for their reliability and, if necessary, to impose additional security constraints on them. In addition, allocation of reactive power can also be optimized.

The "optimal power flow" problem can be stated as follows [13]:

$$\min_{[u]} f([x],[u]),$$

subject to the equality constraints of Eq. (15), with $[u]$ being the vector of control or decision variables and $[x]$ being the vector of dependent variables. Typical control variables are power plant outputs P_i to be scheduled for economic dispatch, but also voltage magnitudes at power plants and substations as well as transformer tap settings which are being controlled to achieve a well-balanced voltage profile throughout the system (problem of reactive power flow). For optimal real and reactive power flow, the objective function f is that of Eq. (1). For the optimization of reactive power flow alone, the objective function is system losses. Without inequality constraints, the optimum is found as the extremum of

$$\mathcal{L} = f([x],[u]) + [\lambda]^T [g([x],[u])] \qquad (18)$$

(superscript T for transposition), which gives the necessary conditions

$$\left[\frac{\partial f}{\partial x}\right] + \left[\frac{\partial g}{\partial x}\right]^{T} \left[\lambda\right] = 0 , \tag{19}$$

$$\left[\frac{\partial f}{\partial u}\right] + \left[\frac{\partial g}{\partial u}\right]^{T} \left[\lambda\right] = 0 , \tag{20}$$

in addition to Eq. (15). If Newton's method is used for the power flow solution, then the Jacobian matrix $[\partial g/\partial x]$ is automatically obtained in factored form (lower and upper triangular matrix) as a by-product of Eq. (16). The same matrix in transposed form appears again in Eq. (19); therefore, Eq. (19) can be solved very fast for $[\lambda]$ with one repeat solution [14]. For a typical 500-node problem this repeat solution takes about 0.5s on a CDC 6400. Inserting $[\lambda]$ thus obtained into Eq. (20) will give the reduced gradient

$$\left[\nabla f\right] = \left[\frac{\partial f}{\partial u}\right] + \left[\frac{\partial g}{\partial u}\right]^{T} \left[\lambda\right], \tag{21}$$

which is the first-order sensitivity of the objective function with respect to the control vector $[u]$, with the equality constraints of Eq. (15) rigidly observed.

This technique for finding the reduced gradient has been used in expanding an existing power flow solution program into an optimization program at Bonneville Power Administration [13]. The basic algorithm is as follows:

1. Guess the control vector $[u]$.

2. Solve the power flow for $[x]$ by Newton's method, with $[u]$ as a fixed parameter. This also yields the Jacobian matrix in factored form which is computationally equivalent to the inverse or transposed inverse [10]. Coming from step 1, 3-4 iterations are required and 1-2 when returning from step 6.

3. Solve Eq. (19) for $[\lambda]$ by a repeat solution with the factored matrix from step 2.

4. Insert $[\lambda]$ from step 3 into Eq. (21) to obtain the reduced gradient $[\nabla f]$.

5. If $[\nabla f]$ is sufficiently small, the minimum has been reached. Otherwise:

6. Make adjustments on $[u]$, either with a straightforward optimum gradient technique,

$$[u^{new}] = [u^{old}] - \alpha \cdot [\nabla f] \tag{22}$$

or some other technique, and return to step 2.

The critical part is step 6. Eq. (22) is one of several possible correction formulas. In Eq. (22), α must be chosen to reach the lowest possible value along the direction of steepest descent $-[\nabla f]$. A combination of first-order adjustments of the type of Eq. (22) and second-order adjustments using approximate values for the diagonal elements of the Hessian matrix was finally adopted [13].

To make the optimization realistic, inequality constraints on some parameters of the form

$$u_{i\,min} \leq u_i \leq u_{i\,max} \tag{23a}$$

$$x_{i\,min} \leq x_i \leq x_{i\,max} \tag{23b}$$

and perhaps functional constraints

$$h_i([x],[u]) \leq 0$$

must be observed. Eq. (23a) is easily fulfilled by assuring that the adjustment algorithm in step 6 sets u_i to the limit value if it falls outside the permissible region. Its component in the reduced gradient must still be computed in the following adjustment cycles to permit the variable to back off the limits again. Eq. (23b) concerns primarily voltage magnitudes V_i on nodes without voltage control. The introduction of penalty terms of the form

$$w_i = k\,(x_i - x_{i\,max})^2 \quad \text{for } x_i > x_{i\,max},$$
$$w_i = 0 \ \text{otherwise,} \tag{24}$$

into the objective function was found to be satisfactory for this problem. The factor k is chosen after the first power flow solution to make the sum of the penalties a certain percentage of the objective function. The percentage factor is given by the engineer, which permits him to place more or less emphasis on the constraints of Eq. (23b). The factor k is not changed thereafter and Eq. (23b) will accordingly be violated somewhat when the solution is reached. From an engineering standpoint, a statement of the form $V_i \leq 1.0$ is not meant to be rigid anyhow and a solution at $V_i = 1.01$ may be perfectly acceptable.

Fig. 4 shows the result of optimizing the reactive power flow in a normally loaded power system with 328 nodes and 80 components in the control vector [u]. Minimization of system losses was achieved with a slight increase in penalties. Increasing the factor k in Eq. (24), that is putting more emphasis on the inequality constraints Eq. (23b), did not change the solution significantly. The engineer can often define such constraints only roughly as desirable goals; strict observance is not warranted in such cases and would prevent a solution in most cases. Reactive power flow optimization for a light load case (night schedule) in a system with 419 nodes and 122 control variables is shown in Fig. 5. Here, the optimization process reduced primarily the penalty terms, which reflects the physical nature of light load power flows where losses cannot be influenced much and where the major problem is to keep the voltages within acceptable limits.

Optimal power flow solution techniques should also be applicable for optimizations over a period of time, since the component λ_{P_i} of the vector $[\lambda]$ associated with the power equation for P_i in Eq. (18) is directly related to λ and the loss factor in Eq. (11). If the objective function contains no penalty terms, then [9]

$$\lambda_{P_i} = \lambda \cdot \left(1 - \frac{\partial P_{LOSS}}{\partial P_i}\right) \tag{25}$$

More research is required, however, before optimization over a period of time becomes practical enough.

Conclusion

Traditional dispatch techniques for the optimal operation of power systems have been used for many years. Newer techniques based on nonlinear programming permit overall optimizations, with the power flow solution obtained simultaneously. One such technique, which was found to be practical enough, is described. Other similar techniques have been reported [15] and continued research should lead the way to on-line applications in the near future.

References

1. F. Noakes, A. Arismunandar, "Bibliography on optimum operation of power systems: 1919-1959, "AIEE Trans. Power Apparatus and Systems, Vol. 81, pp. 864-871, Feb. 1963.

2. M. J. Steinberg and Th. H. Smith, "The theory of incremental rates and their practical application to load division," El. Engg., Vol. 53, pp. 432-445, 571-584, March/April 1934.

3. L. K. Kirchmayer, Economic Operation of Power Plants. New York: J. Wiley Sons, 1958.

4. K. Theilsiefje, "Über hinreichende Bedingungen für ein Minimum der Brennstoffkosten im optimalen Verbundbetrieb," ETZ-A, Vol. 82, pp. 175-179, March 1961.

5. E. E. George, "Intrasystem transmission losses," <u>AIEE Trans.</u>, Vol. 62, pp. 153-158, 1943.

6. W. S. Meyer and V. D. Albertson, "Improved loss formula computation by optimally ordered elimination techniques," <u>IEEE Trans. Power Apparatus and Systems</u>, Vol. PAS-90, pp. 62-69, Jan./Feb. 1971.

7. IEEE Committee Report, "Present practices in the economic operation of power systems," <u>IEEE Trans. Power Apparatus and Systems,</u> Vol. PAS-90, pp. 1768-1775, July/Aug. 1971.

8. W. F. Tinney and C. E. Hart, "Power flow solution by Newton's method," <u>IEEE Trans. Power Apparatus and Systems</u>, Vol. PAS-86, pp. 1449-1460, November 1967.

9. H. W. Dommel, W. F. Tinney and W. L. Powell, "Further developments in Newton's method for power system applications," <u>IEEE Paper No. 70 CP-161-PWR,</u> Winter Power Meeting, New York, Jan. 1970.

10. W. F. Tinney and J. W. Walker, "Direct solution of sparse network equations by optimally ordered triangular factorization," <u>Proc. IEEE,</u> Vol. 55, pp. 1801-1809, Nov. 1967.

11. E. C. Ogbuobiri, W. F. Tinney and J. W. Walker, "Sparsity-directed decomposition for Gaussian elimination on matrices," <u>IEEE Trans. Power Apparatus and Systems,</u> Vol. PAS-89, January 1970.

12. N. Sato and W. F. Tinney, "Techniques for exploiting the sparsity of the network admittance matrix," <u>IEEE Trans. Power Apparatus and Systems,</u> Vol. 82, pp. 944-950, Dec. 1963.

13. H. W. Dommel and W. F. Tinney, "Optimal power flow solutions," <u>IEEE Trans. Power Apparatus and Systems,</u> Vol. PAS-87, pp. 1866-1876, Oct. 1968.

14. H. Dommel, "Digitale Rechenverfahren für elektrische Netze," <u>Archiv für Elektrotechnik,</u> Vol. 48, pp. 41-68, 118-132, Feb./April 1963.

15. A. M. Sasson, F. Aboytes, R. Cardenas, F. Gomez and F. Viloria, "A comparison of power systems static optimization techniques," <u>Proc. 7th Power Industry Computer Application Conf.,</u> pp. 329-337, Boston, Mass., May 23-26, 1971.

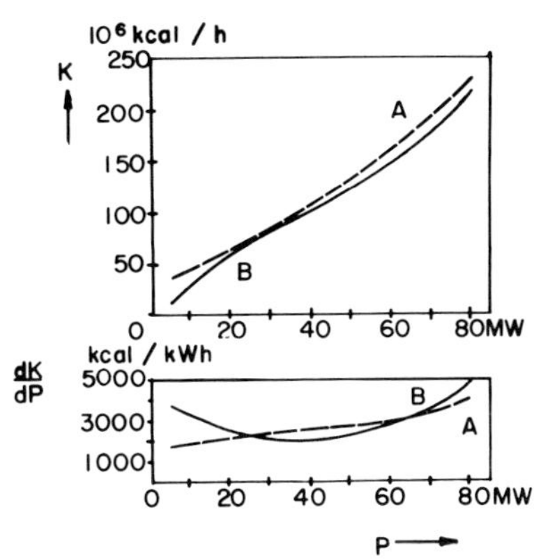

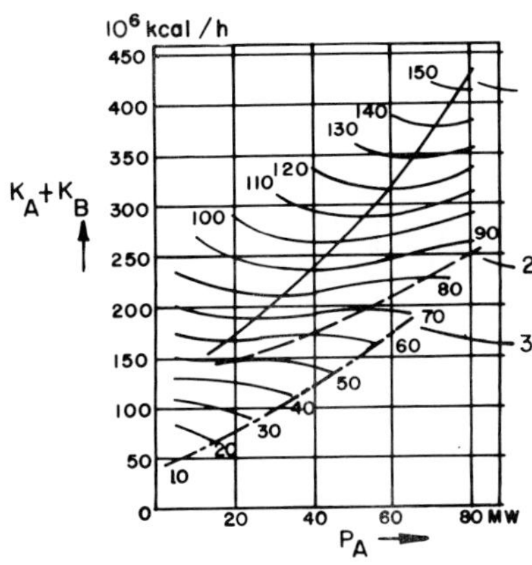

Fig. 1. Costs and incremental costs for units A, B.

Fig. 2. Total costs as a function of P_A (P_{LOAD} = 10 – 160 MW as parameter)
1 = minima in interior
2 = maxima in interior
3 = minima at boundary.

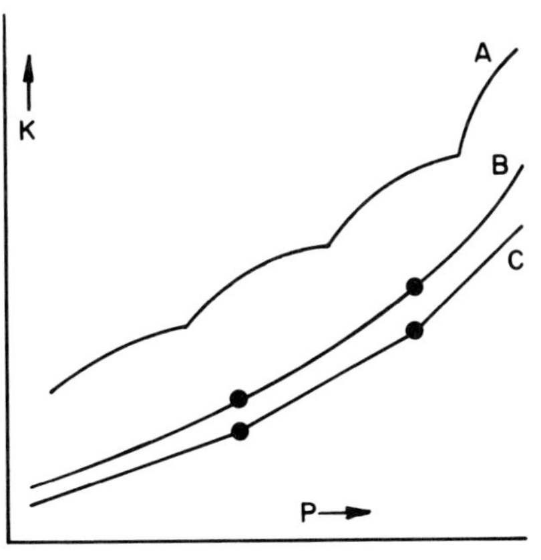

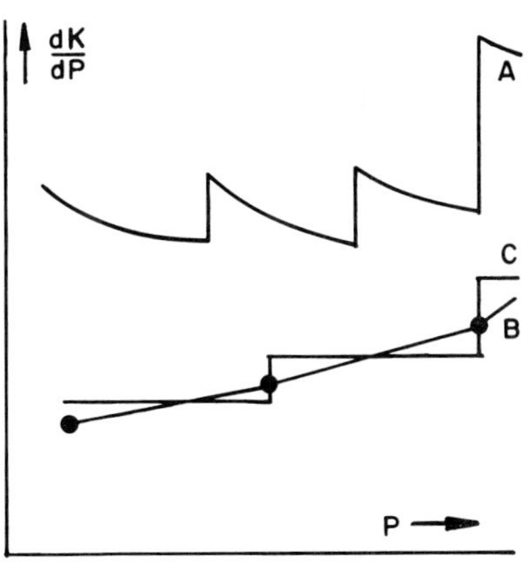

Fig. 3a. Cost curves.

Fig. 3b. Incremental cost curves.

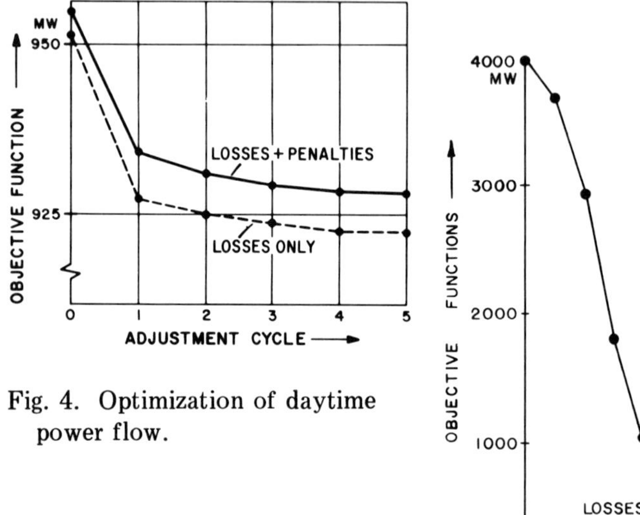

Fig. 4. Optimization of daytime power flow.

Fig. 5. Optimization of nighttime power flow.

WATER POLLUTION CONTROL

Glenn W. Graves

University of California
Los Angeles, California

In an earlier paper [3], "Mathematical Programming for
Regional Water Quality Management," the model presented
dealt with the difficult question of the best treatment
plant configuration in a very limited manner. Flow of effluent was restricted to flow direct to a discharge point
or flow to a single plant and then from the plant to a discharge point. There was no flow between plants and hence
no way to achieve, say, primary treatment for a large percentage of the effluent at huge, cheap primary plants and
subsequent treatment for a much smaller percentage at smaller more expensive secondary and tertiary plants. Since
pipe juncture nodes can be considered plants without treatment, the flow pattern was restricted to single juncture
points. This resulted in practice in manual pipe consoli-
dation and weakened results.
The solution procedure for the original model required
forcing a solution using only the regional plants. This
forces the plants to operate at sufficiently high levels as
to utilize their economies of scale. When the other treatment methods are then introduced, they are compelled to
compete against the efficiently operating plants. Where
economically justified they then force out and close the
non-competitive plants. This is computationally wasteful
as the plants must first be opened and driven down and
closed. There are also difficult problems with infinite
derivatives in the cost function with the initial very low
or zero flows. These problems can be handled much more effectively using discrete integer variables in the formulation of the model. This possibility was not exploited ori-
ginally because of the limitations of the solution techniques for large mixed integer and continuous variable

nonlinear programming problems. Recent research has great-
ly expanded our capability to solve integer programming
problems. This extended capability coupled with new and
more powerful decomposition techniques changes the picture
radically. It now seems attractive to switch to the mixed
integer, continuous variable formulation.

Mixed Integer Regional Treatment Model

In this model we can use a much more extensive and
realistic supporting flow network. This is illustrated by
Figure 1.

The nodes in the network are of three types. A node
is a polluter or source, a sink or a discharge point, or an
intermediate node in the network consisting of a pipe juncture or treatment plant. The sources are fixed at the present sites of the identified major polluters. The discharge nodes are the presently used discharge sections of
the estuary plus any additional potentially attractive discharge sections. The intermediate nodes consist of any
existing treatment plants or pipe junctures plus any additional potentially attractive sites. Combining existing
and new sites into an optimal total network is particularly
nice in this approach because the fixed or capital costs
can be separated from the variable or operating costs. This
permits a true comparison of the marginal cost of a new
plant to an existing plant. The arcs in the network are
pipes. They can consist of existing pipes and potentially
attractive additional pipes. Again the fixed and variable
costs can be separated. Also it is possible to have discrete alternate capacities and enforce a choice of pipe
size. Similarly, we can enforce a choice between alternate
plant capacities at a given site or between alternate plant
sites.

In the mathematical statement of the model we use the
following:

<u>Notation</u>

<u>Physical Quantities</u>

A   - matrix of transfer coefficients (mg/ℓ per lb/day)

C   - vector of D.O. goals (mg/ℓ)

r_i - % removal of B.O.D. at node i

q_{ij} - flow from node i to node j (MGD)

J_i - concentration leaving node i (lb/MG)

m_i - mass of B.O.D. leaving node i (lb/D)

Costs
---

f_i - capital cost of plant i of a given capacity ($)

v_i - operating costs of plant i (($/MGD)/% removal)

f_{ij} - capital cost of pipe between node i and node j of given capacity ($)

v_{ij} - operating cost of pipe between node i and node j of given capacity ($/MGD)

p_i - cost of additional treatment at source ($/% removal)

Index Sets

$R = \{j \mid j$ is a section or reach of the estuary$\}$
$S = \{i \mid$ node i is a source$\}$
$D = \{i \mid$ node i is a discharge point$\}$
$I = \{i \mid$ node i is not a source or sink$\}$
$N = S \cup D \cup I$ set of nodes; $a \subset N \times N$ set of arcs
$a(x) = \{y \in N \mid (x,y) \in a\}$ nodes "after" x
$b(x) = \{y \in N \mid (y,x) \in a\}$ nodes "before" x

Integer Variables

$$z_i = \begin{cases} 0 & \text{plant i is closed} \\ 1 & \text{plant i is open} \end{cases}$$

$$z_{ij} = \begin{cases} 0 & \text{pipe } (i,j) \text{ is closed} \\ 1 & \text{pipe } (i,j) \text{ is open} \end{cases}$$

The mathematical model is:
Subject to
Conservation of Flow

$$\sum_{j \in a(i)} q_{ij} - \sum_{k \in b(i)} q_{ki} = 0 \qquad i \in I$$

$$\sum_{j \in a(i)} q_{ij} = \overline{q}_i \qquad i \in S \ (\overline{q} \text{ current discharge MG/D})$$

Conservation of Concentration

$$J_i = \frac{(\sum_{k \in b(i)} q_{ki} J_k) \cdot (1 - r_i)}{(\sum_{k \in b(i)} q_{ki})} \qquad i \in I \cup D$$

$$J_i = \frac{\overline{m}_i}{\overline{q}_i}(1-r_i) \qquad i \in S \quad (\overline{m}_i \text{ current discharge lb})$$

<u>Capacity at Plant</u> (ℓ_i - lower, μ_i - upper capacity)

$$z_i \cdot \ell_i \leq \sum_{k \in b(i)} q_{ki} \leq z_i \mu_i$$

(Note that when the plant is closed or not chosen for the network all flow is shut off and when it is opened it functions in a restricted range. By breaking up the range nonconvexity can be eliminated and more truly global optimum achieved. This would require merely adding some:

<u>Alternative Constraints (Plants)</u>

$$\sum_{t \in E_i} z_t = 1$$

These constraints would force a choice of only one of the plants in the set E_i.)

<u>Capacity of Pipes</u> (ℓ_{ij} - lower, μ_{ij} - upper capacity)

$$z_{ij} \cdot \ell_{ij} \leq q_{ij} \leq z_{ij} \cdot \mu_{ij}$$

(Again the pipes can be eliminated or forced to function in a restricted range. By breaking up the range nonconvexity can be eliminated. Again we would employ:

<u>Alternative Constraints (Pipes)</u>

$$\sum_{(t,s) \in E_{ij}} z_{ts} = 1$$

These constraints would force a choice of one of the alternatives in the set E_{ij}.)

The quality goals are enforced by introducing the auxillary variables

$$m_t = J_t \cdot \left(\sum_{k \in b(t)} q_{kt} \right) \qquad t \in D$$

$$\overline{m}_t = \text{present discharge in lbs}$$

and the following:

<u>Quality Constraints</u>

$$\sum_{t \in D} a_{it}(m_t - \overline{m}_t) \leq -c_i \qquad i \in R$$

The objective is to minimize the total cost function

$$\text{minimize } T(z_t, z_{ts}, q, r) = PL(z_t, q) + PI(z_{ts}, q) + PO(r)$$

which is broken down into:

Plant Costs

$$PL(z_t, q, r) = \sum_{t \in I} z_t \cdot (f_t + v_t(r) \cdot \sum_{k \in b(t)} q_{kt})$$

Pipe Costs

$$PI(z_{ts}, q) = \sum_{(t,s) \in A} z_{ts} \cdot (f_{ts} + v_{ts} q_{ts})$$

Polluter Costs

$$PO(r) = \sum_{t \in S} r_t \cdot p_t$$

Solution Technique

The foregoing complex mixed integer continuous variable nonlinear programming model presents a truly formidable computational challenge. Although a direct implicit enumeration of the integer variables coupled with the bounding off effect of solving the remaining nonlinear programming problems might be attempted, the outcome of such a venture would be extremely doubtful. Recent advances in Bender's type (see [1]) decomposition offer a much more promising avenue of attack. A Bender's type decomposition applicable to a general nonlinear programming problem

Subject to

$$g^i(y_1, y_2) \le 0$$
$$g^k(y_2) \le 0$$
$$\min g^m(y_1, y_2)$$

is possible. In this type of decomposition a subset of the variables, say, the y_2 vector can be used to parameterize a relaxed

Sub-Problem

Subject to

$$g^i(y_1, \bar{y}_2) \le 0$$
$$\min g^m(y_1, \bar{y}_2)$$

and the parameters y_2 need not be continuous variables. In the present model the integer variables z_t and z_{ts} can be used in this way to yield a supporting network design and a tractable nonlinear programming subproblem.

Whenever a fixed choice of parameters $\bar{y}_2$ satisfying the remaining constraints

$$g^k(\overline{y}_2) \leq 0$$

permits a feasible solution to the subproblem, we obtain a feasible solution to the complete problem. In order to avoid a random search through plausible choices of $\overline{y}_2$, we employ a class of lower bound functions D. In a stepwise procedure we generate members of this class. By ensuring that at least each lower bound function already generated indicates a better possible solution than the best known incumbent feasible solution we guide the choice of the parameter variables. In point of fact since it is a necessary condition that each lower bound function be less than the incumbent when there exists a better solution, the lower bound functions provide us with a convergence criteria. Neglecting infeasible subproblems for the moment we use the following line of reasoning.

<u>Convergence</u>

Generate stepwise a class D of lower bound functions such that for any

$$c^t(y_2) \in D$$

<u>and any y_1</u> such that y_1, y_2 is feasible

$$c^t(y_2) \leq g^m(y_1, y_2).$$

Now attempt to solve the

<u>Master Problem</u>

$$g^k(y_2) \leq 0$$

$$c^t(y_2) \leq g^m(y_1^c, y_2^c) - \epsilon, \quad t = 1, \ldots, n$$

(where y_1^c, y_2^c is the best known incumbent solution).

If a feasible solution y_2 is obtained we test it in the subproblem in an attempt to obtain an improved incumbent. If a feasible solution is <u>not</u> obtained in the master problem, then for at least one <u>lower</u> bound function,

$$g^m(y_1, y_2) \geq c^t(y_2) > g^m(y_1^c, y_2^c) - \epsilon$$

and $g^m(y_1^c, y_2^c)$ is an ϵ-optimal solution. The identical line of reasoning is employed for infeasible subproblems where we simply use a violated subproblem constraint instead of the extremal function. Convergence in a finite number of steps is assured by showing that at most a finite number of $c^t(y_2)$ can be generated between bounded reductions in infeasibility or the extremal function.

The principal difference between this line of reasoning and the standard Bender's approach is that no attempt is made to obtain the greatest lower bound possible with the finite subset of D already generated. This would of course be possible by simply introducing an auxillary variable y_0 and a master problem of the form

$$\text{Subject to} \qquad g^k(y_2) \leq 0$$
$$c^t(y_2) - y_0 \leq 0$$
$$\min y_0.$$

There seems no compelling reason to uniformly depress the available lower bound functions as low as possible. In fact this may simply increase the instability and slow down the convergence.

In order to generate a class of lower bound functions of interest here we require the two following expansions of the functions.

Expansion 1 (with respect to y_1)

$$g^i(y_1, y_2) = g^i(y_1^0, y_2) + \nabla g^i(y_1^0, y_2) \cdot \Delta y_1 \tag{1}$$
$$+ \tfrac{1}{2} \Delta y_1^T H^i(\tilde{y}_1, y_2) \Delta y_1.$$

Now in order to eliminate the gradients and utilize information furnished by the dual variables in the subproblem, we must make:

Expansion 2 (with respect to y_2)

$$\nabla g^i(y_1^0, y_2) \cdot \Delta y_1 = \nabla g^i(y_1^0, y_2^0) \cdot \Delta y_1 \tag{2}$$
$$+ \Delta y_2^T \overline{H}^i(y_1^0, \tilde{y}_2) \cdot \Delta y_1$$

where

$$\overline{H}^i(y_1^0, \tilde{y}_2) = \frac{\partial^2 g^i(y_1, y_2)}{\partial y_{1k} y_{2j}}$$

is a matrix of mixed partial derivatives. Combining the error terms

$$R^i(\Delta y_1, \Delta y_2) = \tfrac{1}{2} \Delta y_1^T H^i(\tilde{y}_1, y_2) \cdot \Delta y_1 \tag{3}$$
$$+ \Delta y_2^T \overline{H}^i(y_1^0, \tilde{y}_2) \cdot \Delta y_1$$

yields the desired representation of each function

$$g^i(y_1, y_2) = g^i(y_1^0, y_2) + \nabla g^i(y_1^0, y_2^0) \cdot \Delta y_1 + R^i(\Delta y_1, \Delta y_2). \tag{4}$$

Using this representation of our functions and the following duality relation,

$$\sum_i x_i \nabla g^i(y_1^o, y_2^o) = \nabla g^m(y_1^o, y_2^o) \tag{5}$$

which is derived as equation (14) of the appendix to "Optimization of a Reverse Osmosis System Using Nonlinear Programming" (see [4]), yields our class of lower bound functions. The argument employed in establishing this class of lower bound functions is similar to the argument used to establish the "Weak Duality Theorem" of linear programming. Any ϵ-feasible solution of the subproblems satisfies

$$g^i(y_1, y_2) = g^i(y_1^o, y_2) + \nabla g^i(y_1^o, y_2^o) \cdot \Delta y_1 + R^i(\Delta y_1, \Delta y_2) \leq \epsilon_1 \tag{6}$$

or

$$\nabla g^i(y_1^o, y_2^o) \cdot \Delta y_1 \leq -g^i(y_1^o, y_2) - R^i(\Delta y_1, \Delta y_2) + \epsilon_1. \tag{7}$$

Using (5) and (7) we demonstrate that as a function of y_2, for any value of y_1

$$\begin{aligned}
g^m(y_1, y_2) &= g^m(y_1^o, y_2) + \nabla g^m(y_1^o, y_2^o) \cdot \Delta y_1 + R^m(\Delta y_1, \Delta y_2) \\
&= g^m(y_1^o, y_2) + [\sum_i x_i \nabla g^i(y_1^o, y_2^o)] \cdot \Delta y_1 + R^m(\Delta y_1, \Delta y_2) \\
&= g^m(y_1^o, y_2) + \sum_i x_i [\nabla g^i(y_1^o, y_2^o) \cdot \Delta y_1] + R^m(\Delta y_1, \Delta y_2) \\
&\geq g^m(y_1^o, y_2) + \sum_i (-x_i) \cdot g^i(y_1^o, y_2) \\
&\qquad + \sum_i (-x_i) \cdot R^i(\Delta y_1, \Delta y_2) + R^m(\Delta y_1, \Delta y_2) \\
&\qquad + \sum_i x_i \cdot \epsilon_1.
\end{aligned} \tag{8}$$

For convenience we restate the conditions for a stationary point of the nonlinear algorithm presented in [4].

<u>Local Linear Stationary Point</u>

$\exists$ a set of dual variables $x_p \ni$

   1) $\sum_p (-x_p) > B_1$ (insufficient resolution)

$\exists$ a w and H $\ni$

   2) $\sum_{p \in H} (-x_p) \cdot g^p(y^o) > -g^w(y^o) - \epsilon$ (inconsistency)

$\exists$ a $y_o \in F \ni$

   3) $\sum_{i=1}^{m-1} (-x_i) \cdot g^i(y^o) > -\epsilon$ (optimality)

Now choose $\epsilon_1 = \epsilon/B_1$ and assume that condition 1) of a local linear stationary point is not violated. Then

$$\sum_i x_i \geq -B_1 \quad \text{and} \quad \sum_i x_i \cdot \epsilon_1 \geq -\epsilon$$

and when the error expression

$$\sum_i (-x_i) \cdot R^i(\Delta y_1, \Delta y_2) + R^m(\Delta y_1, \Delta y_2) \geq 0 \qquad (9)$$

we have our lower bound function

$$g^m(y_1^o, y_2) + \sum_i (-x_i) \cdot g^i(y_1^o, y_2) - \epsilon \leq g^m(y_1, y_2). \qquad (10)$$

(Note that in the case where the functions are separable in y_1 and y_2 as in our model, the convexity of the subproblem in the continuous variables is sufficient to ensure condition (9) and the validity of the decomposition. However, we do not have convexity in the subproblems and a global optimum in our model.)

The decomposition procedure then consists of alternately solving the relaxed subproblems and the following

Master Problem

Find a feasible solution to

1) $g^k(y_2) \leq 0$

2) $g^m(y_1^o, y_2) + \sum_i (-x_i) \cdot g^i(y_1^o, y_2) - \epsilon \leq g^m(y_1^c, y_2^c) - 2\epsilon$

(incumbent)

3) $g^w(y_1^o, y_2) + \sum_{p \in H} (-x_p) g^p(y_1^o, y_2) \leq -\epsilon$

where a constraint of type 2) is added when an optimal solution is obtained for the subproblem and a constraint of type 3) is added when the subproblem is inconsistent.

Convergence to an ϵ-optimal solution in a finite number of steps using this decomposition technique is easily demonstrated. Assume we return from a subproblem without achieving a bounded gain in the incumbent solution when we obtained a feasible solution. Then

$$g^m(y_1^o, y_2) \geq g^m(y_1^c, y_2^c) \qquad (11)$$

and

$$\sum_i (-x_i) \cdot g^i(y_1^o, y_2) > -\epsilon \qquad (12)$$

if this set of dual variables is a repeat we contradict

$$g^m(y_1^o, y_2) + \sum_i (-x_i) g^i(y_1^o, y_2) \leq g^m(y_1^c, y_2^c) - \epsilon \qquad (13)$$

which has been established from the master problem. Since there are only a finite number of dual solutions between bounded gains in a finite number of steps we exceed any lower bound or fail to solve the master. When we fail to solve the master we have established from one of the lower bound functions that we are within ϵ of the optimum.

When we return from the subproblems with an inconsistent solution we have

$$\sum_{p\in H} (-x_p) \cdot g^p(y_1^o, y_2) > -g^w(y_1^o, y_2) - \epsilon. \tag{14}$$

Again if w and H are a repeat we contradict

$$g^w(y_1^o, y_2) + \sum_{p\in H} (-x_p) g^p(y_1^o, y_2) \leq -\epsilon \tag{15}$$

which has been established from the master. So in at most a finite number of steps we resolve the inconsistency or demonstrate that the complete problem is inconsistent.

This type of decomposition has recently been employed in a large warehouse location problem (see [2]) and proved quite successful. Full scale computations on semi-realistic data for the Delaware using our earlier model [3] which is similar to the subproblems has yielded very good results. In the earlier model we solved 80-constraint and 2,000 variable problems in 15-20 minutes of computation on a 360-91 computer. The solution value of $2.3 million dollars for 3 mg/ℓ goals compared very favorably with the $8.3 million dollar cost of the Uniform Treatment at the source policy presently required by the government. These types of regional optimization models would thus seem to offer great promise for regional water quality control.

References

1. Geoffrion, A. M., "Generalized Benders Decomposition," _Journal of Optimization Theory and Application_, forthcoming 1972.
2. Geoffrion, A. M., and G. W. Graves, "Multicommodity Distribution System Design by Benders Decomposition," forthcoming as Working Paper No. 181, Western Management Science Institute, UCLA, December 1971.
3. Graves, G. W., G. B. Hatfield and A. B. Whinston, "Mathematical Programming for Regional Water Quality Management," _Water Resources Research_, 8, 1, February 1972.
4. Hatfield, G. B., and G. W. Graves, "Optimization of a Reverse Osmosis System Using Nonlinear Programming," _Desalination_, 7, 1970.

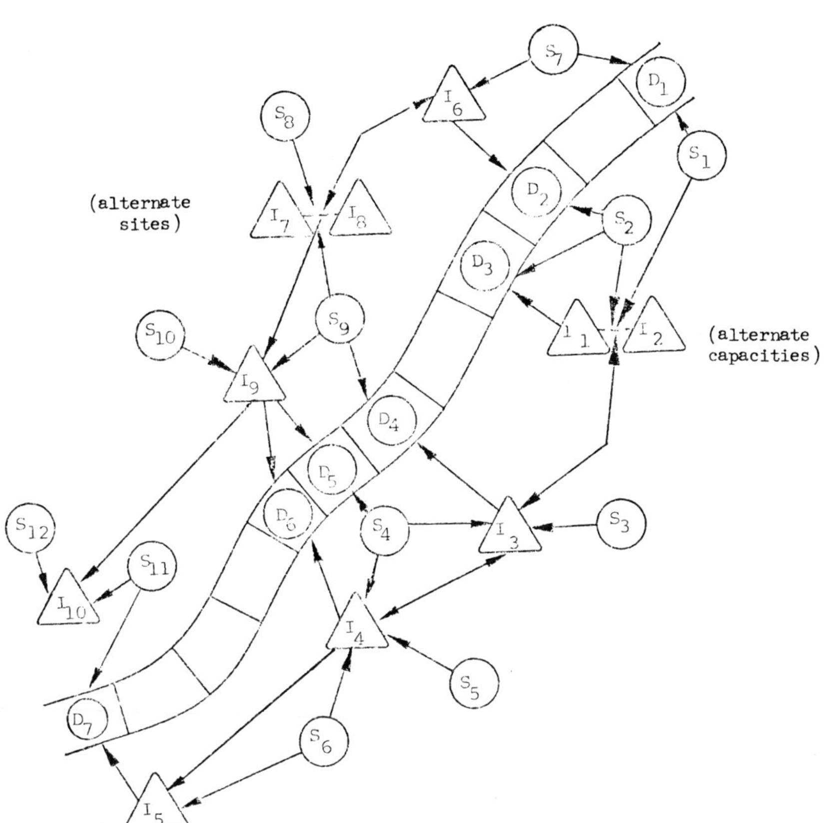

Figure 1. Schematic of General System